Lecture Notes in Physics

AF763614

The Editorial Policy for Proceedings

The series Lecture Notes in Physics reports new developments in physical research and teaching – quickly, informally, and at a high level. The proceedings to be considered for publication in this series should be limited to only a few areas of research, and these should be closely related to each other. The contributions should be of a high standard and should avoid lengthy redraftings of papers already published or about to be published elsewhere. As a whole, the proceedings should aim for a balanced presentation of the theme of the conference including a description of the techniques used and enough motivation for a broad readership. It should not be assumed that the published proceedings must reflect the conference in its entirety. (A listing or abstracts of papers presented at the meeting but not included in the proceedings could be added as an appendix.)
When applying for publication in the series Lecture Notes in Physics the volume's editor(s) should submit sufficient material to enable the series editors and their referees to make a fairly accurate evaluation (e.g. a complete list of speakers and titles of papers to be presented and abstracts). If, based on this information, the proceedings are (tentatively) accepted, the volume's editor(s), whose name(s) will appear on the title pages, should select the papers suitable for publication and have them refereed (as for a journal) when appropriate. As a rule discussions will not be accepted. The series editors and Springer-Verlag will normally not interfere with the detailed editing except in fairly obvious cases or on technical matters.
Final acceptance is expressed by the series editor in charge, in consultation with Springer-Verlag only after receiving the complete manuscript. It might help to send a copy of the authors' manuscripts in advance to the editor in charge to discuss possible revisions with him. As a general rule, the series editor will confirm his tentative acceptance if the final manuscript corresponds to the original concept discussed, if the quality of the contribution meets the requirements of the series, and if the final size of the manuscript does not greatly exceed the number of pages originally agreed upon.
The manuscript should be forwarded to Springer-Verlag shortly after the meeting. In cases of extreme delay (more than six months after the conference) the series editors will check once more the timeliness of the papers. Therefore, the volume's editor(s) should establish strict deadlines, or collect the articles during the conference and have them revised on the spot. If a delay is unavoidable, one should encourage the authors to update their contributions if appropriate. The editors of proceedings are strongly advised to inform contributors about these points at an early stage.
The final manuscript should contain a table of contents and an informative introduction accessible also to readers not particularly familiar with the topic of the conference. The contributions should be in English. The volume's editor(s) should check the contributions for the correct use of language. At Springer-Verlag only the prefaces will be checked by a copy-editor for language and style. Grave linguistic or technical shortcomings may lead to the rejection of contributions by the series editors.
A conference report should not exceed a total of 500 pages. Keeping the size within this bound should be achieved by a stricter selection of articles and not by imposing an upper limit to the length of the individual papers.
Editors receive jointly 30 complimentary copies of their book. They are entitled to purchase further copies of their book at a reduced rate. As a rule no reprints of individual contributions can be supplied. No royalty is paid on Lecture Notes in Physics volumes. Commitment to publish is made by letter of interest rather than by signing a formal contract. Springer-Verlag secures the copyright for each volume.

The Production Process

The books are hardbound, and quality paper appropriate to the needs of the authors is used. Publication time is about ten weeks. More than twenty years of experience guarantee authors the best possible service.
To reach the goal of rapid publication at a low price the technique of photographic reproduction from a camera-ready manuscript was chosen. This process shifts the main responsibility for the technical quality considerably from the publisher to the authors. We therefore urge all authors and editors of proceedings to observe very carefully the essentials for the preparation of camera-ready manuscripts, which we will supply on request. This applies especially to the quality of figures and halftones submitted for publication. In addition, it might be useful to look at some of the volumes already published.
As a special service, we offer free of charge LATEX and TEX macro packages to format the text according to Springer-Verlag's quality requirements. We strongly recommend that you make use of this offer, since the result will be a book of considerably improved technical quality.
To avoid mistakes and time-consuming correspondence during the production period the conference editors should request special instructions from the publisher well before the beginning of the conference. Manuscripts not meeting the technical standard of the series will have to be returned for improvement.

For further information please contact Springer-Verlag, Physics Editorial Department V, Tiergartenstrasse 17, W-6900 Heidelberg, FRG

L. Boi D. Flament J.-M. Salanskis (Eds.)

1830-1930: A Century of Geometry

Epistemology, History and Mathematics

Springer-Verlag
Berlin Heidelberg GmbH

Editors

Luciano Boi
Institut für Philosophie, Wissenschaftstheorie,
Wissenschafts- und Technikgeschichte
Technische Universität, Ernst-Reuter-Platz 7
W-1000 Berlin 10, Fed. Rep. of Germany

Dominique Flament
REHSEIS (UPR 318)
27, rue Damesme
F-75013 Paris, France

Jean-Michel Salanskis
Fondements des Sciences, 3, rue de l'Université
F-67084 Strasbourg, France

ISBN 978-3-662-13890-8 ISBN 978-3-540-47058-8 (eBook)
DOI 10.1007/978-3-540-47058-8

Originally published by Springer-Verlag Berlin Heidelberg New York in 1992
Softcover reprint of the hardcover 1st edition 1992

Typesetting: Camera ready by authors/editors using the TEX macro package from Springer-Verlag Berlin Heidelberg GmbH.

58/3140-543210 - Printed on acid-free paper

Foreword

This volume contains a selection of articles based on lectures presented at the international Conference "1830–1930: Un Siècle de Géométrie: de C.F. Gauss et B. Riemann à H. Poincaré et E. Cartan; épistémologie, histoire et mathématiques", which was held in Paris, 18-23 September 1989, at the Institut Henri Poincaré. The initiative was taken by two research groups: the Séminaire d'Épistémologie des Mathématiques of the École des Hautes Études en Sciences Sociales and the group Recherches Épistémologiques et Historiques sur les Sciences Exactes et sur les Institutions Scientifiques of the Centre National de la Recherche Scientifique.

Since the discovery of non-Euclidean geometries by Gauss, Lobatchevsky and Bolyai at the beginning of the last century, many new geometrical structures and theories have been developed: differential Riemannian geometry, projective geometry, algebraic geometry, symplectic geometry, differential topology, etc. These have deeply modified and extended the concept of space and under the influence of mathematicians like Gauss, Riemann, Grassmann, Poincaré, Lie, Hilbert and Cartan, have supported a fertile process of geometrization which has affected in an essential way mathematics, physics, and our understanding of their relations. From the space-time of Minkowski, Einstein, and Weyl to non-Abelian gauge theories, this process of geometrization has continued at an ever-increasing pace in modern times.

While geometry holds a leading position in modern mathematics and physics, epistemology and the philosophy and history of science still seem to concentrate upon the problem of foundations, formal logic and classical analysis, or the epistemological status of quantum mechanics. But this is obviously insufficient. We therefore urgently need to revisit the philosophical problems concerning the status of geometry and to renew our present epistemological trends with some of the rich reflexions developed at the beginning of this century.

The conference took risks by gathering together historians, philosophers, mathematicians and physicists to re-examine the recent history of geometry in the light of the contemporary developments of mathematics and physics, but we think that the talks and discussions showed that this mixture generated more light than heat, as we had hoped.

The book is divided into five sections, which roughly correspond to the thematic organisation of the five days of the meeting. In any case, these parts seem to us to reflect faithfully the various aspects of geometry as they were addressed by the contributors.

The Editors

Acknowledgements

This book, and the conference from which it comes, could not have been possible without the financial support of the following French institutions: the Ministry of Research, Ministry of Foreign Affairs, Centre National de la Recherche Scientifique, Fondation Hugot du Collège de France, École des Hautes Études en Sciences Sociales, Union des Assurances Populaires, Université Paris XI. We are also grateful to the director and staff of the Institut Poincaré, especially to Mrs. Nocton, who saw to it that every possible comfort was provided for those attending the colloquium while in the institute. Finally, we want to express our gratitude to Catherine Harcour, who helped us in many ways, both in the organisation of the meeting as well as in the technical preparation of the proceedings.

Contents

Part 1 Pluralization of Geometry: New Foundations and Continuity of Problems

Part 2 Historical and Epistemological Aspects of the Connexion Between Physics and Geometry

Part 3 Formalism and Intuition

Part 4 The Philosophical Problem of Space

Part 5 Some Insights About Modern Physics

1

Pluralization of Geometry : New Foundations and Continuity of Problems

The Birth of Non-Euclidean Geometry

Christian Houzel

R.E.H.S.E.I.S. 49, rue Mirabeau 75016 Paris

Introduction

In this essay, we try to sketch the history of the discovery and the first developments of non-Euclidean Geometry, from Gauss' initial researches to Riemann's memoir of 1854. Another era begins with Beltrami's *Saggio* in 1868 but we shall not speak about it.

We shall see that the so called ***hyperbolic Geometry*** (that of Lobachevsky) was discovered several times independently during a period which extends from 1816 to 1829. The problem raised by the theory of parallels was as old as Euclid's *Elements* and the mathematicians never ceased their attempts to resolve it. From Ibn al-Haytham (Xth century) on, those attempts took in most cases the form of the proof of Euclid's fifth postulate by ***reductio ad absurdum***; in the XVIIIth century they led to the works by Saccheri and Lambert which include a lot of results of hyperbolic Geometry. But those attempts were always part of a program aiming to justify the Euclidean theory of parallels by proving the fifth postulate; they reached this aim by admitting, explicitly or not, another equivalent postulate.

Saccheri, having deduced that non-Euclidean parallels must be ***asymptotic***, concludes by noting that they should have a common perpendicular at their point of intersection at infinity, which is "repugnant to the nature of the straight line" (*repugnans naturae lineae rectae*, Saccheri 1733, prop. 33). Lambert does not conclude; he only says that the non-Euclidean hypothesis seems difficult to refute and it was probably the reason why he did not publish his *Theorie der Parallellinien* (this work, written in 1766, was published posthumously in 1786). Yet Lambert had discovered several interesting properties of the non-Euclidean Geometry, among which the existence of an absolute standard of length and the fact that the deficiency of the sum of the angles of a triangle with respect to 2 right angles is proportional to the area of the triangle; this property was compared by Lambert to that of a spherical triangle (the sides of which are arcs of great circles): the excess of the sum of the angles is proportional to the area of the triangle. It is probably the first explicit comparison between plane Geometry and spherical Geometry; Lambert then remarks that the strange Geometry he was studying would be the Geometry on an ***imaginary*** sphere (Lambert 1786).

Several other attempts were made after Saccheri and Lambert, the most prominent of which were those of Legendre in the various editions of his well known *Élémens de Géometrie* and in a late work of 1833 (Legendre 1800, 1823, 1833). He rediscovered the existence of an absolute unit of length and the proportionality of the deficiency of the sum of angles of a triangle to the area; the importance of his contribution lies mostly in the fact that it served to Lobachevsky as a point of departure for his own researches.

In the first third of the XIXth century, some mathematicians do not try any more to establish the Euclidean Geometry, *a priori* considered as irrefutable; on the contrary they undertake to build new Geometries. What had changed to bring such a revolution in the mathematical thought? We shall try to answer, partially at least, to this type of question at the light of the mathematical context of the time and the commentary left by these mathematicians.

Let us begin by some remarks:

1) To the end of the XVIIIth century, Geometry had lost the place it occupied before as paradigm of rational certainty. The development of analytical methods and their growing success had shown that other sources of certainty might exist and that rigour was at work in other domains of mathematics. Progressively, pure Geometry was relegated in a marginal position in the mathematical framework of the time, dominated by Analysis; in Lobachevsky's own words

> "with Newton's and Descartes' time, the whole Mathematics, becoming Analytic, walked so rapid steps forward that they left far behind themselves this study without which they already could do and which had ceased to draw on itself that attention which it deserved before." (Lobachevsky 1829).

2). Only the needs of engineers gave again some life to Geometry, with Monge, L.Carnot and Poncelet. This Geometry, practised by engineers, took a new direction with the *Géométrie de position* and projective Geometry; the creation of these new Geometries was independant, but contemporaneous to that of the non-Euclidean Geometry. At the same epoch, the analogy between plane Geometry and spherical Geometry is recognised and systematically worked out, for instance by Gergonne. In his *New Elements of Geometry*, Lobachevsky states plane Geometry and spherical Geometry side by side, in a sort of counterpoint, underlining the properties which are analogous or identical (Lobachevsky 1835-1838).

3) Newtonian Physics made use of Geometry and that use gradually modified the conception of Geometry itself. Greek Geometry was the science of geometrical *figures*; the concept of space was unknown to it. With the physical use of Geometry, this part of mathematics tends to become the science of space. About the end of the XVIIIth century, that evolution was accomplished; J.Bolyai's essay is termed *scientia spatii absolute vera*.
   By the way, we shall see that Lobachevsky and Bolyai both build their non-Euclidean Geometry in 3 dimensions. On the contrary, the previous attempts to prove the postulate of parallels were exclusively developed in the framework of *plane* Geometry.

4) Philosophical positions conscious or not, coming from the Enlightment, led some mathematicians to abandon the Platonician view of Geometry, based on eternal essences known by intellectual intuition. They tended to replace it by a more empiricist view, based on some mechanist materialism. It is perfectly clear in the case of Lobachevsky and it may be also said, with some nuance, in the case of Gauss.
   The current interpretation of Kantian philosophy had an influence in a different di-

rection; nevertheless, the critical philosophy had opened new horizons to the reflexion on science and I think that it helped the liberation of minds at the beginning of the XIXth century.

Gauss and His Correspondants

After his own testimony, Gauss was interested in the foundations of Geometry at least since the age of 17 in 1794. He studied in Göttingen, where this interest must have been reinforced by his professor A.Kästner (1719-1800). Kästner occupied the chair of mathematics at the university of Göttingen from 1755 to 1800; his favourite subject was precisely the foundations of Geometry and the theory of parallels. Among his other students, several have studied the theory of parallels. For instance Georg Klügel (1732-1812) who wrote, as a thesis, a *Conatuum praecipuorum theoria parallelarum demonstrandi recensio* (1763) with the analysis of about thirty attempts of proofs; Klügel mentions that the fifth postulate might well be undecidable without some recourse to experiment. Farkas Bolyai (1775-1856), Janos' father was another student of Kästner and he proposed a proof for the fifth postulate, which Gauss criticised. Still another of Kästner's students was Bartels (1769-1836), who taught mathematics at the university of Kazan; Lobachevsky was one of his students. At the date July 27, 1797, one can read in Gauss' *Tagebuch* "Plani possibilitatem demonstravi "; later, in September 1799 "In principiis Geometriae egregios progressus fecimus ". On December 16 of the same year 1799, he wrote to his friend F.Bolyai

> "Zwar bin ich auf manches gekommen, was bei den meisten schon für einen Beweis gelten würde, aber was in meinen Augen so gut wie NICHTS beweist, z.B. wenn man beweisen könnte, dass ein geradliniges Dreieck möglich sei, dessen Inhalt grösser wäre als eine jede gegebene Fläche, so bin ich im Stande die ganze Geometrie völlig strenge zu beweisen. "

And he explains that the existence of a triangle with an arbitrary large area cannot be taken as an axiom of Geometry, for the contrary might as well be true.

In 1804, F.Bolyai had sent his own proof of the parallel postulate to Gauss; it contained an attempt to prove the existence of equidistant straight lines, but Gauss pointed out Bolyai's error. In a notebook of 1805, containing however things of an unknown date, Gauss considers a Geometry in which the locus of the points at a given distance from a given straight line is not necessarily a straight line. A note by Schumacher tells us that, by 1808, Gauss had proved that the non-Euclidean hypothesis implied the existence of an absolute unit of length; we know that Lambert and Legendre had come before to the same result. Gauss himself comments about Legendre's arguments in a letter dated of April 11, 1816 and addressed to a professor at Marburg named Gerling (1788-1864); he says that the existence of an absolute unit of length, for instance the length of the side of an equilateral triangle with a given angle, is by no means absurd.

It is certain that at this time (1816), Gauss had acquired the conviction that a consistent Geometry, different of the Euclidean one, was possible; he called this Geometry "anti-Euclidean". His pupil Friedrich Ludwig Wachter (1792-1817) worked on anti-Euclidean

Geometry; he was teaching in the Gymnasium of Dantzig. In a letter to Gauss (December 12, 1816), he says that the Geometry of a sphere of infinite radius in anti-Euclidean Geometry coincides with the usual Euclidean plane Geometry; this fact was then rediscovered by Lobachevsky (who called ***horosphere* - *predel'naja sfera*** - the sphere of infinite radius) and by J.Bolyai.

We know more on Gauss' point of view from a letter he wrote to Olbers (April 28, 1817):

> "Ich komme immer mehr zu der Überzeugung, dass die Nothwendigkeit unserer Geometrie nicht bewiesen werden kann, wenigstens nicht vom *menschlichen* Verstande noch für den menschlichen Verstand. Vielleicht kommen wir in einem andern Leben zu andern Einsichten in das Wesen des Raums, die uns jetzt unerreichbar sind. Bis dahin müsste man die Geometrie nicht mit der Arithmetik, die rein a priori steht, sondern etwa mit der Mechanik in gleichen Rang setzen. "

So, for Gauss, the nature of space was not *a priori* knowable as the nature of number was and Geometry, as a science of physical space, partook with Mechanics some empirical character.

At that time, some other people began to think a non- Euclidean Geometry possible. In a letter to Gerling (August 25, 1818), Gauss expresses his gladness to see the latter talking of such a possibility; but he adds "Aber die Wespen deren Nest Sie aufstören, werden Ihnen um den Kopf fliegen. " We will see other testimonies of this kind of wisdom, almost shyness, on the part of Gauss.

Through the same Gerling, a professor of Law recently called to Marburg and named Ferdinand Karl Schweikart (1780-1857) sent to Gauss a leaflet where he stated the principal properties of an *astral Geometry*. This Schweikart had published in 1807 a *Theorie der Parallellinien* with an attempt to prove the fifth postulate; then he went to the university of Kharkov where he had the opportunity to think anew to the problem and to convince himself that his proof was wrong and that a Geometry without the fifth postulate was possible. In his astral Geometry the sum of the angles of a triangle is less than 2 right angles and the deficiency is an increasing function of the area of the triangle. The height of an isosceles rectangle triangle remains bounded and Schweikart names the least upper bound the *Constant* of the Geometry; the Euclidean case corresponds to the limit case of a Constant infinite. Gauss answers (March 16, 1819) approving Schweikart's work and saying that he himself has developed that Geometry as far as to be able to solve any problem in it if the Constant C is given; the deficiency of the sum of the angles of a triangle is proportional to the area (as Lambert and Legendre had proved before) and the area remains bounded, the least upper bound being

$$\frac{\pi \mathrm{CC}}{(\mathrm{loghyp}(1+\sqrt{2}))^2},$$

attained by a triangle with *asymptotic* sides and vertices rejected to infinity. In a letter to Olbers (May 3, 1827), Gauss expresses how he agrees with Fourier about a several mathematical points; in particular about the impossibility to base the Geometry *a priori*. We can read Fourier's point of view about the foundations of Geometry in a discussion held at the *École Normale* (25 pluviôse an III); Fourier took *distance* as a primitive notion and he then defined the sphere, the plane (as the locus of points equidistant from two given points), the circle and the straight line. We may remark that Lobachevsky's view of the principles of Geometry seems very close to Fourier's (Lobachevsky 1829, 1835- 38). Gauss alludes to the publication of his researches on the foundations of Geometry in a letter to Bessel (January 27, 1829); but he adds

> "vielleicht wird diess auch bei meinen Lebzeiten nie geschehen, da ich das Geschrei der Böotier scheue, wenn ich meine Ansicht ganz ausprechen wollte."

In another letter, to Schumacher (July 12, 1831), he communicates the formula for the length of a circle in the non-Euclidean Geometry:

$$\frac{\pi}{2}k(e^{\frac{r}{k}} - e^{\frac{-r}{k}}),$$

where r is the radius and k is a constant of which we know, after experiment, that it must be very large (*ungeheuer gross*); in the Euclidean Geometry, this constant is infinite. From that period, we have some fragments written by Gauss on the beginnings of non-Euclidean Geometry. Gauss says that a straight (half-) line AM is *parallel* to another one BN (situated in the same plane) when AM and BN do not meet but any straight (half-)line AQ between AB and AM does meet BN. The first properties he proves are the following ones: 1) the parallelism does not depend on the position of the points B and A on their respective lines; 2) it is *symmetric* in the sense that BN is also parallel to AM; 3) when two (half-)lines (in the same plane) are parallel to a third one, they are parallel between them. Then Gauss defines *corresponding points* A and B on two parallel (half-)lines AM and BN as points such that the angles MAB and ABN be equal; the perpendicular to AB at the middle of AB is then parallel to AM and BN. If A' and B' are corresponding points, respectively on AM and BN, one has $AA' = BB'$; when three (half-)lines AM, BN and CN are parallel, with A corresponding to both B and C, then B and C are corresponding points. Gauss calls *Trope* the locus of the corresponding points to a given point A on all the parallels to a given straight (half-)line AM (in the plane); this curve was called *horocycle (predel'naja kruga, predel'naja krivaja)* by Lobachevsky.

Taurinus

Schweikart encouraged his nephew Franz Adolf Taurinus (1794-1874) to pursue his researches. Taurinus was the first to publish positive results about the non-Euclidean Geometry; his letters to Gauss are lost but we have a letter from Gauss to Taurinus (November 8, 1824) where the term *non-Euclidean Geometry* appears for the first time. Gauss talks about his own researches and mentions the absolute unit of length; he adds "Aber mir deucht, wir wissen, trotz der nichtssagenden Wort-Weisheit der Metaphysiker eigentlicht zu wenig oder gar nichts über das wahre Wesen des Raums ", thus echoing what he wrote to Olbers some years before. Taurinus published a *Theorie der Parallellinien* in 1825 and the next year a corrected and completed version. The existence of an absolute unit of length in the non-Euclidean Geometry seemed to him a sufficient reason to reject it. Nevertheless, he developed the theory up to a point where he obtained the trigonometrical formulae, the length and the area of a circle, the area and the volume of a sphere. To get the trigonometrical formulae, he uses the analogy, rediscovered by him, between spherical Geometry and non-Euclidean Geometry: one obtains the non-Euclidean Geometry by considering a sphere with a purely imaginary radius (cf. Lambert) and Taurinus qualifies it by the term *logarithmisch- sphärischen Geometrie* . On a real sphere of radius k, for a triangle ABC, with sides (arcs of great circles) a, b, c (a opposed to A etc.) one has the fundamental formula of trigonometry

$$\cos\frac{a}{k} = \cos\frac{b}{k}\cos\frac{c}{k} + \sin\frac{b}{k}\sin\frac{c}{k}\cos A .$$

When k is replaced by ik ($i = \sqrt{-1}$), this formula becomes

$$\mathrm{ch}\frac{a}{k} = \mathrm{ch}\frac{b}{k}\mathrm{ch}\frac{c}{k} - \mathrm{sh}\frac{b}{k}\mathrm{sh}\frac{c}{k}\cos A$$

with hyperbolic functions instead of the circular ones, and a change of sign.

Let k tend to ∞; at the limit, one obtains $a^2 = b^2 + c^2 - 2bc\cos A$ which the usual formula for plane trigonometry in the Euclidean Geometry. In the case of an equilateral triangle, $a = b = c$ and $\cos A = \frac{\mathrm{ch}(a/k)}{1+\mathrm{ch}(a/k)}$ which is always $> \frac{1}{2}$ so that the angle A is $< \frac{\pi}{3}$; the limit $k = \infty$ gives $\cos A = \frac{1}{2}$ and $A = \frac{\pi}{3}$ (Euclidean Geometry). In the spherical trigonometry, the other formulae are deduced by duality, using the so called *polar triangle* (Chemla 1988); this gives

$$\cos A = -\cos B\cos C + \sin B\sin C\cos\frac{a}{k}$$

which becomes, after replacing k by ik,

$$\cos A = -\cos B\cos C + \sin B\sin C\,\mathrm{ch}\frac{a}{k} .$$

When $A = 0$, that is when the vertex A is rejected to infinity and the sides BA and CA are *parallel*, the first member equals 1; let suppose in addition that $C = \frac{\pi}{2}$ so that $\cos C = 0$ and $\sin C = 1$. The angle B takes a value β given by $1 = \sin\beta\,\mathrm{ch}\frac{a}{k}$; this angle was later called *angle of parallelism* and Lobachevsky noted it by $F(a)$ or $\Pi(a)$.

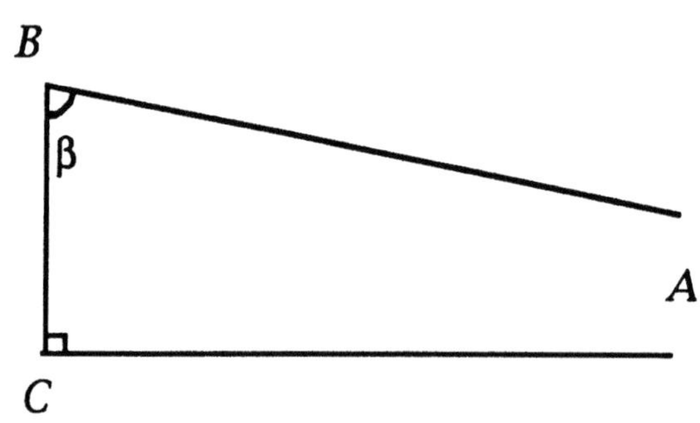

For $\beta = \frac{\pi}{4}$, by adding the symmetric triangle $A'BC$ with respect to BC, one obtains the limit case of the rectangular isosceles triangle considered by Schweikart; so Schweikart's Constant C (not to confuse with the angle C) is given by $\mathrm{ch}\frac{C}{k} = \sqrt{2}$ or $C = k\log(1 + \sqrt{2})$. Taurinus explains that the area of a triangle is proportional to the deficiency of the sum of its angles with respect to π and bounded by a least upper bound πk^2 (cf. Gauss' answer to Schweikart's work). The length of a circle of radius r is $2\pi k \mathrm{sh}\frac{r}{k}$ and its area is $2\pi k^2(\mathrm{ch}\frac{r}{k} - 1)$; the area of a sphere of radius r is $4\pi k^2 \mathrm{sh}^2\frac{r}{k}$ and its volume is $2\pi k^3(\mathrm{sh}\frac{r}{k}\mathrm{ch}\frac{r}{k} - \frac{r}{k})$.

Gauss never expressed any opinion about that work by Taurinus, who was deeply discouraged by this attitude; he decided to burn up his book, which became a bibliographic rarity.

Janos Bolyai

F.Bolyai's son Janos was born in 1802; he received an education in the Royal College for Engineers in Vienna, where he had the opportunity to discuss the foundations of the parallel theory with his friend Carl Szász. Then he went to Temesvár in Transilvania, as a young officer in the Austrian army; there he discovered the non-Euclidean Geometry not later than 1823. From that year dates a letter to his father (November 3) where he announces that he *has created a new universe from nothing*; an abstract of his work was sent by him in 1825 and a first version of the complete work in 1829. The definitive version was finally published in 1832 as an *Appendix* to his father's *Tentamen juventutem studiosam in elementa matheseos purae, elementaris et sublimioris, methodo intuitiva, evidentiaque huic propria introducendi* (J.Bolyai 1832); it had only 26 pages.

The definition of parallels and the first propositions concerning them in Bolyai's paper are almost the same as in Gauss' fragments; they are established by similar methods, principally based on the use of the so called *Pasch axiom.* In addition Bolyai proves the transitivity of the parallelism for straight lines not in the same plane, but in the 3-dimensional space (§7), and the same property for corresponding points on parallel lines (§10). The locus of the points of space corresponding to a given point $\mathfrak{A}$ on the lines of the pencil of parallels to a given $\mathfrak{AM}$ is a surface called an F surface by Bolyai (Lobachevsky's horosphere) and the intersection of an F surface with a plane through $\mathfrak{AM}$ is called an L line (Lobachevsky's horocycle). The rotation of L around $\mathfrak{AM}$ generates F, so that $\mathfrak{AM}$ is called an axis for F (§11); every straight line parallel to $\mathfrak{AM}$ is as well an axis for F (§12). Bolyai proves an analogue of Euclid's fifth axiom for *planes* and *dihedral angles* with parallel edges (§9); then the fifth axiom is true in an F surface for L lines and angles of such lines and the Geometry of an F surface is the usual Euclidean Geometry when L lines are taken as straight lines (§21; cf. what has been said of Wachter's work).

Given two parallel lines $\mathfrak{AM}$ and $\mathfrak{BN}$, the sum of the interior angles made by the tranversal $\mathfrak{AB}$ is either $2R$ or $< 2R$, where R denotes a right angle. If the first cas occurs for *one* pair of parallel lines, it occurs for every pair (§13); the corresponding Geometry is the usual Euclidean Geometry and Bolyai calls it the system Σ. In the second case, one has a new system of Geometry, called S by Bolyai (§14-15); it corresponds to Saccheri's hypothesis of the *acute angle* and Saccheri had proved that this hypothesis is true for every triangle if it is true for only one (Saccheri 1733, prop.7; this fact was rediscovered by Legendre to whom it is generally attributed). In the system Σ an F surface is a plane perpendicular to its axis and an L line is a straight line (§16). On the contrary, in the system S an F surface and an L line are curved but they have a property of *uniformity* (a transitive group of deplacements, §17); the intersection of an F surface by a plane *not containing* any axis is a circle (§18) and the tangent to an L line in a point $\mathfrak{B}$ is the perpendicular to the axis $\mathfrak{BN}$ passing through $\mathfrak{B}$ (§19). Bolyai then considers *equidistant* L lines $\mathfrak{AB}$, $\mathfrak{CD}$ and the ratio X of corresponding arcs on such lines (between the two axes $\mathfrak{AM}$ and $\mathfrak{BN}$ common to both lines) as a function of the distance x (§22-24); it is easily seen that X is an exponential function of x, in Bolyai's terms $Y = X^{\frac{y}{x}}$. In the system Σ, X is a constant equal to 1, but in the system S, $X > 1$ is variable and Bolyai deduces from it that, if $\mathfrak{EF}$ is a third parallel, one can always find such an x that the arc $\mathfrak{CDF}$ be equal to $\mathfrak{AB}$ so that the strips between $\mathfrak{AM}$, $\mathfrak{BN}$ and $\mathfrak{AM}$, $\mathfrak{EP}$ respectively are *congruent* "quod singulare quidem est, sed absurditatem ipsius S evidenter non probat" as Bolyai adds.

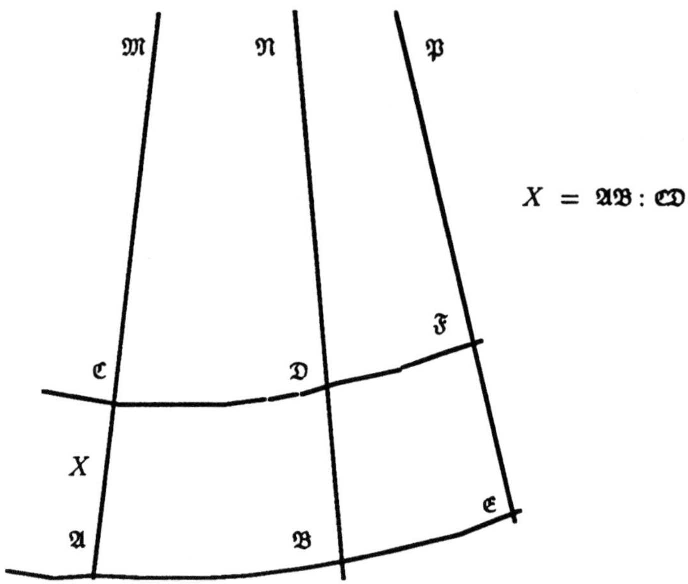

The next theorem is quite remarkable and serves as a base for Bolyai's trigonometry: the sines of the angles of a triangle are proportional to the lengths of the circles with respective radii the opposite sides (§25). Bolyai establishes this result for the case of a right angle triangle $\mathfrak{ABC}$ (with a right angle in $\mathfrak{B}$) by associating to it an L form triangle $\mathfrak{EDC}$ in the F surface through C with axis the perpendicular $\mathfrak{EAM}$ to the plane of the triangle in the point $\mathfrak{A}$. The circle with radius $\mathfrak{AC}$ is the intersection of this F surface with the plane of the triangle so its length is 2π.arc $\mathfrak{EC}$; on the other hand $\sin \mathfrak{CAB} = \sin \mathfrak{DEC} = \mathfrak{DC} : \mathfrak{EC}$ for the Geometry of the F surface is the Euclidean one. From this result, Bolyai deduces the fact that the sines of the angles of a *spherical* triangle are proportional to the sines of the opposite sides so that the spherical trigonometry is the same for the system S as for the system Σ (§26). Then Bolyai is able to give an expression for the angle of parallelism u corresponding to a distance y; with his notations $Y = \cot.\frac{1}{2}u$ (where cot. stands for *cotangent*; §27- 29). He also proves that the product i (not to confuse with $\sqrt{-1}$) of $\tan u$ by the length of the circle of radius y does not depend on y; the value of this product is obtained by letting y tend toward 0 and, using the fact that Y is an exponential with respect to y, one finds that the length of the circle of radius y is $\pi i(Y - Y^{-1})$ where $i = \frac{y}{\text{log.nat.}Y}$ and $Y = e^{y/i}$ (§30). Bolyai's paper continues with the formulae of plane trigonometry for a right angle triangle in the system S (§31), the expression of the element of length of an arbitrary plane curve and of the element of a plane area bounded by a curve (with applications to an L curve and to a circle, §32). He gives also the area and the volume of a sphere. Bolyai proves then that, when i tends to ∞, the limit formulae are those of the system Σ and he says "Num Σ aut S aliquod *reipsa* sit, indecisum manet." The end of the paper is devoted to some geometrical constructions in the system S, the proof that the area of a triangle is proportional to the deficiency of the sum of its angles with respect to $2R$ and a very curious geometrical construction of the quadrature of some circles in the system S (depending on arithmetical properties of the radius).

When F.Bolyai sent to Gauss his son's *Appendix* at the beginning of 1832, Gauss replied that he could not praise that work without praising himself, for he had done such a work (March 6, 1832); he adds that he was stupefied to see that the path taken by J.Bolyai and the results to which he was led coincided almost entirely with his own meditations. Then Gauss makes some remarks about the terminology; he proposes to call *parasphere* the F surface, *paracycle* the L line, *hypercycle* the locus of the points in a plane at a given distance from a right line. F.Bolyai was glad of Gauss' answer, but Janos was deeply depressed by it, for he could not admit that another man had discovered that new universe he was so proud of. After that, J.Bolyai did not publish anything but he did some work on the analogy between spherical and non-Euclidean trigonometry, on the independance of Euclid's axiom of parallels and on the computation of the volume of a tetrahedron in non-Euclidean Geometry (Stäckel 1902). When Lobachevsky published a part of his researches in German in1840, J.Bolyai believed that it was written by Gauss under a pseudonym in order to be harmful to him. He died in 1860.

N. Lobachevsky

Nikolai Ivanovich Lobachevsky was born in 1793; he studied in the University of Kazan, where he entered in 1807. His professor of mathematics was Bartels, a former friend of Gauss and the teaching was a high level one. He became Aspirant (1811), then Adjunct (1814) and Professor (1816) in the same University; since 1827 he was Rector.

He was led to work, as early as 1816, on the foundations of Geometry by the necessities of teaching (Laptev 1951). We have a course *"Geometry "* dated of 1823, which attests that, at this date, he was still convinced of the truth of Euclid's axiom. After stating it, he adds

> "Up to now, one as has not succeeded to find a rigourous proof of that truth. Those which were given may be named only explanations (pojasnenijami), but do not deserve to be considered, in the full sense, mathematical proofs. "

But in 1826, he presented to the Section of Physico-mathematical Sciences of the University of Kazan an *"Exposition succincte des principes de la Géométrie avec une démonstration rigoureuse du théorème des parallèles "* (in French; the text is lost), which constitutes the first part of his *Elements of geometry* of 1829 (*O načalax geometrii*) and where the possibility of the non-Euclidean Geometry is considered. Lobachevsky's point of view on Geometry is apparently a development of d'Alembert's one. He starts with the primitive notions of *solid body* and of *contact* of such bodies; from that he derives the notions of ***surface*** and ***curve***, the dimension being explicitly tied to topological properties. Then comes the notion of distance, which allows to define the sphere (*sfera*, a surface, distinguished from the ball -šar, which is a solid body), the plane, the circle and the right line in the manner of Fourier (see *supra*). The sum of the angles of a triangle cannot be $> \pi$, but it can be $= \pi$ or $< \pi$;

> "Both may be admitted without any contradiction in consequence, whence come two Geometries: the first, *usual*(*upotrebiteljnaja*) up to now on account of its simplicity, in fact concording with every measures; the other, *imaginary* (*voobražajemaja*), more general and therefore difficult in its computations, admits the possibility of the dependance of lines on angles. "

In the imaginary Geometry, Lobachevsky defines parallel straight lines in the same way as Gauss and Bolyai. If the distance of a point to a straight line is a , the parallel from that point to this line makes an angle $F(a)$ with the perpendicular drawn from the point to the line; the angle of parallelism $F(a)$ steadily decreases from $\frac{\pi}{2}$ (when $a = 0$) to 0 (when $a = \infty$) and Lobachevsky defines a continuation of this function for negative values of a by the law $F(-a) = \pi - F(a)$. He introduces the horocycle and the horosphere and asserts that the Geometry of the horosphere is the usual Euclidean plane Geometry; in particular, one can develop the ordinary trigonometry on that surface.

To get the imaginary trigonometry, Lobachevsky considers a right angle triangle ABC (with a right angle in C). He calls $F(a')$ and $F(b')$ the angles respectively opposite to A and to B and he proves the formulae

$$F(c + a') + F(b') = F(a),$$
$$F(c - a') = F(b') + F(a),$$
$$F(b' - a) + F(b' + a) = \frac{\pi}{2}$$

by considering the perpendiculars to AB at the points D, D' of AB such that $AD = AD' = a'$ (they are parallel to CA) and the parallel to them drawn from B; for the last formula (with the hypothesis $b' \geq a$), one draws the perpendiculars NO and IK respectively to BC and CA , with N and I respectively on BC and CA produced, such that $CN = b' - a$ and $AI = a'$ (they are parallel to BA and one also draws the parallel to them from C). So one has $2F(a) = F(c - a') + F(c + a')$, $2F(b') = F(c - a') - F(c + a')$. To each length a, Lobachevsky associates another one α by $F(\alpha) + F(a) = \frac{\pi}{2}$; from the preceding he gets $2F(\beta') = F(a' - c) + F(a' + c)$ and he concludes that there exist anothe right angle triangle with hypotenuse a', one side β' and the opposite angle $F(c)$. The other side of this triangle is b and the opposite angle is $F(\alpha)$, but this results from another construction, using the third dimension and a spherical triangle associated to the plane one. It is obtained by drawing BB' perpendicular to the plane ABC, CC' and AA' parallel to BB' and by considering a sphere with center in A; the straight lines AA', AB and AC meet this sphere at the vertices of the spherical triangle, which Lobachevsky calls (with an abuse) A', B and C. The spherical triangle has a right angle in B , its hypotenuse $A'C$ is $F(b)$ and the sides of the right angle are $BA' = F(c)$ and $BC = F(a')$; the angle C is $F(a)$ and, as the sum of the dihedral angles of three planes meeting two by two along parallel lines is π (this is equivalent to saying that the Geometry of a horosphere is Euclidean), the other angle A' is $\frac{\pi}{2} - F(b')$. Lobachevsky then introduces a third rigth angle triangle LMN, on a horosphere with axis on the line CA; the vertex N is on CA, $A'L$ is parallel to CA and the parallel to CA through the right angle M passes through the orthogonal projection D of A' on AB. As the radius AC of the sphere tends to ∞ the triangles $A'BC$ and LMN differ less and less; using the former relations for the plane triangle with hypotenuse a' and the Euclidean trigonometry for LMN, Lobachevsky gets $F(b) - F(c) = F(a' + \alpha)$ and $\lim \frac{F(a'+\alpha)}{F(a')} = \text{tang}\frac{1}{2}F(\alpha)$. From this equation one sees that the function $\alpha \mapsto \text{tang}\frac{1}{2}F(\alpha)$ statisfies the functional equation of the exponential; Lobachevsky chooses the unit of length in such a way that $\text{tang}\frac{1}{2}F(\alpha) = e^{-\alpha}$. At this point, Lobachevsky is able to write the relations between the sides and the angles of a plane right angle triangle:

$$\sin F(c) = \sin F(a)\sin F(b) = \text{tang}F(a')\text{tang}F(b'), \cos F(b) = \cos F(c)\cos F(a')$$

etc. From these relations, he deduces the formulae of the spherical trigonometry, which are the same as in the Euclidean Geometry (cf. Bolyai's work), then the formulae of the plane non-Euclidean trigonometry:

$$\text{tang}F(a)\sin A = \text{tang}F(b)\sin B, \cos A\cos F(b)\cos F(c) + \frac{\sin F(b)\sin F(c)}{\sin F(a)} = 1$$

etc. For a very small triangle, one gets approximately $b\sin A = a\sin B$, $a^2 = b^2 + c^2 - 2bc\cos A$, etc., the formulae of Euclidean trigonometry.

The second part of the *Elements of Geometry* was published in 1830 and it does not come from the paper of 1826. Lobachevsky begins by saying

> "The theory of parallels explained by us supposes that the lines and the angles are in a dependance which, as we shall see later, nobody is in a position to prove whether it is found in nature or not. "

In order to test the fit of his theory to natural observations, he considers a right angle triangle with vertices at a fixed star with parallax $2p$ and two positions of the Earth diametrally opposite on its orbit; if the diameter of the orbit is a, one has $F(a) > \frac{\pi}{2} - 2p$ and Lobachevsky deduces from this inequality that $a < \text{tang}2p$. According to the data published by Assas-Montdardier in the *Connaissance des temps pour l'an 1831* (Paris 1828), the parallax of 29 Eridan is 2", that of Rigel 1"43 and that of Sirius 1"24; the largest parallax 2" gives $a < 0.000009696$ so that the distance Earth-Sun is very small with respect to the natural unit of the imaginary Geometry. Let 2ω be the deficiency of the sum of the angles of the triangle with respect to π; the angle at the fixed star is $2p - 2\omega$ and the imaginary trigonometry gives $\sin^2(p - \omega) = \sin 2p - \cos^2 p \cot^2 F(\frac{a}{2})$ whence $\cot^2 F(\frac{a}{2}) < \frac{\sin^2 p}{\cos 2p}$ for any parallax p. Replacing $\cot^2 F(\frac{a}{2})$ by $\frac{\sin^2 p'}{\cos 2p'}$ where $p' < p$ is another parallax, Lobachevsky finally gets the inequality $\omega < 2p\sin^2\frac{x}{2}$ with $\sin x = \frac{\sin p'}{\sin p}\sqrt{\frac{\cos 2p}{\cos 2p'}}$; with $p = 1$" (29 Eridan) and $p' = 0$"62 this gives $2\omega < 0$"43; for an isosceles right angle triangle with two sides equal to a the deficiency should be < 0"0000037273 (Lobachevsky's erroneously writes 0"000372). Lobachevsky concludes

> "After that one may imagine, so small is that difference on which our theory of parallels is based, that the precision of all the computations of the usual Geometry

justifies the admitted principles of the latter and allows to regard them as rigourously proved. Nevertheless one cannot escape to the seduction of Mr.Laplace's opinion that the stars visible to us and the Milky Way belong to just one assembly of celestial astral bodies, similar to those we perceive as weakly scintillating spots in the constellations of Orion, Andromede, the Capricorn and others. So, not to mention the fact that in imagination the space may be produced without limit, Nature itself shows us such distances, with respect to which even the distances from our Earth to the fixed stars vanishes to smallness. "

The remainder of Lobachevsky's paper is devoted to the computation of lengths of curves, areas of curved surfaces and volumes in the non-Euclidean Geometry. He begins by writing the equation of a straight line (linear with respect to $\cos F(x)$ and $\sin F(x)\cos F(y)$ where x and y are the cartesian coordinates). The length of an arc of circle is the common limit of the lengths of polygons inscribed and circumscribed; using the non-Euclidean trigonometry one gets $\alpha\cot F(r) = \frac{1}{2}\alpha(e^r - e^{-r})$ where r is the radius and α is the angle of the arc (from the center). On the other hand, this length is $\alpha.s$ where s is the radius of the circle measured along a horocycle in a horosphere; so $s = \cot F(r)$. For an arbitrary plane curve, the element of arc is

$$ds = \sqrt{dy^2 + \frac{dx^2}{\sin^2 F(y)}} = \sqrt{dr^2 + d\phi^2 \cot^2 F(r)}$$

(x, y cartesian coordinates; r, ϕ polar coordinates). Two arcs s, s' of horocycles limited by two common axes and distant of t along the axes are related by $s' = se^{-t}$.

Lobachevsky then studies the area of a triangle and finds it proportional to the deficiency of the sum of the angles. The area of the circle, common limit of the areas of polygons inscribed and circumscribed, is $\pi(e^{\frac{r}{2}} - e^{-\frac{r}{2}}) = 4\pi\cot^2 F(\frac{r}{2})$. For an arbitrary curve, the element of area under the curve and between two infinitely near ordinates of abscisses x and $x + dx$ is $dS = dx\cot F(y)$; Lobachevsky applies that to a segment of a circle, with the limit case of a horocycle. He also computes plane areas in other systems of coordinates.

The area of a curved surface is defined as the limit of those of polyhedra obtained by triangulating the surface. The element of area is written

$$d^2S = dxdy\frac{1}{\sin Z}\sqrt{(\frac{\partial z}{\partial x})^2 + \frac{1}{\sin^2 Y}(\frac{\partial z}{\partial y})^2 + \frac{1}{\sin^2 Y \sin^2 Z}}$$

with $Y = F(y)$ and $Z = F(z)$. It is applied to the area of an equatorial strip on a sphere; a change in the variable of integration leads to a curious formula for the periods of elliptic integral of the two first kinds and to the integral.

$$\frac{\pi R}{4\sin R} = \int_0^\infty \frac{(e^{2x} - 1)e^x x dx}{e^{4x} + 2e^{2x}\cos 2R + 1}$$

The computations of volumes are extremely clever. Lobachevsky begins by the volume limited two equidistant horospheres, at a distance c, and a conical surface generated by segments of common axes; if the area cut out of one horosphere by the conical surface is equal to S, the volume is $\frac{1}{2}S(1 - e^{-2c})$. From this he deduces the volume limited by four planes intersecting along parallels and a cylindrical surface with base a horocycle in one of the planes and generators perpendicular to this plane; then the volume of a pyramid limited by three planes intersecting along parallels and a right angle triangle in a plane orthogonal to one of these lines. The volume of a right circular cone with vertex at infinity is $\pi\log\frac{e^b+e^{-b}}{2}$ if b is the radius of the base; this allows to compute the volume of a "conical envelop" of infinite height, which serves to compute the volume of a right circular cone of finite height x (equal to $\pi(c\cos A - x)$ if c is the length of one generator and A the angle at the vertex between the height and the generator) or, more generally, of any solid of revolution. For the ball of radius c, the volume is $\frac{\pi}{2}(e^{2c} - e^{-2c} - 4c)$. For arbitrary solids, the element of volume is

$$d^3P = \frac{dxdydz}{\sin Y \sin^2 Z} = \frac{1}{4} d\omega d\theta dr \cos\theta (e^r - e^{-r})^2$$

where $Y = F(y)$, $Z = F(z)$ and r, θ, ω are polar coordinates. The computation of the volumes of pyramids or cones in various systems of coordinates leads Lobachevsky to several transformations or exact values of definite integrals; some of them are reduced to *Lobachevsky's function* $\phi(x) = -\int dx \log\cos x$.

Lobachevsky's paper is written in a very concise style and it is difficult to understand. An anonymous recension of it appeared in 1834 in *The Son of Fatherland (Syn ote čestva)*, a journal of Saint-Petersbourg; the author makes fun of Lobachevsky, while admitting that he did not understand anything. This recension is an example of the "clamour of the Beotians" feared by Gauss; it came probably from the circle of Ostrogradsky, who formulated a completely negative judgement about Lobachevsky's work in 1832. But Lobachevsky was not discouraged; on the contrary, he published a new paper *Imaginary Geometry (Voobražajemaja geometrija)* where he starts from a system of three fundamental relations between the angles and the sides of a right angle triangle in the imaginary Geometry to deduce the whole of non-Euclidean trigonometry, where the sum of the angles of a triangle is $< \pi$. He notes that the equations of this trigonometry can be identified with those of spherical trigonometry with imaginary sides; this analytical treatment seems to give a firm ground to the non-Euclidean Geometry and Lobachevsky say's (in the French text of the paper, published in 1837 in Crelle's Journal)

> "Dans la théorie rien ne s'oppose à admettre que la somme des angles d'un triangle rectiligne soit moindre que deux angles droits ; ... L'hypothèse de la somme des angles d'un triangle moindre que deux angles droits ne peut avoir d'application que dans l'analyse, puisque les mesures directes ne nous montrent pas dans la somme des angles d'un triangle la moindre déviation de deux angles droits. "

It is precisely to analytical applications that the remainder of the paper and its continuation *Application of the imaginary Geometry to some integrals (Primenenije voobražajemoj geometrii k nekotorym integralam*, 1836) are devoted; Lobachevsky systematically explores the way to transform or to compute definite integrals interpreted as areas or volumes in the non-Euclidean Geometry, using various systems of coordinates. In the second paper, he shows that all these integrals may be computed by purely analytical methods, but it is much more complicated without the aid of Geometry; the concordance of geometrical and analytical results give a new support to the non- Euclidean Geometry.

From 1835 to 1838, Lobachevsky published in the *Scientific Memoirs of the University of Kazan* a complete exposition of his Geometry under the title *New Elements of*

Geometry with a complete theory of Parallels (Novyje načala geometrii s polnoj teoriej paralleljnyx). In the introduction, he explains Legendre's and L.Bertrand's works and criticise them. He states his own point of view:

> "In the nature we properly know only the movement, without which the sensory impressions are impossible. So, every other concepts, for instance, the Geometrical ones, are artificially produced by our intellect, taken from the properties of movement; and that is why the space in itself, separately, for us does not exist. In consequence there cannot be any contradiction in our intellect when we admit that certain forces in nature follow one Geometry and others another particular one of their own. "

He explains that idea by the example of Newton's law in 1 r^2 and its relation to the area $4\pi r^2$ of the sphere (spherical propagation); the link, so established, between Geometry and Dynamics is very remarkable for that time. The six first chapters of the *New Elements* contain a detailed explanation of the absolute Geometry on the base of *solid bodies* and *contact* as primitive ideas; then comes the theory of parallels in the Euclidean and in the non-Euclidean Geometry (chapters VII-VIII) and the trigonometry, plane and spherical (chapters IX-XI). The last two chapters contain numerical resolutions of triangle in the plane Euclidean and in the spherical Geometries. Lobachevsky's presentation is *not explicitly axiomatic*; he starts from ideas drawn from sensory intuition about solid bodies and the first notions have a marked topological character.

In order to make known his work, Lobachevsky published in 1840 a rather clear explanation of his new Geometry in booklet written in German and titled *Geometrische Untersuchungen zur Theorie der Parallellinien*. Gauss read this work and mentioned it in letters to Gerling (February 4 and 8, 1844); he was so interested as to learn russian in order to be able to read the other works of Lobachevsky. Another mention of Lobachevsky by Gauss is found in some letters to Schumacher; Gauss is very laudatory, speaking of Lobachevsky's masterly way in true geometrical spirit (*...eine meisterhafte Art in ächt geometrischen Geiste*, November 28, 1846). Lobachevsky's work became known to the world through these letters when they were published in the Correspondance between Gauss and Schumacher (1863); Gauss proposed Lobachevsky as a member of the Scientific Society of Göttingen but he never publicly express his approval.

Lobachevsky's last work on Geometry, *Pangeometry*, was published in 1855, simultaneously in Russian and in French. It contains a new exposition of the non-Euclidean Geometry, several results beeing proved by new and simpler methods. Lobachevsky died in 1856.

Under the influence of Gauss' appreciation in his correspondance with Schumacher, R.Baltzer (1818-1887) mentions, in the second edition of his *Elemente der Mathematik* (1867), the possibility of a non-Euclidean Geometry and admits as an experimental fact the equality of the sum of the angles of a triangle to 2 right angles. The same Baltzer drew J.Hoüel's attention to Lobachevsky's work; so Hoüel (1823-1886) published a French translation of Lobachevsky's *Geometrische Untersuchungen* in 1866, with an abstract of the Correspondance between Gauss and Schumacher. The next year (1867), Hoüel published a French translation of Bolyai's *Appendix* preceded by a notice on Bolyai father and son written down by F.Schmidt (1827-1901), who also published his researches about the Bolyais in the *Archiv der Math. und Physik* (1867); Hoüel defended the non-Euclidean Geometry, for instance in his *Essai critique sur les principes fondamentaux de la Géométrie*(1867) and by other translations of texts related to that subject in the following years. In Italy, Lobachevsky and the Bolyais were known at the same period through the work of A.Forti (1867, 1869) and of G.Battaglini (1826- 1894), who published Italian translations of Lobachevsky *Pangeometry* (1867) and of Bolyai's *Appendix* (1868). So, by 1868, the minds were ready to accept the non-Euclidean Geometry; the understanding of its link with the new Differential Geometry was the work of E.Beltrami (1835-1900).

Differential Geometry

We shall be very brief on that theme, developed independantly of the non-Euclidean Geometry during the period we are considering. It apparently comes from problems of Geodesy. Let us first recall that the spherical Geometry was studied from Antiquity because of the necessity of Astronomy; the spherical trigonometry preceded the plane one in the Middle Ages. The proportionality of the area of a spherical triangle to the excess of the sum of its angles with respect to π was proved by A.Girard in 1629. In 1787, Legendre, who was in charge of a geodetical survey, published a result about the comparison of the angles of a spherical triangle and those of a plane triangle with the same sides; the principal part of the difference is $\frac{\sigma}{3R^2}$ where σ is the area of the triangle and R is the radius of the sphere and the error is of order 4 with respect to the diameter of the triangle divided by R. Gauss was also led to the Geometry of surfaces by his work on Geodesy in 1812-1816 and in 1821-1825; his first interest concerned the geodesics on ellipsoids of revolution and the problem of conformal mapping of one surface on another. Before 1813 he had found the idea of the *Gauss map* or *spherical image* of a surface; it is defined by associating, to each point of a given surface, the unit normal to the surface at this point. The *total curvature* of a piece P of the surface is the ratio of the oriented area of the spherical image of P to the area of P; when P is a small disk around a point A, with radius tending to 0, the limit of the total curvature of P is the curvature k (*Krümmungsmaß*; this term was introduced only in 1822) at the point A and it is equal to the product of the principal radii of curvature at A. O.Rodrigues had similar results by 1815 (Rodrigues 1815). In a private note of 1816, Gauss mentions without proof the invariance of the total curvature by isometry (*theorema egregium*). The proof was probably deduced from the first form of the Gauss-Bonnet theorem, saying that the sum of the angles of a geodesic triangle is equal to π plus the oriented area of the spherical image of the triangle; Gauss has written down a proof of that type in the first draft of his *Disquisitiones generales*, dating from 1825 and this paper also contains a generalisation of Legendre's result (comparison of the angles of a triangle to those of a plane triangle) to the case of a geodesic triangle on an arbitrary surface (Gauss made observations to evaluate the corrections in the case of a triangle with vertices at Brocken, Hohehagen and Inselsberg; the sides are approximately 69km, 85km and 107km). In 1822, Gauss found an expression of the curvature in terms of the coefficient of ds^2 in conformal coordinates

$$k = -\frac{1}{m^2}(\frac{\partial^2}{\partial u^2}\log m + \frac{\partial^2}{\partial v^2}\log m) \quad \text{when } ds^2 = m^2(du^2 + dv^2)$$

In 1825, he found the expression of the curvature in terms of the coefficients of ds^2 in polar geodesic coordinates and then in arbitrary coordinates. This formula gave Gauss a new demonstration of his *theorema egregium*, which we can read in the *Disquisitiones generales circa superficies curvas* of 1827. The extension of the Gauss-Bonnet theorem to the case of a piece of the surface bounded by curves which are not necessarily geodesic was obtained by O.Bonnet in 1848.

The next important step in Differential Geometry is marked by Riemann's Habilitation Schrift *Ueber die Hypothesen, welche der Geometrie zu Grunde liegen* of 1854. Riemann's aim is to explain the concept of space which he considers a special case of the more general concept of *many times extended magnitude*; he says that this concept has not yet been studied apart in Gauss' paper on complex numbers and in the philosophical researches of Herbart. The properties distinguishing the space from other thinkable 3 times extended magnitudes can only come from experience (*nur aus der Erfahrung entnommen werden können*); the facts sufficient to determine the metrical relationships of space are not *necessary* but only of an *empiric certainty "sie sind Hypothesen "*. Riemann distinguishes *discrete manifolds* and *continuous manifolds*; as examples of the latter in current life, he gives the manifold constituted by the determinations of the *localities* of the sensory objects and that constituted by the determinations of *colours*. Other examples come from higher mathematics, as the Riemann surfaces in the theory of functions. One important idea of Riemann is to separate extensive relationships from metrical relationships; so he first considers manifolds without any metric, he defines their dimensions and explains how to locate a point in them by numerical coordinates. Riemann also mentions the possibility of infinite dimensional manifolds, for instance that constituted by the determinations of a function in a given domain (a functional space). The metrical relationships are introduced with the hypothesis that the length of a line is independent of its position; so lines may be measured by one of them. An additional hypothesis leads to a ds homogeneous of degree 1 with respect to the differentials dx_i of the coordinates and the simplest form is $ds^2 =$ quadratic form in the dx_i (with coefficients functions of the position). As there are $n\frac{n+1}{2}$ coefficients (where n is the dimension) and only n coordinates, $\frac{n(n-1)}{2}$ relations are necessary in order that the ds^2 could be reduced to the *flat* (*eben*) form $\sum dx_i^2$. To study the problem, Riemann introduces, around a given point, a system of geodesic coordinates such that *at this point* ds^2 has the flat form; at any other point in the neighbourhood, the difference between ds^2 and the flat form has a principal part quadratic with respect to the $x_i dx_j - x_j dx_i$ so of order 4, a property of *infinitesimal flatness*. Riemann considers the quotient of this principal part by the square of the area of the triangle with vertices $(0,0,\ldots)$, $(x_1, x_2, \ldots)$ and $(dx_1, dx_2, \ldots)$; this quotient only depends on the direction of surface generated by $(x_1, x_2, \ldots)$ and $(dx_1, dx_2, \ldots)$ and it coincides, up to a factor $-\frac{3}{4}$, with the Gaussian curvature of the surface with this direction in the considered manifold. When this *sectional curvature* is always 0, the manifold is flat and reciprocally. Riemann also considers manifolds with *constant* curvature α; their ds is reducible to the form

$$\frac{1}{1+\frac{\alpha}{4}\sum x_i^2}\sqrt{\sum dx_i^2}$$

where α may be positive or negative. In these manifolds, the free moving of figures is possible. When $\alpha > 0$, the manifold is applicable on a sphere.

Applying these notions to the space, Riemann introduces new profound ideas and distinctions. First of all, there is an essential difference between the extensive relationships and the metrical ones; the former are determined by *discrete* parameters, the latter by *continuous* parameters. So the knowledge drawn from experience is never completely certain but it is not inaccurate (*nicht ungenau*) when concerning extensive relationships; on the contrary, when concerning metrical relationships, it remains always inaccurate but the probability that it is almost correct is high. The second distinction concerns the difference between an *unlimited* space and an *infinite* space (*Unbegrenztheit* , *Unendlichkeit*); the first property is extensive, the second metrical and one can conceive a finite unlimited space (for instance any space with positive constant curvature). Riemann ends with the question of the extension of spatial relationships in the infinitely large and in the infinitely small. The foundation of the metrical relationships must be sought in the infinitely small; at that level, either the manifold is discrete, or the metrical relationships are determined by the action of *binding forces* (*der Grund der Massverhältnisse ausserhalb, in darauf wirkenden bindenden Kräften, gesucht werden*). This clearly foreshadows the program accomplished by Einstein's General Relativity.

The precise computation of Riemann's curvature tensor is to be found in the second part of a paper on heat theory sent by Riemann to the French Academy of Sciences in 1861. Riemann's paper was published after his death in 1867 and it had an immediate influence, in particular on Helmholtz (1868a, 1868b) and on Beltrami.

Conclusion

When reading Bolyai and Lobachevsky, one is struck by the similarities of their ways. The importance of the trigonometrical developments must be emphasised; they gave an analytical basis to the new Geometry, guaranteeing, to some extent, the consistence of the construction. This construction is *not axiomatic*; on the contrary, modern axiomatics came from a reflexion made necessary by the existence of the non-Euclidean Geometry.

The Differential Geometry is independant of, but very close to, the non-Euclidean Geometry: they communicate through the spherical Geometry and the interpretation of the sum of the angles of a triangle in terms of an area. It is difficult to think that Gauss had not seen the relation between his researches on the theory of parallels and his intrinsic Geometry of surfaces. And we saw that Bolyai and Lobachevsky made explicit the ds^2 of their plane Geometry. Nevertheless, the time was not yet ripe, in the first half of the XIXth century, to interpret the non-Euclidean Geometry (of a synthetic origin) in terms of the Differential Geometry.

References

Assas-Montdardier (comte d') *Mémoire sur la détermination de la parallaxe et du mouvement propre en déclinaison des étoiles au moyen d'une nouvelle méthode d'occultations artificielles*, Connaissance des temps pour l'an 1831, Paris 1828, p. 120-148.

R.Baltzer *Elemente der Mathematik*, 2nd ed., Leipzig 1867.

G.Battaglini *Sulla geometria immaginaria di Lobatschewsky* , Leipzig, Teubner 1867.

E.Beltrami *Saggio di interpretzione della geometria non euclidea*, Giorn.di Mat. 1868, VI.

J.Bolyai *Appendix scientiam spatii absolute veram exhibens: a veritate aut falsitate Axiomatis XI Euclidei (a priori haud unquam decidenda) independentem; adjecta ad casum falsitatis, quadrature circuli geometrica*, added to the *Tentamen* of F.Bolyai, Maros Vásárhelyini 1832; new edition Budapest 1902; French translation by J.Hoüel *La Science absolue de l'espace, indépendante de la vérité ou de la fausseté de l'Axiome X, précédé d'une notice sur la vie et les travaux de W. et de J.Bolyai par M.Fr.Schmidt*, Mém. de la Soc. des Sc. de Bordeaux, 1867, t.V, p.189-248 and Paris, Gauthier-Villars 1868.

R.Bonola *La Geometria non-Euclidea*, Bologne 1906; English translation *Non-Euclidean Geometry*, New York, Dover 1955.

K.Chemla and S.Pahaut *Trigonométrie sphérique: Euler, Lagrange,...*, *in* Sciences à l'époque de la Révolution Française, Paris, Blanchard 1988, p.149-201.

J. le Rond d'Alembert *Géométrie* in Encyclopédie, Paris 1757, t.VII, p.629-638.

P.Dombrowski *150 Years after Gauss' Disquisitiones generales circa superficies curvas*, Astérisque 1979, t.62.

F.Engel and P.Stäckel *Theorie der parallellinien von Euclid bis auf Gauss*, Leipzig, Teubner 1895.

B.V.Fedorenko *Novyje materialy k biografii N.I.Lobačevskogo*, Leningrad, Naučnoje nasledstvo 1988, t.XII.

J.B.Fourier *Séances des Écoles Normales*, nouvelle éd., Paris 1800, t.I, p.28.

A.Forti *Intorno alla geometria immaginaria o non euclidiana. Considerazioni storico-critiche*, Riv.Bolognese di sci., lett., arti et scuole, 1867, t.II, p.171-184.

A.Forti *Intorno alla vita ed agli scritti di Wolfgang e Giovanni Bolyai di Bolya, matematici ungaresi*, Boll. di Bibliogr. et di Storia delle Sci.Mat. et Fis., 1869, t.I, p.277-299.

C.F.Gauss, *Werke*, vol.8, p.159-268 and p.371 ff.

C.F.Gauss *Mathematisches Tagebuch 1796-1814*, Leipzig 1985.

C.F.Gauss *Briefwechsel zwischen C.F.Gauss und H.C.Schumacher*, Altona 1860-63, t.II, p.268-431 and t.V, p.246.

H.von Helmholtz *Ueber die thatsächlichen Grundlagen der Geometrie*, Verh.d.naturw.-med. Vereins, 1868, t.IV, p.197- 202.

H.von Helmholtz *Ueber die Thatsachen, die der Geometrie zum Grunde liegen*, Gött.Nachr., 1868, t.XV, p.193-221; reproduced in English translation in the *Epistemological writings* of H.von Helmholtz, ed. by R.S.Cohen and Y.Elkana, Dordrecht, Reidel,1977, p.39-58.

L.Hörmander *The cosine theorem on a surface and the notion of curvature*, Lecture given at the Swedish Math.Soc., March 17, 1979.

J.Hoüel *Essai critique sur les principes fondamentaux de la géométrie*, Paris, Gauthier-Villars 1867.

C.Houzel *Histoire de la théorie des parallèles*, *in* Mathématiques et Philosophie de l'Antiquité à l'âge classique, Paris, CNRS 1991, p.163-179.

N.I.Idel'son *Lobačevskij-Astronom*, Moscow, Ist.-mat. issledovanija 1949, t.II, p.137-167.

V.F.Kagan *Lobačevskij*, 2nd ed. Moscow 1948.

E.K.Khil'kevich *Iz istorii rasprostranenija i razvitija idej N.I.Lobačevskogo v 60-70-x godax XIX stoletija*, Moscow, Ist.- mat. issledovanija 1949, t.II, p.168-230.

G.Klügel *Conatuum praecipuorum theoriam parallelarum demonstrandi recensio, quam publico examini submitent A.G.Kästner et auctor respondens G.S.Klügel*, Göttingen 1763.

J.H.Lambert *Theorie der parallellinien*, Magazin für die reine und angew. Math. Leipzig 1786, reproduced in Engel and Stäckel p.135-208.

B.L.Laptev *Teorija paralleljnyx prjamyx v rannix rabotax N.I.Loba čevskogo*, Moscow, Ist.-mat. issledovanija 1951, t.IV, p.201-229.

B.L.Laptev and B.A.Rosenfeld *Geometrija in* Matematika XIX Veka, Moscow 1981, t.II, p.9-114.

A.M.Legendre *Sur les opérations trigonométriques dont les résultats dépendent de la figure de la Terre*, Mém.de l'Acad. des Sc. de Paris, 1787, p.352 ff.

A.M.Legendre *Élémens de Géométrie*, 3rd ed., Paris 1800, t.1, p.21-25; 12th ed., Paris 1823, Note II.

A.M.Legendre *Réflexions sur les différentes manières de démontrer la théorie des parallèles* , Mém. de l'Acad. des Sc., 1833, t.XII, p.366-410

N.I.Lobachevsky *Geometrija*, Sobranije Sočinenij t.2, p.43-107.

N.I.Lobachevsky *O načalax geometrii* , Kazanskij vestnik 1829-30, Sobranije Sočinenij t.1, p.185-26.1.

N.I.Lobachevsky *Voobra žajemaja geometrija* , Učenije zapiski kazanskogo Universiteta, 1835, Sobranije Sočinenij t.3, p.16-70.

N.I.Lobachevsky *Primenenije voobra žajemoj geometrii k nekotorym integralam*, Učenyje zapiski kazanskogo Universiteta, 1836, Sobranije Sočinenij t.3, p.181-294.

N.I.Lobachevsky *Géométrie imaginaire*, J.für die reine und angew.Math., XVII4, p.295-320.

N.I.Lobachevsky *Novyje načala geometrii s polnoj teorijej paralleljnyx* , Učenyje zapiski kazanskogo Universiteta, 1835-1838, Sobranije Sočinenij, t.2, p.147-454

N.I.Lobachevsky *Geometrische Untersuchungen zur Theorie der Parallellinien*, Berlin, Finck 1840; French translation by J.Hoüel *Études géométriques sur la théorie des parallèles, suivie d'un Extrait de la correspondance de Gauss et de Schumacher*, Mém. de la Soc. des Sc. de Bordeaux, 1866, t.IV, p.88-120 and Paris, Gauthier-Villars 1866.

N.I.Lobachevsky *Pangeometrija*, Učenije zapiski kazanskogo Universiteta, 1855, Sobranije Sočinenij t.3, p.435- 524; Italian translation by G.Battaglini *Pangeometria o sunto di geometria fondata sopra una teoria generale e rigorosa delle parallele* , Naples 1867.

L.Magnani *Lobačevskij: le dinamiche epistemiche della pratica geometrica non euclidea* , Materiali filosofici 1977, n.3, p.17-48 .

B.Riemann *Ueber die Hypothesen welche der Geometrie zu Grunde liegen*, Abhandl.der Königl. Gesellsch. der Wissensch. zu Göttingen (1867) 13, Gesamm.Math.Werke 2nd ed. p.272-287.

B.Riemann *Commentatio mathematica, qua respondere tentatur quaestioni ab Illma Academia Parisiensi propositae*, Gesamm.Math.Werke, 2nd ed., p.391-404

O.Rodrigues *Recherches sur la théorie analytique des lignes et rayons de courbure des surfaces*, Bull. de la Soc.Philom. 1815.

G.Saccheri *Euclides vindicatus* , Milan 1733, recent ed. New York, Chelsea 1986.

P.Stäckel *Franz Adolf Taurinus*, Abh. zur Gesch.der Math. 1899, IX, p.397-427.

P.Stäckel *Untersuchungen aus der Absoluten Geometrie aus Johann Bolyais Nachlaß*, Math.und Naturw.Berichte aus Ungarn 1902, t.XVIII, p.280-307.

F.A.Taurinus *Theorie der Parallellinien* , Cologne 1825.

F.A.Taurinus *Geometriae prima elementa* , Cologne 1826.

I.Toth *Spekulationen über die Möglichkeit eines nicht euklidischen Raumes vor Einstein* in Einstein Symposium Berlin, Berlin, Springer 1979, p.46-83.

I.Toth *Wann und von wem wurde die nichteuklidische Geometrie begründet?*, Arch.Intern. Hist.des Sc., 1980,vol.30, p.192-205.

S.A.Yanovskaya *O mirobozzrenii N.I.Lobačevskogo*, Moscow, Ist.-mat. issledovanija 1950-51, t.III, p.30-75 and t.IV, p.173-200.

Riemann's Vision of a New Approach to Geometry

Erhard Scholz

Bergische Universität, Gesamthochschule Wuppertal, Postfach 100127 5600 Wuppertal 1 (RFA)

The General Concept of Manifold and Herbart's "Serial Forms"

How can a mathematician outline a fundamentally new vision of a mathematical discipline? He might turn to the philosophy of mathematics and speak *about* mathematics, i.e. on a metalevel, reflecting his own and other mathematicians' work. Or he might try to sketch the architecture of the new mathematical discipline in question. In the latter case he has to introduce concepts, constructions, and theorems as the central technical building blocks of a mathematical theory. Usually he can draw upon a whole network of results of other scientists, which brings his view closer to tradition and attenuates the novelty of his views. Thus, if an epistemological break is intended, at least some elements of the first, more philosophical approach have to be taken up. The occasion of sharp epistemological turns are rare in the history of mathematics. Riemann's contribution to geometry is a most prominent example.

As is well known, Riemann organized his approach to geometry around the new concept of *manifold (Mannigfaltigkeit)* which for obvious reasons he could not define in a mathematical technical sense. He therefore did it in a semi-philosophical way, drawing consciously and cautiously upon hints by C. F. Gauss who had spoken geometrically about complex numbers (Gauss 1831) and J. F. Herbart who had argued for the use of geometrical imagery in all kind of concept formation, his so-called *serial forms (Reihenformen)*. Vaguely speaking a continuous serial form is produced in the imagination when a class of mental images, or presentations (*Vorstellungen*), undergoes what Herbart called a *graded fusion (abgestufte Verschmelzung)*, i.e. a mental fusion which does not destroy the individual presentations but glues them together, with the result that continuous transition from one to another becomes possible (Herbart 1825, 193*).[1] Riemann did not bother much about the specific ontological theorization which Herbart gave of this prima facie psychological process of concept formation in what Herbart called *synechology (Synechologie)*.

[1] An asterix in a Herbart citation indicates that Riemann read and excerpted the corresponding passage during his studies of Herbartian philosophy. For more details see [Scholz 1982a].

Riemann rather preferred to allude only vaguely to this Herbartian conception (1854, 273). He presupposed the existence of concepts, mathematical or not, which may arise as the result of a "graded fusion" into serial forms.[2] He took up the result and opened it to mathematical consideration, thus forming his concept of ***multiply extended magnitude (mehrfach ausgedehnte Grösse)*** or ***manifold.***

To rephrase his approch briefly: In contrast to a set theoretic approach, Riemann presupposed a concept taken from any field of investigation, thus pre-existing in some epistemic sense in both levels of traditional logic, intension (defined by properties in terms of the field specified) and extension (endowed with a well-determined range of instantiation). For Riemann, the extensional aspect of the concept became of prime importance, with little more scrutiny into the related foundational questions. Leaving some sparse – even if in the light of the later development of set theory important – remarks on finite or discrete manifolds aside, Riemann proceeded immediately to those situations where the particular instances of the concept admit *continuous transitions.* That was to be understood in an intuitive sense, as the concept of continuity came to be mathematically analyzed only after the formal definitions of real numbers had appeared and set theoretic ideas were being formulated, that is not before the 1870/80s.[3] Riemann's approach was somehow parallel to the introduction of Herbart's serial forms; but Riemann specified the idea further by a local successive reconstruction in a quasi-cinematical sense, by 1-parameter, 2-parameter, ..., $(n-1)$-parameter, and finally n-parameter variation of the determination of instances of the concept. In these cases he admitted the obvious, but drastically generalized, geometric terminology of *point* for a particular instance of the general concept (manifold).

Another aspect of this local reconstruction is, according to Riemann, the possibility of introducing n-function systems, the values of which separate (locally) points in the manifold. This leads to coordinate systems which make the manifold accessible to mathematical constructions and further investigations. Riemann even hinted at the possibility of infinite dimensional manifolds at this place.[4] This approach is similar to the modern introduction of topological, differential, Riemannian, etc. manifolds; the role of the topological space, however, being taken in a vague sense by a Herbartian-type of "serial form", backed by mathematical intuition.

---

[2] Riemann emphasized, however, that, according to his opinion and in contrast to Herbart's, such continuous *serial forms* (Herbart) or *extended magnitudes* (Riemann) are much more frequent in "higher mathematics" than in other parts of knowledge.

[3] Cf. [Johnson 1979/1981, 1987], [Moore 1989].

[4] "By an n-time repetition of this process, the determination of position in an n-fold extended manifold is reduced to n numerical determinations, and therefore the determination of position in a given manifold is reduced, whenever this is possible, to a finite number of numerical determinations. There are, however, also manifolds in which the fixing of position requires not a finite number, but either an infinite sequence or a continuous manifold of numerical measurements. Such manifolds form e.g. the possibilities for a function in a given region, the possible shapes of a solid figure, etc." (Riemann 1854, 276 – English translation [Spivak 1970]).

Differential Geometry

Already in his inaugural lecture (1854) Riemann pointed clearly to a basic distinction among mathematical investigations which are possible in manifolds, between those "independent of measurement" (analysis situs) and those assuming metrical structures (differential geometry). Even though Riemann hinted in his talk only very briefly at the first branch of this division (*analysis situs*), restricting himself in this respect to the local reconstruction and the choice of coordinates in $\mathbb{R}^n$, it was already quite clear to him that the topological theory of manifolds was a challenging and promising mathematical subject of its own right. I shall come back to this point in the next section of this article.

The main subject of Riemann's lecture, however, was metrical geometry of manifolds, which he introduced as a profound generalization of Gauss' differential geometry of surfaces. It is true that Gauss had prepared the way for Riemann in the best possible way, by working out the intrinsic nature of the metrical geometry of surfaces in his *Disquisitiones generales circa superficies curvas* (1828);[5] and Riemann referred to it quite clearly (1854, 276). Nevertheless Gauss' concept was severely restricted by sticking to the conceptual framework of euclidean space. His surfaces were always embedded in euclidean 3-space, even if their metrical investigation led to intrinsic aspects. The autonomy of the geometrical object (here: surface) was only intended implicitly; Gauss did not, and could not in his framework, state it explicitly. Thus Gauss did not dare to formulate this autonomy conceptually, because he preferred to accept the delimitations of the "euclidean philosophy" of geometry, at least in his published writings. This restriction allowed generalizations of geometric thought as analogy, imagery, or metaphor in other mathematical contexts only.

That changed completely with Riemann. His concept of manifold was formed exactly to transform imagery and metaphor into strictly mathematical concepts of a generalized geometric framework, thus liberating geometrical thought from the euclidean straightjacket. He introduced the concept of a metrical manifold (*Mannigfaltigkeit mit Maßbestimmung*) in the well-known way by the selection of a positive determinate quadratic differential form,

$$ds^2 = \sum_{i,j} g_{ij}\, dx_i dx_j \qquad (1 \leq i, j \leq n),$$

which enabled him to transfer essential constructions of Gauss' theory of surfaces to the generalized geometry of manifolds.

Most important in this respect was the transfer of the curvature concept to manifolds. In his inaugural lecture Riemann did so by introducing the *sectional curvature* of an infinitesimal surface element, even if only by a description of the procedure for deriving it without giving an explicit general formula. This allowed him to speak about *manifolds of constant curvature* as a connecting link between the general differential geometric theory of manifolds and the theory of physical space.

He mentioned two main results:

[5] In particular he elaborated the central role of the *Theorema egregium* (the intrinsic determination of the curvature of the surface) and the *Theorem aureum* (angle sum of geodetic triangles).

1. Manifolds of constant curvature are exactly those in which free mobility of rigid figures is possible.
2. In a manifold of constant curvature α it is always possible to choose local coordinates such that the metric is given by

$$ds^2 = \frac{\alpha \sum_i dx_i^2}{(1 + \frac{\alpha}{4}\sum_i x_i^2)^2} \quad .$$

I mention Riemann's later work (1861) on heat distribution in a homogeneous body only in passing. There Riemann introduced the famous 4-index symbol in second derivatives of the metric as a criterion of flatness of a metrical (Riemannian) manifold, which was later identified as *curvature tensor.*[6]

Steps towards a Topological Theory of Manifolds

I now want to come back to Riemann's basic distinction between studies of "manifolds with metrics" and of manifolds "independent of measurement" or *analysis situs.* Already in his well known doctoral dissertation on complex function theory (1851) Riemann had dealt with questions of analysis situs. There he had introduced *Riemann surfaces* for multi-valued analytic functions in a complex region and had begun to study the topology of compact oriented surfaces with boundaries in detail. He had used *dissection of the surfaces* into simply connected components, resulting in a complete classification of these by the *number of boundary components* and the *order of connectivity* (*Ordnung des Zusammenhangs*) m. The latter had been introduced by him as alternating sum of the number e of cross cuts and the number f of simply connected components

$$m = e - f = -\chi(F), [\quad \chi(F) \quad \text{the Euler characteristic of F}].$$

Of course, one of Riemann's central arguments in this passage was that this number is independent of the specific choices of the dissection process (1851, 11f.).

Three years after his inaugural lecture, in his great *memoir on abelian functions* Riemann used another approach to the characterization of the the topology of surfaces, building on *boundary relations* of systems of closed curves inside the surface (Riemann 1857). This time he studied closed oriented surfaces exclusively and looked for systems of closed curves, which form a complete boundary of a part of the surface. In this connection he introduced an appropriate equivalence concept of curve systems and showed that the maximum number of closed curves which do not form a complete boundary is independent of the specific choice of curves and gives a good classification of the topological type of the surface. As he could show that in the case of closed (Riemann) surfaces the maximum number of boundary independent closed curves (cycles) is even, he was led to the well known *numerical invariant* $2p$ for these surfaces (p was later called *genus* by Clebsch). And of course he showed how the genus characterization translated to the "order of connectivity" m of (1851): Punctuate the closed surface [Euler characteristic $\chi(F) = 2 - 2p \longrightarrow \chi(F') = 1 - 2p$], so you get a surface with boundary to which application of

---

[6] A recent investigation of this work is given in [Farwell/Knee 1990].

the dissection method shows that $2p$ cross cuts can be used to dissect the surface into one simply connected piece (1857, 93ff.). So in the terminology of (1851) the order of connectivity of the punctured surface is $m = 2p - 1$ [which is indeed $-\chi(F')$].

So, if one adds Riemann's insights of (1851) and (1857) to his inaugural lecture, one sees that he did, in the twodimensional case, take the first step to a topological theory of manifolds, containing the two complementary aspects of

- dissection into simply connected cells and counting the Euler characteristic of the cell complex,
- homological characterization of closed surfaces, counting the first Betti number.

All this has been discussed from different points of view in the historical literature, as has the fact that Riemann started to think about a generalization of his analysis situs methods to higher dimensional manifolds [Bollinger 1972, Pont 1974; Scholz 1980 etc.]. He left a *Nachlass*-fragment on this topic, where he experimented with dissection and bordism ideas in higher dimensional manifolds (*Werke* 479–482).

I should only add that an investigation of the fragmentary manuscript itself (Riemann *Nachlass* **16**, 44^r, 46^{r-v}, 49^{r-v}) gives some circumstancial evidence that Riemann worked on these ideas in the time between his doctoral dissertation and his inaugural lecture, not about or even later than 1857 [Scholz 1982b]. So we can read the fragments on analysis situs as a (hidden) background to Riemann's short and in itself rather vague reference to the topology of manifolds in the lecture of 1854.

First Glances at Other Geometrical Structures

The next point I want to discuss is the surprising fine and differentiated approach to geometric thinking that was opened up by Riemann on the basis of his manifold concept. This view on geometry was in line with the most far-reaching and deep-going changes of geometric thought during the turn towards "modern mathematics" of the late 19th and the early 20th century. These changes concern both semantics and the internal structure of geometry. From the point of view of semantics the most striking feature of 19th century development is the turn away of geometric theories from predominantly or even exclusive reference to physical space (even if perhaps understood in a philosophical *a priori* disguise). On the other hand, a whole range of new reference fields arose inside mathematical knowledge itself, in particular analysis, algebra, and arithmetic. As a companion to and a result of this development the geometrical theories became more abstract and more diverse. And yet they were to be kept together by central organizing ideas.

For Riemann the latter function was taken over by his manifold concept which admitted different enrichments with structural ideas derived from the contextual situations. Again it was his function theoretic work, where he developed most clearly (albeit restricted to the real two- or complex one-dimensional case) some basic ideas for the study of manifolds with structures going beyond those he talked about in his 1854 lecture. Most important, from this point of view, are his investigations of *abelian functions* (1857), which contain fundamentally new ideas on surfaces (or curves, depending on the standpoint) from the *complex analytic* and/or complex *birational* point of view.

Here is not the place to discuss these questions in detail,[7] but I have to mention at this point two general and fundamental insights of Riemann. The first is his analysis of the *meromorphic structure* of a compact surface of genus p. Riemann characterized abelian integrals of the second kind by a set of independent conditions (pole behaviour and real part of the periods) using the later disputed Dirichlet principle. In a second step he derived an estimation of the number of linearly independent meromorphic functions on the surface with simple poles in m chosen points as

$$\mu \geq m - p + 1 .$$

The result was later sharpened by his student Roch to the *Riemann- Roch theorem*:

$$\mu = m - p + \tau + 1 ,$$

(with τ = number of linearly independent abelian differentials of the first kind with zeroes at the m given points). *This was the first result of modern geometry to establish a deep-rooted connection between the topology of a manifold and a more refined structure, here the complex analytic one on a complex compact curve.*

The next point to mention in this context is Riemann's insight that the meromorphic functions on a curve can in fact be expressed as *rational functions* in two of them, say z and t, which, read as inhomogeneous coordinates in in $P(2, \mathbf{C})$, let the curve be represented algebraically as

$$F(z,t) = 0, \qquad F \in \mathbf{C}[z,t].$$

This makes it possible, as Riemann stated clearly, to study any compact complex curve from a purely algebraic birational point of view. In particular, the change of representing coordinates (z,t) to (z',t') is given by rational transformations in both directions. Thus Riemann indicated the way towards a purely algebraic structure linked to his manifold concept. The elaboration of these ideas would, of course, later lead to an adaptation of the underlying concept of manifold to the algebraic-geometric context and a transformation into different types of algebraic varieties. This is a story far away from Riemann's days and closer to the present than most of the other points mentioned here.[8]

Foundations of Geometry

Coming back to the inaugural lecture, it has to be said that Riemann gave only slight indications that the manifold concept could be developed further from the point of view of analytic or even purely algebraic structures.[9] The title of his talk indicated another line of investigation, namely the foundations of geometry. There is no doubt that, after

[7] See e.g. [Dieudonné 1974, 42ff.; Gray 1989, 361ff.; Scholz 1980, 68ff.]

[8] See [Dieudonné 1974] for a first historiographic overview.

[9] Riemann mentioned in passing, however, that "...Such investigations [of analysis situs of manifolds, E.S.] have become a necessity for several parts of mathematics, e.g., for the treatment of many-valued analytic functions, and the dearth of such studies is one of the principal reasons why the celebrated theorem of Abel and the contributions of Lagrange, Pfaff and Jacobi to the general theory of differential equations have remained unfruitful for so long." (1854, 274)

his inaugural lecture became accessible to the wider scientific public with its publication in 1867 (*Göttinger Abhandlungen*), the early reception of Riemann's geometric ideas was eminently important and influential in the debate about the character and interpretation of ***non-euclidean geometry***. That is true, in particular, for Riemann's influence on Beltrami (to be seen by the latter's conceptual progress between his two 1868 papers on non-euclidean geometry), on Helmholtz (even if also Helmholtz's ideas on free mobility as the central "fact" lying at the base of geometry were developed before he had read Riemann) and on Clifford, to give just the three most outstanding examples.

Indeed Riemann's approach, and in particular his discussion of manifolds of constant curvature, can easily be read in the context of the investigations of Bolyai and Lobachevskii, because Riemann outlined a sophisticated conceptual framework for a possible and satisfying interpretation of non-euclidean geometry. Surprisingly ***there are no indications whatsoever that Riemann knew more than superficially of Bolyai's and Lobachevskii's work and maybe even not at all.*** Consequently he did not bother about the intimate potential connection between his and their considerations. I shall try to give the main arguments for this thesis which I have discussed more in detail elsewhere [Scholz 1982b].

It may even be surprising that in Riemann's inaugural lecture the only "modern reformer of geometry" cited by name was Legendre. That fits the observation that Riemann never in his talk (or elsewhere) mentioned the axiom of parallels – not even as a sideremark comparable to that referring to the topological theory of manifolds and its role in complex function theory. That must be startling if one tries to see a conscious reference to non-euclidean geometry in Riemann's title, *Hypotheses which lie at the basis of geometry*. This is the more so, as the last two sections of the second part of the talk (on differential geometry in manifolds) deal with manifolds of constant curvature, so that a sideremark on the different behaviour of parallels, dependent on the curvature, would have had an obvious place and context.

It becomes even clearer that Riemann never bothered about the foundational questions of geometry in the logical sense, when one takes into account a passage in his *Nachlass*, which was apparently written in the years 1852/1853, i.e. some time, although not much, before his inaugural lecture. In this pre-1854 fragment Riemann experimented with the idea of a manifold and dealt with the relationship of the manifold concept to foundational questions of geometry.[10] In particular he pointed out that a treatment of geometry from the manifold point of view would make superfluous all the specifically geometric axioms of Euclid and provide the possibility of reducing the necessary axioms to those "which hold for quantities in general...". As the only example of what could be proved in this framework, Riemann cited Euclid's axiom 9 which states that, given two points, there exists only one straight line incident with them. Again the parallel problem is not even mentioned.

In what follows Riemann stated quite frankly why he was content with this simple example and why he did not see a reason to go further into these foundational questions:

> "But even if it is of interest to grasp the possibility of this mode of treatment of geometry, the execution of the latter would be extremely fruitless, as by this means we

[10] Riemann *Nachlass* (map **16**, folio 40^{r-v}) published in [Scholz 1982b, 228–230] with a correction by E. Neuenschwander.

would not find new theorems, and what seems simple and clear in the presentation in space, would thereby get involved and difficult." (Riemann Nachlass, **16**, 40^r)

This is clear testimony that Riemann did not show much interest in detailed studies of the logical foundations of geometry, precisely because he presumed them to be fruitless from the point of view of new theorems. This position is completely understandable from his point of view, *but it cannot be upheld if one is familiar with the works of Bolyai and/or Lobachevskii.* Their studies of absolute geometry and of horocycle geometry in the non-euclidean case [Gray 1979], to name just two examples, is too obviously incompatible with such a strict verdict of fruitlessness.

On the other hand, Riemann gave a clue why, in his opinion, the study of manifolds *really* mattered. In direct continuation of the quotation just given he went on:

> "Therefore one has taken everywhere the opposite road, and each time one encounters manifolds of several dimensions in geometry [mathematics? E.S.], as in the doctrine of definite integrals in the theory of imaginery quantities, one takes spatial intuition as an aid. It is well known how one gets thus a real overview over the subject and how only thus are precisely the essential points emphasized."

This quotation agrees completely with the main line of thought in Riemann's inaugural lecture, as far as manifolds inside mathematics are concerned. If Riemann had obtained a deeper knowledge of the more recent foundational studies of Bolyai or Lobachevskii during the time which elapsed between the formulation of these remarks and his inaugural lecture, he would have had sufficient reason to mention these new and unexpected aspects. That, however, is not the case.

Manifold Concept and Physical Space

We have to conclude that Riemann's "hypotheses which lie at the basis of geometry" have connotations different from the ones given in the late 1860s and 1870s when the debate on non-euclidean geometry ran high. One of these other connotations has already been mentioned. Riemann very consciously began the introduction of (non-metaphoric) geometric language into other mathematical fields (complex function theory, differential equations etc.) and needed therefore new fundamental concepts and "hypotheses" of geometry. This was the purely mathematical aspect of his enterprise, which was kept in second place in his 1854-talk.

The *main goal* of Riemann's inaugural lecture, on the other hand, was a *reformulation of the conceptual foundations of physical geometry.* This is made completely clear by the talk's architecture and the selection of main topics which culminate in a proposal of how the manifold concept could be used to analyze more deeply the properties of physical space. This goal also illuminates the reasons why Riemann choose manifolds of constant curvature as the only class of examples for a more detailed treatment in his second, differential geometric, part.

Riemann's intention in the last part of his lecture was to outline methodologically how the manifold concept could be used to improve the comprehension of physical space.

Already at the beginning of this discussion he stated that for such an application of the new concept it is essential to know something about conditions "which are sufficient and necessary for determining the metric relations of Space" (taking for granted that physical space can be analyzed as a manifold with Riemannian differential geometric metric). manifolds of constant curvature became important in this context, because in this class of examples such conditions are easily characterized.

- If free movement of rigid bodies is assumed, curvature is constant, and determination of the angle sum in one triangle uniquely determines curvature and metric of the whole manifold.
- If the angle sum of all triangles equals two right angles, all sectional curvatures are zero and space is euclidean (at least locally).
- If neither is the case the determination of the metric is completely open.

After a short excursion into the questions of unboundedness and infiniteness of space, which Riemann himself classified as "idle questions (müssige Fragen) for the explanation of nature" (1854, 285), he came back to the metrical relations of space. He thought them of much greater importance, as "... to discover causal connections one pursues phenomena into the spatially small, just as far as the microscope permits." And "... (u)pon the exactness with which we pursue phenomena into the infinitely small, does our knowledge of their causal connections essentially depend." (*ibid.*)

Here again he discussed the existing empirical evidence from the point of view of a theoretical alternative. The main observation referred to (without mentioning names) was Bessel's recent astronomical measurements (in 1838) of fixed star parallaxes, which had the highest technical standards of the time and had given positive values for some (nearer) stars but zero for most of them. That was good evidence for the existence of large triangles of astronomical scale with angle sum π (if one excludes the case of positive curvature, what Riemann apparently did without further notice). Riemann now drew the widely known conclusion:

- If one assumes free mobility of rigid bodies, then space is euclidean with the best precision available at the time.
- If, however, free mobility does not hold, then only "the total curvature of every measurable portion of Space is not perceptibly different from zero", leaving open the possibility of drastic changes of curvature in the small, which cancel out if integrated over larger regions (1854, 285).

Riemann closed this passage with the warning that one should not take euclidean geometry for granted, however convincing it may seem at the time. His argument in this connection even casts some doubts on the general applicability of his own metrical concept and shows that also his new concepts cannot be considered as a new type of (neo-)Kantian *a priori*. He noted that "... the empirical notions on which the metric determinations of Space are based, the concept of a solid body and that of a light ray, lose their validity in the infinitely small." Therefore one should always be open to a revision of the fundamental concepts of (physical) geometry "... as soon as it permits a simpler way of explaining phenomena." (*ibid.*) This is, from hindsight, a particularly striking remark, as the change to a Lorentzian or semi-Riemannian metric in special and general relativity was a revision of fundamental concepts of this type and – to comment this remark of Riemann even more anachronistically – something similar is being looked for today in the ongoing search for a quantum structure of space(-time).

All this shows clearly, how Riemann wanted to proceed in the elaboration of the "hypotheses which lie at the basis of (physical) geometry". These have to be rethought and perhaps revised time and again, with each fundamentally new piece of evidence about the physical tools of metrical measurement. He finished with a short remark on what was the role of mathematics in this process:

> "Investigations like the one just made, which begin from general concepts, can serve only to insure that this work is not hindered by too restrictive concepts, and that progress in comprehending the connection of things is not obstructed by traditional prejudices." (1854, 286)

This remark leads to our last point, Riemann's philosophical definition of the task of mathematics in the cognition of physical reality.

Riemann's Epistemology of Mathematics

Riemann obviously was neither an empiricist nor a Kantian. He could, in contrast to e.g. Gauss, go beyond the restrictive limits of euclidean geometry so easily, because he had worked out a good understanding of post-Kantian German dialectical philosophy, in particular by his extensive and detailed studies of the philosophy of Johann Friedrich Herbart (1776 – 1841). In fact, Herbart had defended a sideline of philosophy, which was essentially realistic in its methodology and ontology without losing its commitments to dialectics in its epistemology. This brought it closer to the lines of thought of scientists than the mainstream of German idealist philosophy of the time. I cannot go here into much detail,[11] but I want to outline some philosophical aspects underlying Riemann's mathematical work, which were heavily influenced by Herbart and surely of a greater overall importance for his work than just the vague reference to Herbart's "serial forms".

First of all the Herbartian background gave Riemann a post-Kantian view of epistemology. Herbart, in contrast to the idealist dialecticians of the time, saw the role of dialectical development mainly in concept formation and in the methodology of knowledge. He was no dialectician as far as ontology is concerned. We know from Riemann's *Nachlass* that the latter studied exactly intensely, and essentially with agreement, those parts which were constitutive for Herbart's epistemology.[12] That formed the background for Riemann's own developmental dialectical position with respect to epistemology, stated explicitly e.g. in the philosophical fragments published by Weber in Riemann's *Werke* (521–525).

As a consequence, Riemann did not share the restrictive Kantian view of a uniquely determined structure of synthetic knowledge *a priori*, of which mathematics, according

[11] See for this subject [Scholz 1982a].

[12] Riemann *Nachlaß* (16, 59^v, 64^r, 141^r) – see [Scholz 1982a] – and Riemann's selfdescription: "The author [Riemann, E.S.] is a Herbartian in psychology and epistemology [...]; in most cases he cannot agree, however, with Herbart's natural philosophy and the metaphysical disciplines (ontology and synechology) referring to it." (Riemann *Werke*, 508)

to Kant, forms an essential part. For Riemann there was no place for a purely *a priori* deduction of transcendental forms of cognition.[13] But on the other hand he also did not give in to the pitfalls of empiricism. Theoretical knowledge, in particular mathematical theory, insofar as it constitutes a conceptual framework for scientific knowledge, plays, according to Riemann, a role of what I want to call a *relative* or *dialectical a priori* with respect to empirical knowledge.

- This knowledge is *a priori*, because it is never to be derived by induction, generalization, or even straightforward idealization from experience. It is constituted by a deliberate conceptual creation and serves as a theoretical system of reference for empirical investigations and thus plays a formative role for the cognition of the empirical world.
- On the other hand this knowledge is *relative* and *dialectical.* Its structure is not uniquely determined, i.e. there is place for theoretical choices in the process of generation of the concepts, and these choices are done in consideration of the available empirical evidence. Just as little is it stable in time; it is subject to changes during the historical process of refinement of knowledge. *Refinement* (Riemann's term) may be read as a pragmatic expression for a type of conceptual change which overcomes the old structure without destroying completely the latter's validity. It thus shares the characteristic features of dialectical negation (*Aufhebung*), even if presented in less elaborate language.

Both aspects were already inherent in Herbart's epistemology, but they were formulated by Riemann in his epistemological fragments as his own position.[14] Mathematics plays, according to Riemann, an essentially critical role. It has to ensure that cognition of reality "... is not hindered by too restrictive concepts, and that progress in comprehending the connection of things is not obstructed by traditional prejudices..." (full citation above). It goes without saying that for Riemann the critical function of mathematical investigations is not restricted to undermining the validity of the old concepts, but it also establishes new and wider ones.

This point of view enabled Riemann to conceive such a fundamental revision of the conceptual framework of physical space as given in his inaugural lecture. We know from Gauss that the latter had thought about the necessity of going beyond the Kantian-Euclidean standpoint, but had never dared to come out with such a position in the scientific public. Riemann knew of the changes which had taken place on the philosophical terrain during the early century and used these as a positive reference system for his own proposals with respect to physical geometry.

There remains one more point to add. Riemann's clear conceptual orientation, which led him to single out central concepts in the different mathematical fields he worked in

[13]The anti Kantian tendency of Riemann's 1854 talk is discussed more in detail by [Nowak 1989].

[14]"...By the concepts through which we conceive nature, not only are our perceptions complemented in each moment, but also future perceptions are singled out as necessary, or, insofar as the conceptual system is not complete enough for that purpose, determined as probable ..." And a little later: "... The conceptual systems which underlie them now [the exact sciences, E.S.], have been formed by gradual change of older conceptual systems, and the reasons which resulted in new modes of explanation, can be reduced to contradictions or improbabilites, which turned up in the older modes of explanations." (Riemann *Werke*, 521)

(manifolds in geometry, Riemann surface in function theory, Riemann integral in convergence theory of trigonometric series etc.) was also much in line with Herbart's concept of ***philosophical studies of the sciences.*** Herbart, as one of Germany's educational philosophers in the first decades of 19th century, had seen a close connection between philosophy, philosophical studies of the sciences and a type of social reform which was fostered by the state but carried on by a class of (scientifically) educated men. In order to fulfil such a function the sciences should not be pursued in a predominantly technical style, but studied in what he called ***philosophical spirit.*** This lead Herbart to postulate a continually renewed search for and elaboration of central concepts in the different scientific disciplines. In fact the sciences ought to organize themselves around ***central concepts (Hauptbegriffe)*** (Herbart 1807). Philosophy proper should then work out the connections and dissolve possible contradictions between the central concepts of the different scientific fields. By such a systematic but open communication between philosophical studies of science and philosophy proper, science and philosophy would be able to fulfil their *Bildungsauftrag*, their educational goal. Riemann had studied intensely and excerpted these comments of Herbart, as we know from his *Nachlass* [Scholz 1982a, 424ff.]. That seems to have given him a sort of mirror for the self-reflection on the task and method of mathematics.

Riemann's mathematical work is penetrated by such a deep conceptual orientation that it cannot be better characterized than as "philosophical study of mathematics" in the Herbartian sense. And it even throws some light on the relationship between mathematics and physical sciences, as seen by Riemann, when one substitutes mathematics for philosophy in the Herbartian communication network of the sciences and philosophy. In fact, Riemann did in his inaugural lecture, what philosophers should do, according to Herbart, with respect to the scientific disciplines. He investigated the central concept of manifold, to be found in different mathematical and physical sciences, in order to clarify the connections between its different specifications, dissolve possible contradictions and elaborate it, in order to insure the possibility of further progress of scientific knowledge.

So again we find a deep convergence of ideas between Herbart and Riemann on this methodological level, referring to the most general description of the task of mathematical research. This convergence leads us to the claim that Riemann, in his mathematical research, took up orientations for scientific investigations, which had been worked out by Herbart, among others, on a philosophical level and which signify a broader social and cultural influence on the sciences and mathematics in the first half of the nineteenth century.

References

[Beltrami, Eugenio 1868a.] Saggio di interpretazione della Geometria non Euclidea. *Giornale di Mathematica 6*, 284–312. Opere **1**, 374 – 405.

[Beltrami, Eugenio 1868b.] Teoria fondamentale degli spazi di curvatura costante. *Annali di Mathematica* (2) **2**, 232–255. Opere **1**, 262–280.

[Bollinger, Maja 1972.] Geschichtliche Entwicklung des Homologiebegriffs. *Archive for History of Exact Sciences* **9**, 94–170.

[Dieudonné, Jean 1974.] *Cours de géométrie algébrique*, t. **1**. Paris: Presses Universitaires de France.

[Farwell, Ruth; Knee, Christopher 1990.] The missing link: Riemann's "Commentatio", differential geometry and topology. *Historia Mathematica* **17**, 223–255.

[Gauss, Carl Friedrich 1828.] Disquisitiones generales circa superficies curvas. *Commentationes Societatis Gottingensis*, 99–146. Werke **4**, 217–258. German by A. Wangerin, Leipzig 1900.

[Gauss, Carl Friedrich 1831.] Theoria residuorum biquadraticorum, Comment. secund. *Göttingische gelehrte Anzeigen*. Werke **2** (1863), 169–178.

[Gray, Jeremy J. 1979.] *Ideas of Space. Euclidean, Non-Euclidean and Relativistic.* Oxford: Clarendon. ²1989.

[Gray, Jeremy J. 1989.] *Algebraic geometry in the late nineteeth century.* In [Rowe/McCleary 1989, 361–388].

[Herbart, Johann Friedrich 1807.] Über philosophisches Studium. Werke **2**, 227–296.

[Herbart, Johann Friedrich 1825.] *Psychologie als Wissenschaft, Zweiter analytischer Theil.* Werke **6**, 1–339.

[Herbart, Johann Friedrich, Werke.] *Sämtliche Werke in chronologischer Reihenfolge.* Hrs. K. Kehrbach; O. Flügel. Langensalza 1899–1912. Reprint Aalen: Scientia Verlag 1964.

[Johnson, Dale 1979/1981.] The problem of the invariance of dimension in the growth of modern topology, I, II. *Archive for History of Exact Sciences* **20** (1979), 97–188, **25** (1981), 85–267.

[Johnson, Dale 1987.] L. E. J. Brouwer's coming of age as a topologist. In: E. Phillips (ed.). *Studies in the History of Mathematics.* Mathematics Association of America, Studies in Mathematics **26**, 61–97.

[Moore, Gregory H. 1989.] Towards a history of Cantor's continuum problem. In [Rowe/McCleary 1989, 79–121].

[Nowak, Gregory 1989.] Riemann's Habilitationsvortrag and the synthetic *a priori* status of geometry. In [Rowe/McCleary 1989, 17–48].

[Pont, Jean-Claude 1974.] *La topologie algébrique des origines a Poincaré.* Paris: Presses Universitaires de France.

[Riemann, Bernhard 1851.] Grundlagen für eine allgemeine Theorie der Functionen einer veränderlichen complexen Grösse. *Inauguraldissertation* Göttingen. Werke, 3–45.

[Riemann, Bernhard 1854.] Über die Hypothesen, welche der Geometrie zu Grunde liegen. Habilitationsvortrag Göttingen. *Göttinger Abhandlungen* **13** (1867). Werke, 272–287.

[Riemann, Bernhard 1857.] Theorie der abelschen Funktionen. *Journal für Mathematik* **54**. Werke 86–144.

[Riemann, Bernhard 1861.] Commentatio mathematica, qua respondere tentatur quaestioni ab illustrissima Academia Parisiensi proposita: "Trouver quel doit être l'état calorifique d'un corps solide homogène indéfini ...". *Werke*, 391–404.

[Riemann, Bernhard, Werke.] *Gesammelte mathematische Werke und wissenschaftlicher Nachlaß.* Leipzig 1876, 2. erweiterte Aufl. 1892. Neudruck New York: Dover 1955. Neudruck Nendeln: Sändig 1978. Erweiterter Neudruck (Hrsg. Narasimhan) Berlin etc.: Springer 1990.

[Riemann, Bernhard, Nachlaß.] *Codex Ms Riemann*, map **16**. Handschriftenabteilung, Universitätsbibliothek Göttingen.

[Rowe, David; McCleary, John 1989.] *The History of Modern Mathematics.* vol. **1**. Boston etc.: Academic Press.

[Scholz, Erhard 1980.] *Geschichte des Mannigfaltigkeitsbegriffs von Riemann bis Poincaré.* Basel – Boston – Stuttgart: Birkhäuser.

[Scholz, Erhard 1982a.] Herbart's influence on Bernhard Riemann. *Historia Mathematica* **9**, 423–440.

[Scholz, Erhard 1982b.] Riemanns frühe Notizen zum Mannigfaltigkeitsbegriff und zu den Grundlagen der Geometrie. *Archive for History of Exact Sciences* **27**, 213–282.

[Spivak, Michael 1970.] *A Comprehensive Introduction to Differential Geometry.* Vol. **2**. Boston: Publish or Perish.

Poincaré and Klein – Groups and Geometries

J.J. Gray

Faculty of Mathematics, Open University, Milton Keynes, MK7 6AA, England

The theme of this conference which I wish to address is the status of geometry, in particular the introduction of group-theoretic ideas into geometry; I hope that what I shall say will provide an introduction to some of the papers to be presented later. I shall go on to describe how Poincaré came to develop his ideas about non-Euclidean geometry at the very start of his career. More generally, I shall claim that non-Euclidean geometry was a decisive arena for the recognition of the importance that ideas of group theory play in geometry of any kind.

A naive view of geometry would be that it is the study of figures in space. By a sleight of hand this is often reduced to the study of curves in the plane. Within this subject a paradigm place is reserved for the study of lines and circles - the subject matter of Euclid's *Elements*. This mildly interesting topic was defended on two grounds: that it led into the study of physical space, as exemplified by Newton's *Principia*, and that it introduced the mind to the logical treatment of our most elementary intuitions about shape. I quote from the 1860 reprint of Legendre's influential *Eléments de géométrie* :

> "La géométrie est une science qui a pour objet la mesure de l'étendue."

As is well known, Legendre's book was a deliberate attempt to restore to the teaching of mathematics in France something like the classical Euclidean approach to geometry, and his book is written throughout in the style of definitions and numbered theorems.

This state of affairs could be found in other countries, in Great Britain, for example, as Joan Richards has recently described. What interests us today is that it is was not, even by the time of the first edition of Legendre's book (1794), unproblematic. As Christian Houzel described at this conference, the question of the parallel postulate has a long history. The early period has recently been described at length by Jean-Claude Pont [1986]; I too have offered overviews [1987], [1989]. Most remarkably, by the time Legendre's book was re-issued in 1860, there were two convincing accounts of a geometry different from Euclid's, those offered by Lobachevskii and Bolyai. It cannot be stressed too much that their work, while lacking in rigour, posed the gravest threats to the entire structure of contemporary knowledge about, and interest in, geometry[1].

It threatened one's knowledge, because it described an alternative geometry of space, one just as physically plausible as Euclid's and Newton's, but in important ways different.

[1] For the views of Bolyai, Lobachevskii and other early writers such as Taurinus, see Gray [1989] and Bonola [1912], which also reprints some of their original papers.

As you know, Lobachevskii even conducted an empirical investigation to see if space was Euclidean or non-Euclidean; the results were inconclusive, but the idea that such an investigation was necessary was revolutionary. It threatened one's interest in geometry because it raised the question of what the basic objects of one's spatial intuition are. If, after all, they are not to be described as Euclid had originally done, what, one might ask, was the value of teaching everyone Euclidean geometry. This is simultaneously a problem of pedagogy and in the philosophy of mathematics.

On the other hand, as is well-known, the work of Lobachevskii and Bolyai was not favourably received in their lifetimes, and it is worth considering briefly why this was. As Houzel has described, both men assumed that it was possible to discuss the configuration in a plane consisting of a line, l, a point P not on that line, and a line m through the point P and asymptotic to the line l which is such that the perpendicular from P to l makes an acute angle with the line m. The line m was said to be parallel to the line l. The assumption that such a configuration exists, which of course denies the validity of the Euclidean parallel postulate, was not defended by them. What their work accomplished was a thorough trigonometric description of the consequences of making such an assumption; it could not logically amount to a vindication of it.

Foremost amongst their results were the analogues in hyperbolic trigonometry of the formulae of spherical trigonometry. This analogy had been discovered earlier by Taurinus (as Houzel described). Not only is their work independent, however, it offered a derivation of the new formulae where Taurinus had merely relied on the analogy alone. This derivation, which both men put forward using a bold argument in 3 dimensions, was an important step in showing how the theorems of non-Euclidean geometry followed from the assumption that a new definition of parallels was possible. Nonetheless, critics could still argue that what was missing was a geometric interpretation of the formulae. Lobachevskii tried to defend his work by arguing that geometry was necessarily to be written in formulae, since its subject was measurement, but his argument was obscure and indeed only becomes intelligible after the work of Riemann and Beltrami.

There can be no doubt that the impact of the discovery of non-Euclidean geometry, however much it was resisted and delayed, animated the minds of those leading geometers who chose to reconsider the foundations of their subject. Yet the introduction of group-theoretic ideas is still somewhat different, and does not seem to form any obvious next step from the work of Bolyai and Lobachevskii. The relevant context is the other major change in the 19th century study of geometry, the study of projective geometry.

The obvious fact about transformations in projective geometry is that such transformations are obtrusive. That is to say, their use obtrudes or forces itself onto one's awareness in a way that Euclidean ones do not. It is my view that at least in the early work of Poncelet and Chasles the subject was conceived of as the study of projective transformations of Euclidean figures. To speak very schematically, I would say that the study of projective space *per se* is more the work of German mathematicians. Thus, within the algebraic tradition, Möbius, Plücker and especially Hesse, and within the synthetic tradition most notably von Staudt, are those who most clearly saw their subject as the study of a new space. Since that space is patently not physically plausible it need not detain us here, although the nature of projective space is something I shall refer to later. But it seems that it is this German tradition that was the first to make explicit the idea that the study of geometry necessarily involves families of transformations.

There is no doubt that the most influential figure in the creation of group theory in the mid-nineteenth century was Camille Jordan. Following his announcement of some results concerning groups of motions in 1867, he published a lengthy paper on that subject in 1868/9. In his "Mémoire sur les groupes de mouvements" his theme was the application of group theory to the results of Bravais and others on crystal lattices. This paper is widely taken to mark the recognition that group-theoretic ideas matter in the study of Euclidean geometry[2]. However, there are others much more intimately involved with geometry as usually understood. Curiously, they all have almost the same date, the date of the publication of Riemann's "Ueber die Hypothesen, welche der Geometrie zu Grunde liegen": 1867±1. They are three famous papers by Helmholtz: "Uber die thatsächlichen Grundlagen der Geometrie" (1866) , "Ueber die Thatsachen, die der Geometrie zu Grunde liegen" (1868), with its deliberate echo of Riemann's paper, and "The origin and meaning of geometric axioms" (1876, but originally given as a published lecture in 1870); Beltrami's "Saggio di interpetrazione della geometria non-euclidea" (1868), and Houël's *Essai critique sur les principes fondamentaux de la géométrie élémentaire, etc* , (1867).

The date is interesting. To judge from Koenigsberger's biography of Helmholtz, Helmholtz had not known of Riemann's ideas on geometry for very long by 1866. In 1867 Houël translated the first of Helmholtz's papers into French; in 1869 he was to do the same for Beltrami's paper. Beltrami had delayed publication of his paper for a year while he worried that Cremona's criticisms of it might contain a fatal truth. Although aware of Riemann's work, he did not embrace it until his second publication on this theme, when he was able to study n-dimensional space.

So it might seem that the inspiration is Riemann's. That cannot wholly be true. All these writers pass immediately to the idea that it is the motion of figures without alteration that is the basic property of geometry, and that this is a fact obtained from experience. Riemann's much more profound and difficult idea is the unaltered motion of infinitesimal line segments. Helmholtz defended his view by arguing that although it restricted space to have constant curvature, it was precisely in such spaces that there could be a plausible mechanics. Indeed, he was at pains to show that a non-Euclidean space could be experienced, and not merely imagined. Houël argued that

> "Après avoir aquis l'idée de grandeur ou d'étendue par la considération du mouvement, . . . nous constatons que certains corps . . . nous présentent toujours . . . des dimensions . . . que nous jugeons être les mêmes. . . . L'idée du mouvement [géométrique] . . . n'est pas une idée plus complexe que celle de grandeur ou d'étendue. On peut même dire, en toute rigueur, que cette idée est identique avec celle de grandeur . . . .

And Beltrami wrote (I quote from the Houël translation) that "Le critérium fondamental de la Géométrie élémentaire consiste dans la *superposition des figures égales.*" None of these writers shared Riemann's appreciation of the role experience plays in the formation of our geometrical concepts, a topic on which Erhard Scholz has written recently, see his [1982]. They took only the simple and intuitive case of bodies moving unaltered, thus restricting attention to spaces of constant curvature, but obtaining far more obviously a group of transformations associated to a geometry.

---

[2] For discussions of Jordan's paper, see D.E. Rowe [1989], Hawkins [1989], and Scholz [1989].

Riemann's ideas were quite different. He based his account of geometry on a series of novel ideas which Scholz discusses below. Among them is the idea that all that is needed to do geometry on a manifold is an infinitesimal line element which can be moved around invariantly. It follows, as he saw, that such a manifold may well have variable curvature[3], and manifolds of constant curvature will appear only as a special case. Such was the novelty of this point of view, and so great are the technical complications in determining curvature in general, that this special case was singled out by him as worthy of attention. Riemann indicated how the expression for the line element would take a simple form when the manifold had constant curvature. If, moreover, the manifold was only two-dimensional (a surface) and if bodies could be moved around freely, then he indicated that the angle sum of a triangle was known in every case when it was known in one. These brief remarks were as near as he came to discussing non- Euclidean geometry; the names of Bolyai and Lobachevskii are missing from his *Habilitationsvortrag*, and it is likely that he did not know of their work. In any case, his own foundations of geometry lay far deeper than their simple antithesis of Euclidean and non-Euclidean geometry[4].

The restriction to the case of constant curvature is thus one way in which the non-Euclidean geometers differed from Riemann. Curiously, as is often remarked, the restriction to constant curvature is artificial for a mathematician, and was shown to be inadequate for physics with the advent of Einstein's general theory of relativity. There is another reason for suggesting that Riemann's influence was not the only one at work: the interrupted and subterranean tradition in the study of the foundations of geometry which places the idea of motion underneath Euclid's *Elements* , usually in order to shore it up. I cite only two Arabic authors to illustrate my point, Thabit ibn Qurra and ibn al Haytham (for details see Jaouiche, [1986]). I mention this to show that it is not necessary to have read Riemann to have come to the idea of motion as fundamental to geometry.

Nonetheless, there is a sense in which Riemann's example was decisive. It certainly inspired Helmholtz, and the difference between Beltrami's *Saggio* and his second paper on the subject, [1869], is due to Beltrami's awareness of the German's work. In the second paper he dealt confidently with the differential geometry of n dímensions and described in detail the non-Euclidean geometry of 3 dimensions. Moreover, as a footnote in the *Saggio* suggests, the effect of reading Riemann's *Habilitationsvortrag* was to enable Beltrami to overcome his misgivings about the ideas in his *Saggio* and to proceed with its publication[5].

Among mathematicians, if not among philosophers, the publications of the articles by Riemann, Helmholtz, and Beltrami were decisive in establishing the acceptance of non-Euclidean geometry. It is customary among historians to turn next to Klein's pamphlet "Vergleichende Betrachtungen über neuere geometrische Forschungen", better known as the Erlanger Programm of 1872, although its involvement with non-Euclidean geometry is distinctly slight. The sequence of ideas that Klein, then only 23 and newly enrolled as a Professor at Erlangen, chose to present then went as follows. Projective geometry, he began, was the most important development of geometry of the last 50 years. However, a general principle capable of embracing the conceptions of ordinary and projective figures

[3] Riemann generally considered the sectional curvatures at a point.

[4] I am indebted to Erhard Scholz for his discussion of Riemann's ideas.

[5] Houël's active role behind the scenes was also important, and should be illuminated by L. Boi's study of the Beltrami-Houël correspondence .

is required, for during that time geometry has become too broken up. The chief idea is that of a group of transformations, geometric properties are those invariant under the appropriate group. How two such groups may be related was then described, then projective geometry was described in this setting. Then Klein showed how two seemingly different geometries can turn out to be the same in this setting, and gave examples. Various examples of different geometries followed, wherever there was an example of a group acting on a set. Only in Note V did Klein turn to the topic of non-Euclidean geometry, which he had described at length in a paper in the *Mathematische Annalen.* (For whatever reasons he made a point of disassociating himself from the name.) He presented non-Euclidean geometry as a subgeometry of projective geometry.

Even this brief summary makes clear that Klein's message was to present (almost) all geometries as types of projective geometry; he was classifying them. His tool was the recognition that all of them were unified around the idea of groups of transformations acting on spaces. However, although there is much in this famous pamphlet that merits, and is finally acquiring, reconsideration, we can content ourselves with this reminder of what is well-known. It is the first place where the idea that the transformations of a geometry form a group is presented as central to the very definition of a geometry. But Tom Hawkins [1984] has argued convincingly that the initial impact of the *Erlanger Programm* was slight and that it had in its early years, 'but a limited circulation' (to quote Klein himself on the occasion of its English translation.) It was not this pamphlet but the papers Klein published through the 1870s which spread his message. Prominent among these is one which I shall describe only to indicate how, as late as 1878, Klein saw non-Euclidean geometry.

The paper is his long paper "Über die Transformationen siebenter Ordnung der elliptischen Functionen". In it he introduces a group of matrices (with integer entries and determinant 1; strictly speaking this is $PSL(2,\mathbb{Z})$ acting on the upper-half plane, and whose fundamental domain (the term is modern, but the concept is there) gives a Riemann surface of genus 3. The appropriate quotient group, which has order 168, is the group of symmetries of the surface. Klein showed that the Riemann surface is a plane projective curve, and so its symmetry group should have been found by Jordan when he listed the finite subgroups of $Sl(3,\mathbb{C})$, but it had been overlooked. Klein concluded the paper with a thorough description of how the curve looks. It is very easy to say how Klein saw non- Euclidean geometry fitting into this story. He did not. The language of non-Euclidean geometry is nowhere used, and it does not seem to have been considered. This is surprising, when one considers how natural it would have been to see the upper-half plane as representing non-Euclidean space.

The situation with Poincaré is very different. He himself described it once, when he talked to the *Société de Psychologie* in Paris in 1908[6]. Not so many years ago I was fortunate enough to find three unpublished essays he wrote in 1880 which greatly elucidate the matter[7]. We know from the essay that at that time in 1880 Poincaré was interested

[6] That lecture was later printed as an essay, [1909], and forms the centre-piece of Hadamard's book *The psychology of invention in the mathematical field.*

[7] I have discussed them at length in print, [1982] and [1986], and I continue to hope that they will be published in their own right by any French scholar willing to ensure the accuracy of the end-product.

in some geometric transformations. These had to do with the analytic continuation of a complex variable. In the paper he presented to the psychologists he told us that

> "arrivés à Coutances, nous montâmes dans un omnibus [...] [et] l'idée me vint, [...], que les transformations dont j'avais fait l'usage [...] étaient identiques à celles de la géométrie non-euclidienne [...]"

From the first of the supplementary essays we can learn a great deal more, but first a little more background must be described. The original problem Poincaré had in mind had to do with ordinary differential equations in the complex domain. His paper on this topic had been entered for the prize offered by the *Académie des Sciences.* He had obtained a spider's web (in his phrase) of triangles in the unit disc and needed to know if they ever overlapped[8]. Since their sides were generally curved, in fact arcs of circles perpendicular to the unit circle, he had decided to study them by straightening them out, thus obtaining a different figure, still bounded by the unit circle, but in which all the sides were straight. His realisation on boarding the bus was that this second picture was the Kleinian (or projective) one of non-Euclidean geometry and so, necessarily, his original picture must be. But in this picture the transformations mapping one triangle to another preserve angles, and it turned out that the new picture Poincaré had found is a conformal description of non-Euclidean geometry.

The three essays mentioned above are Poincaré's supplements to his prize essay. The first was received by the *Académie* on 28 June 1880. There on p. 11 we read:

> "Il existe des liens étroits entre les considérations qui précèdent et la géométrie non-euclidienne de Lobatchewski. Qu'est-ce en effet qu'une Géométrie? C'est l'étude du *groupe d'opérations* formé par les déplacements que l'on peut faire subir à une figure sans la déformer. Dans la Géométrie euclidienne ce groupe se réduit à des *rotations* et à des *translations*. Dans la pseudogéométrie de Lobatchewski, il est plus compliqué."

This might seem to be the Kleinian view, but I think it is not. There is no mention of projective geometry here, the description is naively metrical. Rather, what is described agrees in spirit with what Beltrami had written, and since that had been published in France in Houël's translation it is surely a much more likely source for Poincaré, who was not a well-read mathematician. This impression is supported by the way in which Poincaré, like Beltrami, referred to Lobachevskii but not Bolyai.

Of interest to us here is the way Poincaré amplified his treatment of non-Euclidean geometry in the second supplement, dated 6 September 1880. Admitting straight away that "Je crains d'avoir manqué de clarté dans mon premier supplément", he began by reviewing all the basic concepts of non-Euclidean geometry in his conformal model. He defined lines as arcs of circles perpendicular to the boundary circles, and angles as the (Euclidian) angle between their tangents. To define the distance between two points he extended the (non-Euclidean) line joining them to the boundary circle and took the logarithm of the cross-ratio of the four points he now had. (This is Klein's definition). Area he defined as a certain integral. The elements of his group of transformations were those of the form

[8] See Poincaré [1912], and Gray [1986].

$$z' = \frac{(az+b)}{(a'z+b')}$$

which preserved the boundary circle. After distinguishing various kinds of non-Euclidean motion he remarked:

> "Les mouvements pseudogéométriques transforment les droites et les cercles pseudogéométriques en droits et en cercles pseudogéométriques, ils conservent les longeurs, les angles, et les surfaces pseudogéométriques. Dóù l'important résultat qui suit : *Il y a entre les longueurs, les angles, et les surfaces pseudogéométriques les mêmes relations qu'entre les longeurs, les angles, et les surfaces géométriques, sauf celles qui sont une conséquence du postulatum d'Euclide.* (Underlined by Poincaré.)[9]

He then proved what he admitted he had earlier not established rigorously, namely that the non-Euclidean disc is decomposable into these attractive triangular patterns. Thereafter he resumed the analysis of his original problems concerning differential equations, and we need not follow him here.

Group theory enters Poincaré's work quite differently from the way in enters Klein's. Both studied discrete subgroups of the group of non-Euclidean transformations. But Poincaré was the one who emphasised the natural metrical geometry of the non-Euclidean plane which therefore imposed in a natural way a metric on the surfaces. Klein preferred the older view of Riemann's, that a surface was given by an equation and so could be regarded as a Riemann surface spread out over the Riemann sphere.

What does this tell us about the status of geometry? It was not long before first Poincaré and then Klein realised that in some way every surface (with a few simple exceptions like the sphere and the torus) was somehow capable of being cut open and realised isometrically as a region of the non-Euclidean plane. This remarkable result, which took much longer to be proved, says that non-Euclidean geometry is the natural geometry on all these surfaces. So it extends to surfaces an appropriate way of thinking metrically. Previously the geometrical language had been algebraic, the curves had been considered as embedded in a projective space. So this is a great extension of one's power to think geometrically about curves and the complex functions defined on them; the whole subject became called geometric function theory. I believe this was a great vindication of the value of thinking about groups when confronted with a problem in geometry, and one much more eloquent than Klein's *Erlanger Programm*. It is comparable in its significance to the work of Lie, and helped incidentally to make the *Erlanger Programm* popular when it was republished in the 1890s.

Poincaré's achievement is all the more striking when one considered the status of complex algebraic curves at the time. Until the 1850s there seems to have been a widespread reluctance to think of an equation $f(x, y) = 0$ as defining a locus in a space of two complex variables. Even when the algebraic treatment of such curves, thought of initially as real curves, required complex points of intersection with other curves, and even complex points of inflection, mathematicians stopped short of considering the locus as consisting of complex points. The first to introduce complex considerations systematically into geometry was von Staudt, in his *Beiträge zur Geometrie der Lage* [1856]. But his synthetic approach did not lend itself to computation. Complex points were re- interpreted, I would

[9] Poincaré, second supplement, p. 2.

say explained away, in terms of various kinds of real involutions. The more analytically or algebraically minded were offered Riemann's approach of thinking of a complex locus as spread out over the complex x-plane, but that too was found to be difficult. Authors like Clebsch and Klein only gradually took the step of admitting complex coordinates into analytic geometry. By 1880 their approach was gaining recognition, but it was far from hallowed by tradition, such was the hold of Euclidean 2- and 3-dimensional space on people's minds. So Poincaré's conception of a complex algebraic curve as also a surface already reflected recent developments in the subject. To endow such surfaces with an intrinsic geometry was therefore a most dramatic step, whether or not that geometry was non-Euclidean[10].

A second point, exemplified not only by Jordan's paper of 1869 but by much of the subsequent work on Riemann surfaces and non-Euclidean geometry, is that henceforth theorems about transformations, and groups of transformations, become the subject matter of geometry, and the study of figures *per se* recedes somewhat. If a geometry has a group, then to study that group is to study geometry. That step takes us a long way from Euclid, although the restriction to constant curvature keeps us (one might say, not too seriously) about the same distance from Cartan.

But there is something else, with which I wish to end. In one of the few major papers in our subject (Nagel, [1939]), Ernest Nagel explored what he believed to be the connection between modern geometry and the rise of modern logic. It was his thesis that projective geometry, with its striking duality between points and lines in the plane, made it difficult for the more critically aware mathematicians to leave geometry resting on an intuitive base. Nagel argued that the example of geometry was vitally important in moving mathematicians away from believing that their subject rested on a careful mode of abstraction, and towards believing that what made mathematics work was its mode of reasoning. In short, that projective geometry pushed Pasch, Peano, and Hilbert towards abstract axiomatics, which, when Nagel was writing, was the language of the mathematical gospel. That this is not the whole story is shown by Toepell's recent interesting study of Hilbert's route to axiomatic geometry (Toepell, [1986]).

I would add that the importance of non-Euclidean geometry was underestimated by Nagel. The idea of groups of transformations can be found under the surface of Hilbert's systems, of course, as it was all along, so to speak, in Euclid's. But the novelty of Hilbert's approach was that it took geometry beyond the setting of algebraic varieties and Riemannian manifolds. The importance of non-Euclidean geometry was its physical and intuitive plausibility. It struck directly at the idea that one's intuition, whether or not it was supported by some sort of experience, could supply a geometrical description of space. What is striking is the essential realism of all the proponents of non-Euclidean geometry I have discussed. All of them take positions that are either explicitly concerned to establish the (possible) reality of non-Euclidean geometry, or else that are compatible with it. Recently, Corinna Mette has published a study [1986] arguing that Poincaré's conventionalism can be understood in the light of his early involvement with the concepts of mathematical invariants. Interesting though her book is, and surely correct in its attempt to draw out Poincaré's philosophical ideas by seeking their roots in his mathematical work, I think the emphasis on invariants is misplaced. Poincaré's work is as good an example as you

[10]Poincaré did not speak of the geometry as intrinsic in the sense of Gauss. The first to put forward that point of view explicitly was Klein, and, in a different way, Segre.

can find of group-theoretic concepts being used to do what invariant theory would have been used for a generation earlier. It may also be that Poincaré's early work contains seeds of his later intuitionism, if it can be so described, and his hostility to the ideas of Zermelo and Couturat on the foundations of mathematics.

The whole enterprise of seeking to understand the later, philosophical Poincaré by reconsidering his early, strictly mathematical work is obviously precarious methodologically. But it is testimony to the importance of non-Euclidean geometry that its discovery continues to raise important historical, epistemological, and mathematical questions. I have tried to show how in Poincaré's hands it became a show-case of the viewpoint according to which one's intuitive ideas rested on a (previously concealed) use of the idea of groups of transformations, and how it led to a new way of introducing geometrical ideas into the study of curves and Riemann surfaces.

References

Beltrami, E. (1868) "Saggio di interpetrazione della geometria non-euclidea", Giornale di matematiche, 6, 248-312, *Opere matematiche*, I, 375-405.

Bonola, R. (1912) *Non-Euclidean geometry ; a critical and historical study of its development*, English translation by H.S. Carslaw of *La Geometria non-euclidea*, (1906), Dover reprint 1955.

Gray, J.J. (1982) "The three supplements to Poincaré's prize essay of 1880 on Fuchsian functions and differential equations", *Archives internationales d'histoire des sciences* , 32, 221-235.

Gray, J.J. (1987) "The discovery of non-Euclidean geometry", *Studies in the history of mathematics*, ed. E.R. Phillips, MAA Studies in mathematics, vol 26, 37-60.

Gray, J.J. (1989) *Ideas of space; Euclidean, non-Euclidean, and relativistic*, Oxford University Press, 2nd edition.

Hawkins, T. (1984) "The *Erlanger Programm* of Felix Klein: Reflections on its Place in the History of Mathematics", *Historia Mathematica* , 11, 442-470.

Hawkins, T. (1989) "Line geometry, differential equations, and the birth of Lie's theory of groups", in D.E. Rowe, J. McCleary (eds) *The history of modern mathematics*, vol 1, 275-330.

Helmholtz, H. von (1866) "Über die thatschlichen Grundlagen der Geometrie" , *Verhandlungen des naturhistorischen- medecinischen Vereins zu Heidelberg*, 4, 197-202.

Helmholtz, H. von (1868) "Ueber die Thatsachen, die der Geometrie zu Grunde liegen", *Nachrichten von der königlichen Gesellschaft der Wissenschaften zu Göttingen*, 193-222.

Helmholtz, H. von (1870,1876) "The origin and meaning of geometrical axioms", *Mind*, 1, 301-321.

Houël, G.J. (1867) *Essai critique sur les principes fondamentaux de la géométrie élémentaire, etc* .

Jaouiche, K. (1986) *La théorie des parallèles en pays d'islam*, Vrin, Paris.

Jordan, C. 1868/9. "Mémoire sur les groupes de mouvements" *Annali di matematiche*, II, 167-215 and 322-345, *Œuvres* , IV, 231-302.

Klein, C.F. (1872) *Vergleichende Betrachtungen ber neuere geometrische Forschungen*, Deichert, Erlangen, *Gesammelte mathematische Abhandlungen*, 1, 460-497.

Klein, C.F. 1879 "Über die Transformationen siebenter Ordnung der elliptischen Functionen", *Mathematische Annalen*, 14, *Gesammelte mathematische Abhandlungen*, III, 90-134.

Koenigsberger, L. *Hermann von Helmholtz*, English translation, abridged, F.A. Welby, Dover reprint 1965.

Legendre, A.M. (1860) *Eléments de géométrie*, 14th edition, Paris
Mette, C. (1986) *Invariantentheorie als Grundlage des Konventionalismus*, die blaue eule, Essen.
Nagel, E. (1939) "The formation of modern conceptions of formal logic in the development of geometry", *Osiris*, 7.
Poincaré , J.H.. (1908) "L'invention mathématique", *Science et méthode* , 43-63.
Pont, J.-C. (1986) *L'aventure des parallèles*, Lang, Berne.
Richards, J. *Mathematical visions*, Academic Press, San Diego, 1988.
Riemann, B. (1854, 1867) "Ueber die Hypothesen, welche der Geometrie zu Grunde liegen" Abhandlungen der königlichen Gesellschaft der Wissenschaften zu Göttingen, 13, 1-20, *Werke*, (1876) 272-287.
Rowe, D.E. (1989) "The early geometrical works of Sophus Lie and Felix Klein", in D.E. Rowe, J. McCleary (eds) *The history of modern mathematics*, vol 1, 209-274.
Scholz, E. (1982) "Herbart's influence on Bernhard Riemann", Historia Mathematica , 9, 413-440.
Scholz, E. (1989) "Crystallographic symmetry concepts and group theory", in D.E. Rowe, J. McCleary (eds) *The history of modern mathematics*, vol 2, 3-28.
Von Staudt, (1856) *Beiträge zur Geometrie der Lage.*
Toepell, M.-M., (1986) "*Über die Entstehung von David Hilberts 'Grundlagen der Geometrie'* " Vandenhoeck and Ruprecht, Göttingen.

Klein, Lie, and the "Erlanger Programm"·

David E. Rowe

Department of Mathematics, Pace University, Pleasantville, New York, USA[1].

Felix Klein's *Erlanger Programm* [Klein 1872b] has long been regarded as a milestone in the history of geometry. First published as a pamphlet in connection with Klein's appointment to a chair in Erlangen, its fame rests on the radically new insight that one can meaningfully classify different geometries by distinguishing their respective transformation groups. In modern terms, this amounts to the following notion: Klein regarded two geometrical spaces M and M'—he referred to *Mannigfaltigkeiten*, but employed this term in a more restricted sense than is done today[2]—with associated transformation groups G and G' as essentially identical if there exists a bijective mapping $f : M \rightarrow M'$ which induces an isomorphism $f^* : G \rightarrow G'$ between the transformation groups. Having located the key nexus of ideas for characterizing a geometry, Klein then formulated the objective of geometrical researches in the following language:

> Es ist eine Mannigfaltigkeit und in derselben eine Transformationsgruppe gegeben; man soll die der Mannigfaltigkeit angehörigen Gebilde hinsichtlich solcher Eigenschaften untersuchen, die durch die Transformationen der Gruppe nicht geändert werden.... [oder] Man entwickle die auf die Gruppe bezügliche Invariantentheorie.[Klein 1872b, 463-464]

These central features of Klein's *Erlanger Programm* are, of course, famous and familiar to everyone with a passing knowledge of the history of geometry. It appears, however, that relatively few are aware of the actual historical context that shaped Klein's ideas or the curious and convoluted manner in which they entered into the mainstream of mathematics. In fact, these questions have only received serious attention by historians of mathematics rather recently (Cf. [Hawkins, 1984, 1989], [Rowe, 1989]). This situation has created a certain folklore surrounding the *Erlanger Programm* that, in my opinion, largely misrepresents the work's significance while overlooking most of what makes it a rich and rewarding classic in the history of mathematics.

A few remarks regarding the origins of the work should be made at the outset. Although it contains many pronouncements, like the one quoted above, that evoke a programmatic ring, one should bear in mind that Klein's essay, "Vergleichende Betrachtungen über neuere geometrische Forschungen," became known as the *Erlanger Programm* for reasons that have nothing to do with its contents, but rather because it was submitted

[1] The research presented herein was partially funded by National Science Foundation grant DIR 8821421. The author would like to thank Thomas Hawkins for several stimulating conversations on the subject of this paper.

[2] See [Scholz 1980, pp. 131-136].

as a “Programm zum Eintritt in die philosophische Fakultät” at Erlangen. Furthermore, the oft-repeated statement that the *Erlanger Programm* formed the subject of Klein's speech before the faculty senate, his “Erlanger Antrittsrede,” also stands in need of correction (see [Rowe, 1985]). A more significant misconception surrounding the *Erlanger Programm* concerns the nature of its influence on contemporary geometers. As [Hawkins 1984] has noted, outside of a handful of Klein's close friends (among them Sophus Lie and Max Noether) very few mathematicians appear to have seen the work prior to the publication of a series of translations, beginning with Gino Fano's Italian translation of 1890, and its republication in *Mathematische Annalen* in 1893.[3] Thus if the *Erlanger Programm* is to be viewed as initiating a new program for geometrical research, Hawkins' analysis strongly suggests that its impact was rather marginal and certainly delayed.

The inner logic of the *Erlanger Programm* would seem to lead naturally to a research program for classifying geometries by means of group theory. This approach had, in fact, already been employed by Klein and Lie in their work on W-curves in the plane [Klein and Lie, 1871], i. e. those curves left invariant by a 1-parameter group of commutative linear transformations. One also finds an allusion to a grandiose conception of a “Galois theory” for geometry in the *Schlussbemerkungen* of the *Erlanger Programm*, although this would seem to have been little more than a vague analogy. As Hawkins has emphasized, back in 1872 neither Klein nor Lie appear to have taken any serious interest in pursuing research on the classification of continuous groups. Indeed, their preliminary (unpublished) work on W-curves in 3-space had come to a standstill simply because they found that the number of cases that needed to be handled became intractable.[4]

Of course, Lie took up the question of classifying continuous groups and their differential invariants little more than a year later, but it was his new insights into contact transformations and their applications to systems of ordinary and partial differential equations that led him to do so.[5] Lie was thoroughly familiar with the contents of the *Erlanger Programm* through numerous and detailed discussions with Klein that took place during September and October of 1872 while Klein composed it. He had also criticized an earlier version, no longer extant, written in December of the previous year. Although on more than one occasion he expressed a high regard for the novel ideas set forth in the final version, Lie never regarded the *Erlanger Programm* as providing an important impulse for group-theoretic research, as the following passage from a later letter to Klein indicates:

> Dass ich Deine Programmschrift sehr schätze, weiss Du. Ich habe das Wort 'gedankenreich' eingeschaltet. Ganz sicher ist sie mir anregend gewesen. Aber anderseits ist zu bemerken, dass die Aufgabe: **alle** Gruppen zu bestimmen nicht in Deiner Schrift gestellt wurde; und wohl aus dem Grunde dass eine solche Aufgabe sich damals Dir wie mir als absurd oder unmöglich stellte. Anderseits giebt Deine Schrift auch nicht Mitteln zur Erledigung meines Problems, jedenfalls nicht andere als solche die früher bekannt waren.[6]

[3] See [Hawkins 1984]; following the Italian translation, translations of the *Erlanger Programm* appeared in French (1891), English (1893), Russian (1896), Hungarian (1897), and Polish (1905).

[4] See [Hawkins 1989, pp. 308-309] and [Rowe 1989, pp. 236-239].

[5] For details, see [Hawkins 1989, 1990].

[6] This passage is quoted in [Hawkins 1989, p. 321].

Lie's assessment of the situation is not only apt, it also suggests that those who have attached such great significance to the group-theoretic implications of the *Erlanger Programm*, viewing it essentially as the first stage in a program for classifying continuous groups, have not fully appreciated its author's intentions. In fact, Klein's principal goal focused on bringing the group concept to bear on geometry as the subject appeared in 1872. He hoped thereby to restore some order to a field that had become notoriously disjointed, with practitioners who sometimes behaved as if geometry had descended to a veritable Tower of Babel condition. Klein addressed this issue in some detail in the first two notes he appended to the *Erlanger Programm*. The first of these dealt with the antagonism between the synthetic and analytic approaches to geometry, a tension he regarded as no longer holding much significance, whereas the second criticized the present-day separation of geometry into subdisciplines.[7]

Thus, Klein's ideas had little to do with extending the frontiers of an already over-extended subject, and what he sought instead was to bring some cohesion into an otherwise chaotic state of affairs. The group concept obviously played an indispensible role in this, but it served primarily as an organizing principle for subject matter already at hand. Indeed, nearly all the examples in the *Erlanger Programm* involved familiar groups acting on well-known spaces, and when this was not case—as in §8, which briefly discussed Cremona transformations, analysis situs, and Graßmann's theory—the fertility of Klein's approach for these more general situations was far less clear.

In short, what makes the *Erlanger Programm* so fascinating is not the group theory but the geometry connected with it. Its interest derives from Klein's virtuoso command of a wealth of striking examples spun together to make a profound point: from a modern perspective, what Klein accomplished was to provide geometers with a clear and (until about 1920) complete conception of what their subject was all about. He never intended, however, that the answer be abstract, axiomatic, or shorn of the rich material that motivated it, but rather that it should differentiate between what was *essential* and *inessential* when one considered a space and a certain group of transformations acting on it. The following passage from the *Erlanger Programm* illustrates, for example, that, for Klein, the important thing was not the nature of the fundamental elements within a geometrical space but rather how different systems of geometry could be structurally related even when their underlying space elements had no connection with one another whatsoever:

> Instead of the points of a line, plane, space, or any manifold under investigation, we may use instead any figure contained within the manifold: a group of points, curve, surface, etc. As there is nothing at all determined at the outset about the number of arbitrary parameters upon which these figures should depend, the number of dimensions of the line, plane, space, etc. is likewise arbitrary and depends only on the choice of space element. *But so long as we base our geometrical investigation on the same group of transformations, the geometrical content remains unchanged.* That is, every theorem resulting from *one* choice of space element will also be a theorem under any other choice; only the arrangement and correlation of the theorems will be changed. The essential thing is thus the group of transformations; the number of dimensions to be assigned to a manifold is only of secondary importance.[8]

[7] [Klein 1872b, pp. 490-491].

[8] [Klein 1872b, pp. 470-471].

The full implications of these remarks can only be understood by referring to the rich panoply of geometrical structures Klein had in mind when he wrote these words. In what follows, I will briefly sketch at least some of the key geometrical concepts that served as source material for Klein's *Erlanger Programm.* A thorough treatment of this subject would necessarily include a discussion of: 1) line geometry according to Plücker and Klein; 2) Klein's elaboration of Cayley's work on the foundations of geometry; and 3) Lie's insight into the connection between line geometry and sphere geometry. Considerations of space, however, require me to confine most of my remarks to topic 1), which in many repects is the least familiar of the three. For further details on Lie's early geometrical work and his collaboration with Klein, the reader should consult [Hawkins 1989] and [Rowe 1989].

Line Geometry according to Plücker and Klein

Plücker's approach to line geometry, as set forth in his *Geometrie des Raumes* [Plücker, 1868,1869] and elaborated by his student Felix Klein,[9] represented the 4-parameter family of lines in $P^3(\mathbb{C})$ with six homogeneous line coordinates (p_{ij}). These (p_{ij}) can be obtained by taking any two points (x_1, x_2, x_3, x_4) and (y_1, y_2, y_3, y_4) on a given line ℓ and forming the six possible 2×2 determinants $\rho(p_{ij}) = x_i y_j - x_j y_i$. The (p_{ij}) thereby satisfy the fundamental relation:

$$P = p_{12}p_{34} + p_{13}p_{42} + p_{14}p_{23} = 0.$$

One also has the possibility of a dual representation by taking any two planes (u_1, u_2, u_3, u_4) and (v_1, v_2, v_3, v_4) that contain ℓ and forming $\sigma(q_{ij}) = u_i v_j - u_j v_i$, where the q_{ij} satisfy the same identity:

$$Q = q_{12}q_{34} + q_{13}q_{42} + q_{14}q_{23} = 0.$$

The coordinates (p_{ij}) and (q_{ij}) represent the same line if

$$\rho p_{ij} = \frac{\partial Q}{\partial q_{ij}} \qquad (\sigma q_{ij} = \frac{\partial P}{\partial p_{ij}}).$$

Moreover, the condition for two lines (p_{ij}) and (p'_{ij}) to intersect is given by:

$$\sum p_{ij} q'_{ij} = \sum q_{ij} p'_{ij} = 0$$

or

$$\sum p_{ij} \frac{\partial P}{\partial p'_{ij}} = \sum q_{ij} \frac{\partial Q}{\partial q'_{ij}} = 0.$$

That the first quantity vanishes is readily seen by choosing two pairs of points x, y and x', y' on each of the two given lines ℓ and ℓ'. Since these four points lie in a plane, the 4×4 determinant $|xyx'y'| = 0$. Expanding this determinant out by its 2×2 minors yields the above identities. These sorts of calculations permeate Plücker's approach to line geometry, reflecting close ties with projective methods, especially invariant theory.

9 The exposition that follows is based primarily on [Klein 1869], [Klein 1871c], and [Klein 1892-93].

The fundamental objects of line geometry are called line complexes. An nth degree line complex is determined by a homogeneous nth degree equation $F(p_{ij}) = 0$. Its local behavior can be studied by fixing a point $x = (x_1, x_2, x_3, x_4)$, so that $F(p_{ij}) = F(x_i y_j - x_j y_i) = 0$ results in an nth degree equation in (y_1, y_2, y_3, y_4); thus, the set of lines through x determine a cone of the nth order. The intersection of two line complexes, $\Omega_m = 0$, $\Omega_n = 0$ results in a so-called *congruence* with order and class equal to mn. It consists of a 2-parameter family of lines with mn lines passing through a point in general position and with mn lines lying in a general plane. Except for special cases, the lines of a congruence envelope a so-called caustic surface (*Brennfläche*) with two components. If one adjoins a third line complex $\Omega_\ell = 0$, then the intersection with the other two yields a ruled surface of order $2\ell mn$, and intersecting with a fourth complex Ω_o results in $2\ell mno$ lines in space.

The theory of first-degree or linear line complexes clearly illustrates the central role played by invariant theory in this field. Such complexes have the form $\Omega_1(p_{ij}) = \sum a_{ij} p_{ij} = 0$, which leads to what is otherwise known as a *Nullsystem* of lines, an object that Möbius had studied in some detail.[10] Following the approach of Plücker and Klein, the expression $A = a_{12}a_{34} + a_{13}a_{42} + a_{14}a_{23}$ is called the *invariant* of the complex. Its geometric significance can be interpreted as follows: if $A = 0$, then the (a_{ij}) satisfy the identity $P = 0$ and therefore can themselves be viewed as line coordinates. By setting $a_{ij} = q'_{ij}$, one has $\sum a_{ij} p_{ij} = \sum q'_{ij} p_{ij} = 0$, and the complex consists of all lines (p_{ij}) that intersect a fixed line (q'_{ij}), a so-called *special linear complex*.

As noted above, the intersection of two linear complexes, $\sum a_{ij} p_{ij} = 0$ and $\sum a'_{ij} p_{ij} = 0$, determines a congruence K of the first order and class. In fact, the lines $\ell = (p_{ij}) \in K$ satisfy the condition:

$$\sum (\lambda a_{ij} + \lambda' a'_{ij}) p_{ij} = 0, \quad \forall \lambda, \lambda' \in \mathbb{C}.$$

This 1-parameter family of linear complexes contains two special complexes. The condition for this being:

$$(\lambda a_{12} + \lambda' a'_{12})(\lambda a_{34} + \lambda' a'_{34}) + (\lambda a_{13} + \lambda' a'_{13})(\lambda a_{42} + \lambda' a'_{42}) +$$
$$(\lambda a_{14} + \lambda' a'_{14})(\lambda a_{23} + \lambda' a'_{23}) = 0$$

or

$$\lambda^2 A + \lambda'^2 A' + \lambda\lambda'(a_{12}a'_{34} + a_{13}a'_{42} + a_{14}a'_{23}) = 0.$$

This leads to a quadratic equation in λ/λ' whose solutions determine the two special complexes. Thus the congruence K consists of all lines ℓ that meet two fixed lines, the *Leitlinien* of K. The *Leitlinien* can be real or conjugate imaginary, and if the equation contains a double root they coincide, in which case K is not a true congruence but rather a special linear complex. In the special case where $A = A' = 0$ and $a_{12}a'_{34} + a_{13}a'_{42} + a_{14}a'_{23} = 0$, the above condition holds automatically and the two complexes $\sum a_{ij} p_{ij} = 0$ and $\sum a'_{ij} p_{ij} = 0$ are both special: $(a_{ij}) = \ell_1$, $(a'_{ij}) = \ell_2$ with $\ell_1 \cap \ell_2 = \{p\} \neq \emptyset$. Here K has degenerated into the union of two congruences: K', the congruence of first order and class zero consisting of all lines through point p, and K'', the congruence of order zero and class one composed of all lines in the plane determined by ℓ_1 and ℓ_2.

---

10 See, for example, [Lie 1896, p. 272].

The expression

$$a_{12}a'_{34} + a_{13}a'_{42} + a_{14}a'_{23} = \sum a'_{ij}\frac{\partial A}{\partial a_{ij}}$$

was introduced by Klein as the *joint invariant* or *simultaneous invariant* of the two linear complexes $\sum a_{ij}p_{ij} = 0$ and $\sum a'_{ij}p_{ij} = 0$, a concept of central importance for line geometry. When it vanishes, the complexes are said to lie in involution with respect to one another, a notion that generalizes the usual concept of involution in projective geometry.

Klein made use of this notion in a bold new setting: he treated the coefficients (a_{ij}) of the linear complex $\sum a_{ij}p_{ij} = 0$ as coordinates in $P^5(\mathbb{C})$. The special linear complexes are then those that satisfy the condition $A = 0$, and therefore constitute a 4-dimensional submanifold $M_4^{(2)}$ (the superscript (2) designates the degree of A). The general linear equation in these new coordinates $\sum a_{ij}B_{ij} = 0$ with fixed (a_{ij}) and variable (B_{ij}) represents a 4-parameter family of linear complexes (B_{ij}) all of which lie in involution with respect to (a_{ij}). Similarly two such equations

$$\sum a_{ij}B_{ij} = 0, \quad \sum a'_{ij}B'_{ij} = 0$$

can be combined into the single equation $\sum(\lambda a_{ij} + \lambda' a'_{ij})B_{ij} = 0$ which asserts that the (B_{ij}) form a 3-parameter family of linear complexes each of which lies in involution with respect to the 1-parameter family $\lambda a_{ij} + \lambda' a'_{ij}$. Finally, three such equations produce two families of 2-parameter linear complexes, each member of the one family lying in involution with respect to all the members of the other. Their mutual relationship can be partially visualized by recalling that the intersection of three linear complexes $(a_{ij}), (a'_{ij}), (a''_{ij})$ yields a ruled surface S of order $2 \cdot 1 \cdot 1 \cdot 1 = 2$, i. e. a hyperboloid. By an argument similar to the one made above, one notes that the family $\lambda a_{ij} + \lambda' a'_{ij} + \lambda'' a''_{ij}$ contains a one-parameter system of *special* linear complexes as does the family (B_{ij}). These two systems of special linear complexes can thus be identified with the two systems of generators for the surface S. The fact that each special complex lies in involution with respect to all the members of the other family—i.e. that their joint invariant vanishes—amounts to the condition that each line in one set of generators intersects all the lines of the other set.

Klein and Lie utilized the notion of involutionary line-complexes in a wide variety of settings. One of the most striking, however, was Klein's reformulation of Dupin's Theorem for line geometry:

> If four complexes lie pairwise in involution with respect to a common line ℓ, and if furthermore any three of the four also lie pairwise in involution with respect to the next [infinitesimally close] nearby line common to all three, then their common ruled surface intersects the *Brennfläche* belonging to any two of the three complexes near the line ℓ in the direction of an asymptotic curve.[11]

To make this a little more precise, it should be noted that this theorem concerns second-degree line complexes. For this purpose, Klein introduced new coordinates $x_1, \ldots, x_6$ which are homogeneous functions of the p_{ij} satisfying the condition

[11] [Klein 1871a, p. 99].

$$x_1^2 + x_2^2 + \ldots + x_6^2 = 0.$$

He further showed that the 1-parameter family of second-degree complexes

$$C(\lambda) = \sum \frac{x_i^2}{\lambda - k_i} = 0$$

has the property that given any line ℓ, there exist $\lambda_1, \lambda_2, \lambda_3, \lambda_4$ such that $\ell \in C(\lambda_i), i = 1, \ldots, 4$, and where $C(\lambda_i)$ and $C(\lambda_j)$ lie in involution with respect to ℓ (meaning that the expression $\sum \frac{\partial A}{\partial x_i} \frac{\partial B}{\partial x_i}$ vanishes). Here the ruled surface S determined by any three of the $C(\lambda_i)$ meets the *Brennfläche* determined by any two of them along an asymptotic curve. If, in particular, $C(\lambda_1) = C(\lambda_2)$, then the congruence they determine consists of the singular lines belonging to this complex, and in this case the *Brennfläche* degenerates into the singularity surface common to all the $C(\lambda_i)$ (the points of $P^3(\mathbb{C})$ where the 2nd-order cone of lines degenerates into two planes) and a second surface of the 12th order and class. Since the singularity surface for families like the $C(\lambda_i)$ is a Kummer surface, Klein's approach can thus be used (at least in principle) to determine the asymptotic curves for Kummer surfaces.

Already back in [Klein 1869, p. 79], Klein had noted the formal analogy between the equation for the $C(\lambda)$ and the the confocal quadric surfaces $\sum_{k=1}^{3} \frac{x_i^2}{a_i - \lambda} = 1$. The latter, of course, provided the classic illustration of Dupin's Theorem: that a system of three mutually orthogonal families of surfaces intersect one another along lines of curvature. Still, the connection remained obscure until the spring of 1870, when Klein was visiting Paris and became aware of the work of Darboux and Moutard on confocal cyclides, surfaces which subsume the confocal quadric surfaces and Dupin cyclides as special cases. Using a system of pentaspheric coordinates, these may be written in the form:

$$\sum_{k=1}^{5} \frac{y_k^2}{\alpha_k - \lambda} = 0.$$

Klein later recalled the impression this work made on him:

> I remember how one day Darboux showed me a manuscript giving a detailed treatment of the theory of cyclides, and how he added the remark that he had obtained the same formula as I had earlier only with five variables instead of six. There was no doubt that here there must exist a transfer principle [*Übertragungsprincip*] connecting line geometry with metric geometry.....[12]

Since the confocal cyclides were intimately tied to Dupin's Theorem, Klein naturally sought to frame a suitable analogy for line geometry. Before he was able to do this, however, Lie made the seminal discovery of a contact transformation that maps lines to spheres and in so doing sends the asymptotic curves of one surface over into the lines of curvature of the image surface. The immediate payoff came when Lie found that he could use the results of Darboux and Moutard, who had found the lines of curvature for their generalized cyclides, in order to determine the asymptotic curves of a Kummer

[12] Quoted from F. Klein, "Über Lie's und meine Arbeiten aus den Jahren 1870-72," *Lie Nachlass, Universitetsbibliotheket, Oslo.*

surface. This was precisely the result that Klein eventually generalized in his new version of Dupin's Theorem.

Klein's search for a suitable line-geometric formulation of Dupin's Theorem also led him to the insight that "line geometry is equivalent to metric geometry in four variables,"[13] an idea he spelled out soon afterward in an important paper, "*Über Liniengeometrie und metrische Geometrie*" [Klein 1871c]. Although written in October 1871, this work already contained many of the fundamental leitmotifs and examples that Klein took up one year later in the *Erlanger Programm*.

The construction ran as follows: Klein considered those linear transformations $T : P^5(\mathbb{C}) \to P^5(\mathbb{C})$ where the coordinates (p_{ij}) are subject to the single condition that $M_4^{(2)} = \{(p_{ij}) : P = 0\}$ remains invariant. He then showed that these mappings T send lines to lines and points to either points or planes. Thus when restricted to $M_4^{(2)}$ they are identical with the totality of all collineations and dual transformations on the lines in space. He concluded that: "All of line geometry can thereby be reduced to the following problem: one investigates the projective properties of the $M_4^{(2)}$ in $P^5(\mathbb{C})$; then one translates the results into the language of line geometry."[14] Naturally, this construction could also be interpreted, following Klein, in terms of the 5-dimensional space of all linear complexes where the submanifold $M_4^{(2)}$ represents the special linear complexes.

To see how he made the connection with metric geometry, it may be helpful to begin with the following example cited in the *Erlanger Programm*.[15] Consider the stereographic projection $f : S \to P^2(\mathbb{C})$ from a fixed point O on a 2nd-degree surface S in $P^3(\mathbb{C})$. This mapping is one-to-one except for two points O' and O'' in the range whose preimages are the two generators of S that pass through O. Now when the group of linear transformations of $P^2(\mathbb{C})$ that leaves O' and O'' fixed is pulled back by f^{-1}, one obtains the group of linear transformations of S that fix O, i. e., the restrictions to S of those linear transformations of $P^3(\mathbb{C})$ that leave both S and O invariant. Furthermore, since any two points of $P^2(\mathbb{C})$ are projectively equivalent, one can take O' and O'' to be the circular points at infinity. But, by a well-known theorem of Cayley, the linear transformations of $P^2(\mathbb{C})$ that leave these points fixed are precisely the transformations of plane Euclidean geometry. Klein's conclusion: the study of projective invariants on a 2nd-degree surface with a single point held fixed is equivalent to studying ordinary Euclidean geometry in the plane.

Klein employed essentially the same argument to prove the general case in section 2 of "*Über Liniengeometrie und metrische Geometrie.*" There he observed that if $M_{n-1}^{(2)}$ is a hypersurface in $P^n(\mathbb{C})$ given by a 2nd-degree equation, then one can exhibit a mapping $f : M_{n-1}^{(2)} \to P^{n-1}(\mathbb{C})$ analogous to the one given above. Thus, in the domain space, it has a singular point O, the projection point, and elsewhere it is one-to-one. In the range space $P^{n-1}(\mathbb{C})$, on the other hand, the singular set will be a hypersurface $M_{n-3}^{(1,2)}$ determined by one linear and one quadratic equation. This is the same type of hypersurface that arises in the ordinary metric geometry of $P^{n-1}(\mathbb{C})$, and since any two such hypersurfaces are projectively equivalent, it follows that one can replace this $M_{n-3}^{(1,2)}$ by

$$S^{n-3} = \{(x_1, \ldots, x_n) | x_1^2 + \ldots + x_n^2 = 0, x_n = 0\},$$

[13] First announced in [Klein 1871a, p. 99], but proven in [Klein 1871c].

[14] [Klein 1871c, p. 112].

[15] [Klein 1872b, pp. 469-470].

the $(n-3)$-dimensional imaginary sphere at infinity, which is simply the higher-dimensional analogue of the spherical-circle at infinity. Klein then showed that the group of linear transformations of $P^n(\mathbb{C})$ leaving $M^{(2)}_{n-1}$ invariant corresponds to the group of all similarity mappings and transformations by reciprocal radii acting on $P^{n-1}(\mathbb{C})$. When one further restricts the group of linear transformations acting on $P^n(\mathbb{C})$ to those which leave O as well as $M^{(2)}_{n-1}$ fixed, then one obtains exactly those linear transformations of $P^{n-1}(\mathbb{C})$ that leave S^{n-3} invariant, i. e. the motions of $(n-1)$-dimensional Euclidean geometry. The case when $n = 5$ establishes an isomorphism bewteen line geometry and the metric geometry of 4-space.

This background at least provides a glimpse of how Klein's work in line geometry informed his broad-ranging conception of geometry as set forth in the *Erlanger Programm*. Here, as elsewhere, he certainly had many important predecessors who laid out a good deal of the critical groundwork for him, and Klein was, indeed, someone who knew how to stand on the shoulders of giants—among them Plücker, Cayley, Lie, Riemann, and Grassmann. Still, it must be admitted that none of these distinguished figures ever developed a conception of geometry that even approaches the bold and sweeping vision of the *Erlanger Programm*.

References

Hawkins, Thomas, 1984. "The *Erlanger Programm* of Felix Klein : Reflections on its Place in the History of Mathematics," *Historia Mathematica*, 11(1984), 442–470.

Hawkins, T., 1989. "Line Geometry, Differential Equations, and the Birth of Lie's Theory of Groups," D. Rowe and J. McCleary, eds., *The History of Modern Mathematics*, vol. 1, (Boston: Academic Press, 1989), pp. 275-327.

Hawkins, T., 1990. "Jacobi and the Birth of Lie's Theory of Groups." Preprint.

Klein, Felix, 1868. "Ueber die Transformation der allgemeinen Gleichungn des zweiten Grades zwischen Linien-Coordinaten auf eine canonische Form," *GMA*, vol. 1, pp. 3-49.

Klein, F., 1869. "Zur Theorie der Linienkomplexe des ersten und zweiten Grades," *GMA*, vol. 1, pp. 53–80.

Klein, F., 1871a. "Über einen Satz aus der Theorie der Linienkomplexe, welcher dem Dupinschen Theorem analog ist," *GMA*, vol. 1, pp. 98-105.

Klein, F., 1871b. "Über die sogenannte Nicht-Euklidische Geometrie," (*Erster Aufsatz*), *GMA*, vol. 1, pp. 254–305.

Klein, F., 1871c. "Über Liniengeometrie und metrische Geometrie," *GMA*, vol. 1, pp. 106-126.

Klein, F., 1872a. "Über die sogenannte Nicht-Euklidische Geometrie," (*Zweiter Aufsatz*), *GMA*, vol. 1, pp. 311–343.

Klein, F., 1872b. *Vergleichende Betrachtungen über neuere geometrische Forschungen.* Programm zum Eintritt in die philosophische Facultät und den Senat der k. Friedrich-Alexanders-Universität zu Erlangen. (Erlangen: Deichert, 1872), *GMA*, vol. 1, pp. 460–497.

Klein, F., 1892-93. *Vorlesungen über höhere Geometrie*, 3rd. ed. revised and edited by Wilhelm Blaschke. (New York: Chelsea, 1949).

Klein, F., 1921, 1922, 1923. *Gesammelte Mathematische Abhandlungen GMA*, 3 vols. (Berlin/-Heidelberg/New York: Springer, 1921–23, reprinted 1973).

Klein, F., 1926. *Vorlesungen über die Entwicklung der Mathematik im 19. Jahrhundert*, vol. 1 (Berlin: Springer, 1926).

Klein, F. and Lie, S.,1871. "Über diejenigen ebenen Kurven, welche durch ein geschlossenes System von einfach unendlich vielen vertauschbaren linearen Transformationen in sich übergehen," *GMA*, vol. 1, pp. 424–459.

Lie, Sophus, 1871a. "Über diejenige Theorie eines Raumes mit beliebigvielen Dimensionen, die der Krümmungstheorie des gewöhnlichen Raumes entspricht," *Göttinger Nachrichten*, May 1871 (*GA*, vol. 1, pp. 215–228).

Lie, S., 1871b. "Über Komplexe, insbesondere Linien- und Kugelkomplexe, mit Anwendung auf die Theorie partieller Differentialgleichungen," *GA*, vol. II, pp. 1–121.

Lie, S., 1896.*Geometrie der Berührungstransformationen*, (Leipzig: Teubner, 1896).

Lie, S., 1924-1960.*Gesammelte Abhandlungen*, *GA*, 7 vols. (Leipzig: Teubner and Oslo: Haschehoug, 1924–60).

Lie, S. and Klein, F. 1870. "Über dieHaupttangentenkurven der Kümmerschen Fläche vierten Grades mit 16 Knotenpunkten," *GMA*, vol. 1, pp. 90–97.

Plücker, Julius, 1868, 1869. *Neue Geometrie des Raumes gegründet auf die Betrachtung der geraden Linie als Raumelement*,, vol. 1 (A. Clebsch, ed.), vol. 2 (F. Klein, ed.), (Leipzig: Teubner, 1868–69).

Rowe, David, 1985. "Felix Klein's *Erlanger Antrittrede*: A Transcription with English Translation and Commentary, *Historia Mathematica*, 12(1985), 123–141.

Rowe, D., 1988."Der Briefwechsel Sophus Lie – Felix Klein, eine Einsicht in ihre persönlichen und wissenschaftlichen Beziehungen," *NTM. Schriftenreihe für Geschichte der Naturwissenschaften, Technik und Medizin*, 25(1)(1988), 37–47.

Rowe, D., 1989. "The Early Geometrical Works of Sophus Lie and Felix Klein," D. Rowe and J. McCleary, eds., *The History of Modern Mathematics*, vol. 1, (Boston: Academic Press, 1989), pp. 209-273.

Scholz, Erhard, 1980.*Geschichte des Mannigfaltigkeitsbegriffs von Riemann bis Poincaré*. (Boston: Birkhäuser, 1980).

Tobies, Renate, 1981.*Felix Klein, Biographien hervorragender Naturwissenschaftler, Techniker und Mediziner*, vol. 50. Leipzig : BSB B. G. Teubner, 1981.

Wussing, Hans, 1984. *The Genesis of the Abstract Group Concept*, trans. by Abe Shenitzer (Cambridge, Mass./London: MIT Press, 1984).

Ziegler, Renatus , 1985. *Die Geschichte der geometrischen Mechanik im 19. Jahrhundert* (Stuttgart : Steiner, 1985).

Apparent Contours from Monge to Todd

Bernard Teissier

D. M. I. Ecole Normale Supérieure. 45, rue d'Ulm 75230 Paris cedex 05 (France).

The way we perceive the shape of a smooth object $A \in \mathbb{R}^3$ with our eye is through its apparent contour, the curve on the object where the lines passing through our eye and tangent to the boundary surface $S = \partial A$ of A touches S. One may say that this fact has been at the origin of a certain quantity of experimental Mathematics since the invention of perspective. One may also speculate that the origin of the mathematical study of apparent contours is, stated in modern terms, a problem of measure of complexity; first remark that the apparent contour depends only upon the bounding surface S; it is also the apparent contour of S. Remark also that we may measure the complexity of an algebraic surface by the degree of its equation. Now the following is a natural problem: if S is an algebraic surface of degree m, how complicated is its apparent contour; what is its degree (as a space curve)? Intersecting everything with a plane containing the origin 0 (the eye) and not tangent to S reduces the problem to the following: given a nonsingular algebraic curve C of degree m in $\mathbb{R}^2$, how many of the lines tangent to C pass through a given point 0?

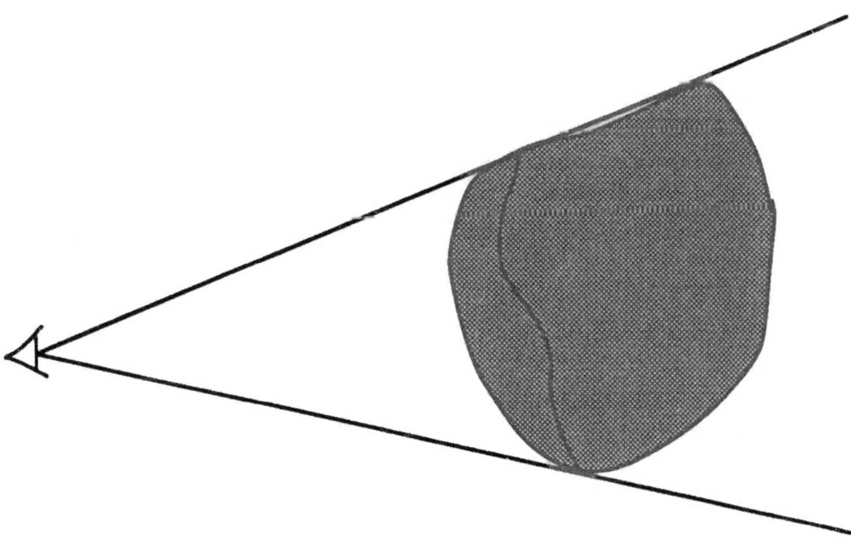

This question would be quite natural for a mathematician like Monge, who used extensively the concept of apparent contour, in particular in his work on fortifications, and provided what one may call "geometric algorithms" to draw apparent contours to special surfaces in descriptive geometry. According to Salmon ([14]), however, the first to study this question was Wallis, who said that the number of these tangents, for a general point 0 of the plane, was at most m^2.

Later Poncelet, for whom the natural setting of this problem was complex projective geometry, considered, following Monge, the *polar curve* (the terminology is his): Let

$$f(X,Y,X) = 0,$$

where f is a homogeneous polynomial of degree m, be an equation for C. The points of C where the tangent goes though the point of $\mathbb{P}^2(\mathbb{C})$ with coordinates (ξ, η, ζ) are on C and on the curve of degree $m-1$ with equation

$$P_{(\xi,\eta,\zeta)}(f) = \xi \frac{\partial f}{\partial X} + \eta \frac{\partial f}{\partial Y} + \zeta \frac{\partial f}{\partial Z} = 0$$

obtained by *polarizing* the polynomial f with respect to the point (ξ, η, ζ). If C is non singular, the points we seek are all the intersection points of C and $P_{(\xi,\eta,\zeta)}(C)$. By Bezout's theorem, the number of these points counted with multiplicity is $m(m-1)$, for every point (ξ, η, ζ), so the number of real tangents to the real part of C passing through a point $0 \in \mathbb{R}^2$ is at most that number.

Monge and Poncelet almost never wrote equations; their arguments were "synthetic", and a bit difficult to follow for modern geometers. The first to write equations in our style are Bobillier ([1]) and Plücker ([8], [9]). They used, perhaps for the first time, the projective coordinates of a point of projective space. They also considered the family of polar curves as parametrized linearly by the coordinates (ξ, η, ζ), and began the study of what is one of the first examples of a *linear system* of curves of degree > 3 with base points (when the curve has singularities, all the polar curves go through the singular points).

In fact a number of important ideas are introduced at that time. First and foremost is the use of the complex projective plane as a natural setting for geometry, an idea due mostly to Poncelet. Second, the idea of the group of projective transformations, of course not stated so explicitely or in these terms, but certainly present in the background of the work of Monge and Poncelet. Then there is projective duality, very dear to Poncelet and viewed perhaps as an extra transformation of the projective plane.

A line in the projective space $\mathbb{P}^2$ is by definition a point in the dual projective space $\check{\mathbb{P}}^2$.

Poncelet saw that given a nondegenerate conic Q, since the polar curve of Q with respect to any point is a line, we get an isomorphism between $\mathbb{P}^2$ and its dual $\check{\mathbb{P}}^2$; he insisted that duality was defined with respect to a non degenerate conic. This insistence was partly a reply to Gergonne's idea of a sort of metamathematical principle of duality. We shall, however, refrain from identifying in this way, as was done at that time, $\mathbb{P}^2$ and $\check{\mathbb{P}}^2$.

The collection of points of $\check{\mathbb{P}}^2$ corresponding to the lines in $\mathbb{P}^2$ tangent to an algebraic curve C is an algebraic curve $\check{C} \subset \check{\mathbb{P}}^2$. A point x in $\mathbb{P}^2$ corresponds to a line $\check{x}$ in $\check{\mathbb{P}}^2$; each point of this line represents a line in $\mathbb{P}^2$ which contains x, and the lines through x tangent to C correspond to the intersection points in $\check{\mathbb{P}}^2$ of the curve $\check{C}$ and the line $\check{x}$. So the class of the curve C, defined as the number of lines tangent to C at non singular points and passing through a given general point of $\mathbb{P}^2$, is the degree $\check{m}$ of $\check{C}$; as we saw above, it is equal to $m(m-1)$ if C has no singularities. It is geometrically obvious that $\check{\check{C}} = C$; this is called biduality (it is wrong if we do geometry over a field of positive characteristic). If the curve $\check{C}$ had no singularities as well, the computation of degrees would give $m(m-1)(m^2-m-1) = m$, which holds only for $m = 2$. So if $m > 2$ the dual of a non singular curve has singularities ; for a general non singular curve, double points (a.k.a. nodes) corresponding to double tangents of C and cusps corresponding to its inflexion points.

To understand biduality better, it becomes important to find the class of a projective plane curve with singularities, at least when these singularities are the simplest: nodes and cusps. This was done by Plücker and the formula for a curve with δ nodes and κ cusps is

$$\check{m} = m(m-1) - 2\delta - 3\kappa$$

One said that" a node decreases the class by two, and a cusp by three"
This is perhaps the first example of a search of numerical invariants of singularities.

In fact, as I mentionned earlier, from the beginning the theory was extended to surfaces in $\mathbb{P}^3$; given such a surface S defined by the homogeneous polynomial $F(x,y,z,t) = 0$, a point $(\xi : \eta : \zeta : \tau) \in \mathbb{P}^3$ defines a *polar surface* $P_{(\xi:\eta:\zeta:\tau)}(S)$; the intersection of S and $P(S)$ is the apparent contour of S from the point $(\xi : \eta : \zeta : \tau)$. Now the intersection of the polar surfaces of two points lying on a line ℓ is a curve, called the *polar curve* of S with respect to ℓ. The dual surface $\check{S} \subset \check{\mathbb{P}}^3$ is the closure of the locus of points in $\check{\mathbb{P}}^3$ corresponding to the tangent planes to S at non singular points of S. The projective dual of ℓ is a line $\check{\ell}$ in $\check{\mathbb{P}}^3$, and the number of points of intersection of $\check{\ell}$ with $\check{S}$ is the number of hyperplanes tangent to S at non singular points and containing ℓ. If ℓ is a general line, the points of intersection of $\check{\ell}$ with $\check{S}$ are all simple.

This is a modernized account of what was known about the apparent contours in the middle of the last century.

To summarize, the problem of estimating the complexity of the apparent contour, together with the concept of duality intoduced by Poncelet, led to the construction of the dual curve of a given algebraic curve and the computation of its degree for curves and surfaces with simple singularities. The equation of the dual curve or surface are called "tangential equations" of the original curve or surface.

This current of thought, rather dominant in "pure " geometry in France and Germany, seems to have been for a while fairly independant of the other developments of the time, including the work of Gauss and Riemann. Think that the second edition of Poncelet's magnum opus "Traité des propriétés projectives des figures" was published in 1863. I will not describe the work of Riemann, but quote Dieudonné in [4]:

> "Riemann's Memoir on abelian functions avoided the geometric language, and it is only in 1863 that Roch and Clebsch begin to link Riemann's results with the projective Geometry of plane algebraic curves. Their first successes quickly attract followers, and around 1870 an active scool of algebraic Geometry develops, around Brill and Max Noether in Germany, Smith and Cayley in England, Halphen in France, Zeuthen in Denmark, and the first generation of the Italian geometers, Cremona, C. Segre and Bertini. The main theme of their researches will be the mutual adaptation of the algebraic Geometry of the beginning of the century and the new ideas of Riemann."

Now if we follow the thread of apparent contours, we have to move to England where Cayley and H.J.S. Smith continue to think about the general Plücker formulas for plane curves and for surfaces, and the diminution of class that a singular point imposes on a surface. Cayley proposed the idea that each singular point of a plane curve should be "equivalent to certain numbers of cusps and nodes, as far as Plücker formulas are concerned. This caused H.J.S. Smith to write a very interesting paper containing in particular results on the contact of the polar curve with the given one near one of its singular points. It is amusing to note also that the famous "rational double points" which were so carefully studied in the 1960's and 70's appear in Salmon indexed not by their "Coxeter number" which comes from their connexion with simple Lie groups, and is equal to their "Milnor number" which is a topological invariant, but by the *diminution of class* which their presence on a projective surface would impose. For example the modern E_8 is Salmon's U_{10}.

Salmon himself made extensive computations of the numerical characters of the dual surface of a general algebraic surface of degree m lying in $\mathbb{P}^3$.

In the meantime, the identification of compact Riemann surfaces with non singular projective algebraic curves and their topological classification by the genus , as well as the birational classification by the same genus had been perfected, and by the end of the last century, the *ordre du jour* was the extension of this classification to projective surfaces. So it is not very surprising that the next important step for us comes from the search of invariants of projective surfaces.

In the theory of algebraic curves, an important formula states that given an algebraic map $f: C \to C'$ between algebraic curves, which is of degree $\deg f = d$ (meaning that for a general point $c' \in C'$, $f^{-1}(c')$ consists of d points, and is ramified at the points $x_i \in C$, $1 \leq i \leq r$, which means that near x_i, in suitable local coordinates on C and C', the map f is of the form $t \mapsto t^{e_i+1}$ with $e_i \in \mathrm{N}$, $e_i \geq 1$. The integer e_i is the *ramification index* of f at x_i. Then we have the Riemann-Hurwitz formula relating the genus of C and the genus of C' via d and the ramification indices:

$$2g(C) - 2 = d(2g(C') - 2) + \sum_i e_i$$

If we apply this formula to the case $C' = \mathbb{P}^1$, knowing that any compact algebraic curve is a finite ramified covering of $\mathbb{P}^1$, we find that we can calculate the genus of C from any linear system of points (such a system is always made of the fibers of a map $C \to \mathbb{P}^1$) if we know its degree and its singularities: we get

$$2g(C) = 2 - 2d + \sum e_i$$

The ramification points x_i can be computed as the so-called *jacobian divisor* of the linear system, which consists of the singular points, properly counted, of the singular members of the linear system. In particular if C is a plane curve and the linear system is the system of its plane sections by lines through a general point $x = (\xi : \eta : \zeta)$ of $\mathbb{P}^2$, the map f is the projection from C to $\mathbb{P}^1$ from x; its degree is the degree m of C and its ramification points are exactly the points where the line from x is tangent to C. Since x is general, these are simple tangency points, so the e_i are equal to 1, and their number is equal to the class $\check{m}$ of C; the formula gives

$$2g(C) - 2 = -2m + \check{m} \ ,$$

thus computing the genus from the degree and the class.

This is the first example of the relation between the "characteristic classes" (in this case only the genus) and the polar classes; in this case the curve itself, of degree m and the degree of the polar locus, or apparent contour from x, i.e. in this case the class $\check{m}$. There is a similar construction for surfaces, which gives the *Zeuthen-Segre invariant* :

Consider a linear system of curves on a non singular algebraic surface S, without base points, i.e. defined as the fibers of an algebraic map $f: S \to \mathbb{P}^1$. Let us assume that the general fiber $F = f^{-1}(t)$ (for "generic" t), which is non singular, is of genus g, and that there are σ singular fibers, each having a single ordinary double point as singularity. A computation of topological Euler-Poincaré characteristics, nowadays quite standard, gives

$$\chi(S) = 2\chi(F) + \sigma = 4 - 4g + \sigma$$

so that $Z(S) = \sigma - 4g = \chi(S) - 4$ does not depend upon the choice of the pencil of curves; it is the *Zeuthen-Segre invariant* of the surface S. We may in fact also allow pencils with base points; if there are b of these, then $Z(S) = \sigma - 4g - b$. Such pencils correspond to maps $\tilde{f}: \tilde{S} \to \mathbb{P}^1$, where $\tilde{S} \to S$ is a blowing up map determined by the structure of the base points.

Now let us take a surface S of degree m and a pencil of hyperplanes in $\mathbb{P}^3$; we can view it as the pencil of hyperplanes containing a line (a copy of $\mathbb{P}^1$) $\ell \subset \mathbb{P}^3$, and if we choose another line $\mathbb{P}^1 \subset \mathbb{P}^3$ not meeting ℓ, the map which to a point $s \in S \setminus \ell$ associates the intersection point with $\mathbb{P}^1$ of the plane determined by s and ℓ extends to a map $\tilde{S} \to \mathbb{P}^1$, where $\tilde{S} \to S$ is the blowing up of $S \cap \ell$ in S. The singular fibers correspond to planes in the pencil that are tangent to S; if ℓ is general, they are simply tangent, and their number is the class $\check{m}$ of the surface S. The Zeuthen-Segre formula gives, taking into account the fact that we have blown up m points on S, the equality $Z(S) = \chi(S) - 4 = \check{m} - 4g - m$, where now g is the genus of a general plane section of S.

We may interpret all this in two ways; first, as Zeuthen-Segre and Severi did, as creating an invariant of the algebraic surface S, which is in fact $\chi(S) - 4$, in terms of the singular curves of a linear system on S, and second as saying that, given S, if the general curve of a pencil has high genus, then there must be many degenerate fibers.

Severi and others tried to generalize this construction of invariants via the singularities of elements of linear systems, but it was Todd who in 1936 found the right formulation, which is based on the apparently less general case of linear projections. The basic idea is to generalize the notion of apparent contour, considering what Todd calls the "Polar loci" of a projective variety $X \subset \mathbb{P}^n$. Then it turns out that certain formal linear combinations of the intersections of general polar loci of X with general linear sections (of various dimensions) of X are invariants of X, i.e. do not depend upon the projective embedding of X and the choices of polar loci and linear sections.

More precisely, given a non singular $d-1$-dimensional variety X in $\mathbb{P}^{N-1}$, for a linear subspace $L \subset \mathbb{P}^{N-1}$ of dimension $N - d + k - 2$, i.e. of codimension $d - k + 1$, let us set

$$P_k(X; L) = \{x \in X / \dim(T_{X,x} \cap L) \geq k - 1\}$$

This is the *Polar variety* of X associated to L; if L is general, it is either empty or purely of codimension k in X. We see that this construction is a direct generalization of the apparent contour. The eye 0 is replaced by the linear subspace L!

Todd shows that the following formal linear combinations of varieties

$$X_k = \sum_{j=0}^{j=k} (-1)^j \binom{d-k+j+1}{j} P_{k-j}(X;L) \cap H_j$$

where H_j is a linear subspace of codimension j, are independant of all the choices made and of the embedding of X in a projective space, provided that the $L's$ and the $H_j's$ have been chosen general enough.

The linear combination is at first sight a rather awkward object to deal with. The idea is that X_i represents a generalized variety of codimension i in X and what we should remember is that any numerical character $e(Y)$ associated to algebraic varieties Y, and which is *additive* in the sense that $e(Y_1 \cup Y_2) = e(Y_1) + e(Y_2)$ whenever Y_1 and Y_2 have the same dimension, can be extended by linearity to such a generalized variety. In particular, given a partition $i_1, \ldots, i_k$ of $d-1$, we obtain that the intersection numbers

$$(X_{i_1} \cdots X_{i_k})$$

which are well defined since the intersection of the corresponding varieties (each one assumed to be a general representative obtained by taking general and independant linear spaces) is of dimension zero, depend only upon the structure of X as an algebraic variety.

In fact, Todd preferred to introduce an equivalence relation between varieties, called nowadays *rational equivalence* and instead of considering "general linear spaces" and so on considered the equivalence classes of the corresponding objects, which (by definition of rational equivalence) turn out to be independant of the choices of linear spaces and so on. This relation, however, considers as equivalent objects which are geometrically very different, while the geometry (the topology, and in fact much more) of the polar variety corresponding to a general linear space is perfectly well defined and contains a lot of information.

In any case, one of the main results of Todd is that the numbers $(X_{i_1} \cdots X_{i_k})$ depend only upon X, that they are independent invariants (there is no relation between them valid for all X's) and finally that the *arithmetic genus* of X is a linear function of these invariants. The arithmetic genus is the Euler characteristic in coherent cohomology $\chi(X, \mathcal{O}_X)$ of the sheaf of algebraic functions on X; by Serre duality (see [5]), it is equal to $(-1)^d \chi(X, \omega_X)$ where $d = \dim X$ and ω_X is the sheaf of germs of holomorphic d-forms on X. Thus, since the Euler-Poincaré characteristic of a sheaf is the stable, or computable form of the dimension of the space of global sections, the arithmetic genus is a generalization of the analytic definition of the genus of a non singular curve (dimension of the space of regular differential forms) to an arbitrary algebraic variety. The expression for the arithmetic genus is not simple; it involves what is now called the *Todd class* (see [5]) of the tangent bundle.

On the other hand, the topological Euler-Poincaré characteristic of X can be computed, in the same way we used for the Zeuthen-Segre invariant, to show the equality

$$\chi(X) = \deg X_d = \sum_{j=0}^{d} (j+1)(P_{d-j}(X).H_j)$$

where $(a.b)$ denotes the intersection number, in this case since we intersect with a linear space of complementary dimension, it is just the degree of the projective variety $P_j(X)$.

So Todd's results give a rather complete generalization of the genus formula, both in its analytic and its topological aspects. The connection between these two aspects goes much deeper and leads to the Hirzebruch-Riemann-Roch theorem. One must first realize, after Nakano, Hirzebruch, Serre, Gamkrelidze (see [5]), that the invariants X_k of Todd (or rather their cohomology classes) coincide with the *Chern classes* of the tangent bundle of X, which provides a good reason for them to be invariant!.

The Chern classes are cohomology classes expressing the obstruction to the existence of k-uples of everywhere linearly independant sections of the tangent bundle. The best known of them is the Euler class, expressing the obstruction to finding an everywhere non zero section of the tangent bundle, that is, finding an everywhere non zero vector field; it is represented by finite collections of points of X, the zeroes of a general vector field. All these finite collections of points are rationnally equivalent and their number, counted with multiplicities, is the Euler-Poincaré characteristic of X.

Then we remark that the formula for the topological Euler characteristic can be inverted to compute the degrees of the polar varieties from the topological Euler characteristics of general plane sections (of all dimensions) of X. Using the expression for the arithmetic genus which I mentionned, one expresses in the end the arithmetic genus as a linear function of the Euler-Poincaré characteristics of X, its polar varieties, and their plane sections.

Following this thread of the determination of the geometry of an object from its apparent contours, we have been led not only to the determination of such a fundamental topological invariant as the Euler-Poincaré characteristic, but also to an extension to arbitrary dimensions of the connexion between the geometry of a variety and the theory of algebraic functions on this variety discovered by Riemann. In recent years this geometric viewpoint has been extended to singular projective varieties, and to the study of an analytic space embedded in $\mathbf{C}^N$ in the vicinity of one of its singular points (see [6], [7], [18]). There is much left to understand, and in particular what form a similar theory would take for real algebraic or semi-algebraic varieties in $\mathbf{R}^N$.

Finally I remark that the analogous problem of showing that the collection of its orthogonal projections to subspaces $\mathbf{R}^i \subset \mathbf{R}^d$ determines a convex body $K \subset \mathbf{R}^d$ has been completely solved only in the last fifty years or so. The problem of doing so effectively (up to some approximation) is not yet satisfactorily solved.

References

This is only a sample

[1] Bobillier, *Annales de mathématiques pures et appliquées,* t. 18, 1827-28, p.253.

[2] A. Cayley, *Quarterly Journal,* t. VII (= Collected papers, Vol. 5, p.520).

[3] A. Cayley, A memoir on the theory of reciprocal surfaces, *Phil. Trans. Royal Soc. London,* **119**, 201-229 (= Collected papers, Vol. 6, 329-358).

[4] J. Dieudonné, *Cours de Géométrie algébrique* , Vol. 1 , P.U.F., Paris.

[5] F. Hirzebruch, *Topological methods in algebraic Geometry,* Grundlehren der Math. Wiss., No. 131, Springer 1966.

[6] D.T. Lê et B. Teissier, Variétés polaires locales et classes de Chern des variétés singulières, *Annals of Math.,* 2, **114**, (1981), 457-491.

[7] D.T. Lê et B. Teissier, Cycles évanescents, sections planes, et conditions de Whitney, II, *Proc. A.M.S. Symp. in pure Math.* Vol. 40 (1983), part 2, 65-103.

[8] Plücker, solution d'une question fondamentale concernant la théorie générale des courbes, *Journal de Crelle,* t. 12, 1834, 105-108.

[9] Plücker, sur les points singuliers des courbes, *Journal de Liouville,* t. 2, 1837, 11-15.

[10] J.-V. Poncelet, *Applications d'Analyse et de Géométrie* Gauthier-Villars, Paris 1864.

[11] J.-V. Poncelet, *traité des propriétés projectives des figures,* deuxième édition, Gauthier-Villars, Paris 1866.

[12] J.-V. Poncelet, Théorie générale des polaires réciproques, *Annales de Math. pures et appliquées,* t.8, 1817-18, p. 201.

[13] F. Severi, *Commentarii math. Helv.* **4**, 1932, p.268.

[14] G. Salmon, *Traité de Géométrie analytique* (Courbes planes), suivi de *Étude sur les points singuliers* par G. Halphen, Gauthier-Villars, Paris 1884.

[15] J.A. Todd, The geometrical invariants of algebraic loci. *Proc. London Mat. Soc.* , **43**, 1937, 127-138.

[16] J.A. Todd, The arithmetical invariants of algebraic loci. *Proc. London Mat. Soc.* , **43**, 1937, 190-225.

[17] B. Teissier, Quelques points de l'histoire des variétés polaires, de Poncelet à nos jours. *Séminaire d'analyse de Clermont-Ferrand,* 1989.

[18] B. Teissier, Sur la classification des singularités des espaces analytiques complexes, *Proc. Int. Congress of Math.,* Vol. 1, 763-781, PWN, Varsovie 1984.

2

Historical and Epistemological Aspects of the Connexion between Physics and Geometry

L'Espace : Concept Abstrait et/ou physique ; la Géométrie entre Formalisation Mathématique et Etude de la Nature

Luciano Boi

Associé au *Séminaire d'Epistémologie des Mathématiques*, Ecole des Hautes Etudes en Sciences Sociales, 54 bd Raspail, 75006 Paris ; et Boursier de recherche de l'*Alexander von Humboldt-Stiftung*, c/o Institut für Philosophie, Wissenschaftstheorie, Wissenschafts- und Technikgeschichte, Technische Universität, Ernst-Reuter-Platz 7, 1000 Berlin 10[1]

So z. B. lässt sich eine vollkommen in sich abgeschlossene mathematische Theorie zusammenstellen, welche von den für die einzelnen Punkte geltenden Elementargesetzen bis zu den Vorgäng in dem uns wirklich gegebenen continuirlich erfüllten Raume fortschreitet, ohne zu scheiden, ob es sich um die Schwerkraft, oder die Electricität, oder den Magnetismus, oder das Gleichgewicht der Wärme handelt. (B. Riemann, *Pädagogische Seminar*, 1850).

The hypothesis that space is not homoloidal (flat), and again that its geometrical character may change with the time, may or may not be destined to play a great part in the physics of the future ; yet we cannot refuse to consider them as possible explanations of physical phenomena [...]. (W.K. Clifford, *The Common Sense of the Exact Sciences*, 1885).

[...] Hauptaufgabe der Mathematik ist, unbekannte Dinge aus bekannten Eigenschaften zu bestimmen. (S. Lie, *Theorie der Transformationsgruppen*, 1893).

Nowhere do mathematics, natural sciences, and

[1] Ce texte est une version révisée de la conférence présentée au Colloque. Je tiens à remercier ici les participants au *Séminaire d'Epistémologie des Mathématiques* à l'Ecole des Hautes Etudes en Sciences Sociales, avec qui j'ai pu discuter à plusieurs reprises de la problématique développée ici.
La version finale de ce texte a été rédigée lors d'un séjour de recherche à la Technische Universität de Berlin en été 1991. J'aimerais exprimer ici ma reconnaisance envers l'*Alexander von Humboldt-Stiftung* qui a rendu possible un tel séjour.

philosophy permeate one another so intimately as in the problem of space.
(H. Weyl, *Philosophy of Mathematics and Natural Science*, 1949).

1. Introduction

Notre propos ici, consistera à s'interroger sur le statut de la géométrie telle qu'elle s'est développée de la seconde moitié du XIXe siècle au début du XXe, c'est-à-dire, pour mieux nous situer historiquement, entre les travaux de Riemann, Clifford, Beltrami, Helmholtz, Klein, Lie et Poincaré, et ceux de Hilbert, Cartan et Weyl. Au cours de cette période, la géométrie connaît la transformation sans doute la plus fondamentale de son histoire, qui investit aussi bien ses méthodes que ses concepts. Les rapports de cette science aux autres branches des mathématiques, notamment l'algèbre et l'analyse, en seront profondément changés, de même que ses rapports à la physique et aux autres sciences exactes de la nature comme, par exemple, l'astronomie. Dès lors on ne parlera plus de la géométrie ou de l'espace, mais *des* géométries et *des* espaces. La reconnaissance sur le plan mathématique d'une pluralité de géométries a constitué un fait historique capital. Pour la première fois, avec la découverte des géométries non-euclidiennes, la conception d'une géométrie unique et d'un espace absolu se trouve complètement remise en question au profit d'une autre, radicalement différente : en tant que science des formes pures, la géométrie appartient aux mathématiques au même titre que l'arithmétique et l'algèbre ; tandis qu'en tant que science des formes réelles, elle est intimement liée à la physique[2].

[2] Cette idée de la double nature de la géométrie on la retrouve, par exemple, chez Grassmann, dans l'introduction à sa *théorie de l'extension (Die lineale Ausdehnungslehre)* de 1844. Il écrit à ce propos : "Die Stellung der Geometrie zur Formenlehre hängt von dem Verhältniss ab, in welchem die Anschauung des Raumes zum reinen Denken steht. Wenngleich wir nun sagten, es trete jene Anschauung dem Denken als selbständig gegebenes gegenüber, so ist damit doch nicht behauptet, dass die Anschauung des Raumes uns erst aus der Betrachtung der räumlichen Dinge würde ; sondern sie ist eine Grundanschauung, die mit dem Geöffnetsein unseres Sinnes für die sinnliche Welt uns mitgegeben ist, und die uns eben so ursprünglich anhaftet, wie der Leib der Seele. Auf gleiche Weise verhält es sich mit der Zeit und mit der auf die Anschauungen der Zeit und des Raumes gegründeten Bewegung, weshalb man auch die reine Bewegungslehre (Phorometrie) mit gleichem Rechte wie die Geometrie den mathematischen Wissenschaften beigezählt hat. Aus der Anschauung der Bewegung fliesst vermittelst des Gegensatzes von Ursache und Wirkung der Begriff der bewegenden Kraft, so dass also Geometrie, Phorometrie und Mechanik als Anwendungen der Formenlehre auf die Grundanschauungen der sinnlichen Welt erscheinen." ; *Hermann Grassmanns Gesammelte Mathematische und Physikalische Werke*, Bd. I, Teil I : *Die Ausdehnungslehre von 1844 und die Geometrische Analyse*, Unter Mitwirkung von E. Study u. herausgegeben von F. Engel, Leipzig, 1894 (Einleitung, p. 24).
La même idée, exprimée autrement, a été affirmée par Gauss : "Nach meiner innigsten Überzeugung hat die Raumlehre in unserm Wissen a priori eine ganz andere Stellung, wie die reine Grössenlehre ; es geht unserer Kenntniss von jener durchaus diejenige vollständige Überzeugung von ihrer Nothwendigkeit (also auch von ihrer absoluten Wahrheit) ab, die der letztern eigen ist ; wir müssen in Demuth zugeben, dass, wenn die Zahl bloss unsers Geistes Product ist, der Raum auch ausser unserm Geiste eine Realität hat, der wir a priori ihre Gesetze nicht vollständig vorschreiben können..." (*Lettre de Gauss à Bessel*, Göttingen, le 9 avril 1830).

L'existence de plusieurs géométries qui se réalisent également au point de vue de la physique, a été montrée de façon décisive grâce à la théorie de la relativité d'Einstein, bien que déjà notamment Riemann et Clifford eurent admis qu'une géométrie différente de celle euclidienne pouvait s'appliquer à notre espace physique. Mais pour arriver à admettre cela, il a fallu d'abord critiquer profondément la conception qu'on s'était fait de l'espace, qui n'était plus à penser, ni comme le lieu où peuvent se construire les figures, ni comme celui où se meuvent les corps. Il était à la fois le concept mathématique primordial de la géométrie dont il héritait la forme et la structure, et un concept réel ayant un contenu physique. Nous nous bornerons ici à considérer les parties de la géométrie plus directement liées aux développements mathématiques des géométries non-euclidiennes.

On peut dire, pour schématiser, que la géométrie au XIXe siècle s'est développée selon deux directions et conceptions différentes : la première, privilégiant sa nature théorique, s'est préoccupée avant tout de la développer mathématiquement, et cela par l'approfondissement de ses méthodes, une définition plus rigoureuse de ses objets, et la découverte de nouvelles propriétés formelles ; en revanche, la deuxième conception, en plus de reconnaître à la géométrie le statut d'une science mathématique, lui a attribué celui d'une science physique, c'est-à-dire que les concepts abstraits de la géométrie auraient un contenu physique réel. Puisque d'après cette conception la géométrie permet de relier l'analyse mathématique à la physique, elle est par conséquent à l'origine de la constitution de la physique-mathématique. On reviendra sur ce point dans la dernière partie de notre analyse. Remarquons encore ceci, tandis que dans le premier cas la géométrie est pensée comme une science qui n'a aucun rapport de nécessité avec l'expérience, dans le second, en revanche, on affirme que la géométrie dépend dans une certaine mesure de l'expérience. Il y a toutefois une affinité importante entre ces deux conceptions : elles partagent l'idée que l'élaboration mathématique de la géométrie n'a besoin pour être justifiée d'aucun recours à l'expérience, du moment que ses concepts abstraits ou ses structures formelles n'ont pas de fondement empirique. L'expérience intervient uniquement au moment de déterminer sa validité. Ces deux conceptions ne sont pas, non plus, en contradiction : la géométrie peut être développée, soit en demeurant toujours à l'intérieur du discours mathématique, soit en travaillant à la frontière des mathématiques et de la physique, c'est-à-dire en essayant de comprendre la signification et la portée qu'elle peut avoir pour la physique. Même les développements les plus récents de la géométrie ont renforcé l'idée que cette science possède un statut spécial. D'une part, elle apparaît de plus en plus étroitement liée à l'algèbre, l'analyse et l'arithmétique, jusqu'à en être inséparable, et de ce point de vue la géométrie aujourd'hui n'est que l'étude de structures mathématiques abstraites, des "objets" qui forment de telles structures et des propriétés formelles qui caractérisent ces mêmes objets[3]. Par ailleurs, on comprend de plus en plus

[3] "We wish to explain what we mean by contemporary differential geometry as contrasted to classical differential geometry. Classical differential geometry [...] is a study of curves and surfaces in 3-dimensional Euclidean space whose principal method is differential calculus. [...] Contemporary differential geometry is the study of a geometric object given on a differentiable manifold. We shall again not try to define "geometric objects". It is sufficient to understand, for example, a connection, a linear connection, a Riemannian metric, or a Kählerian metric, and so on. Given such a structure Γ on a differentiable manifold M, the first object is to study the properties of Γ. Then, there are problems such as the study of the group of automorphisms of $A(\Gamma)$ of the structure Γ and its relation to properties of Γ, or the study of relationship between properties of Γ and the topological properties of the manifold M." ; K. Nomizu,

clairement que certains concepts géométriques expriment de façon non contingente un contenu physique et que, peut-être, la vraie signification des phénomènes physiques réside entièrement dans les concepts géométriques qui le représentent. On pourrait à la limite conclure que les phénomènes physiques n'ont au fond qu'une signification géométrique[4].

2. Le Concept de Variété et la Constitution Mathématique de l'Espace

Riemann a été sans doute le premier à expliciter mathématiquement la double nature de l'espace. Il expose sa nouvelle conception de la géométrie dans le mémoire *Über die Hypothesen, welche der Geometrie zu Grunde liegen*, présenté à Göttingen en 1854 pour l'obtention de son *Habilitation*. Ce travail contient à la fois, des idées mathématiques nouvelles, une valeur philosophique remarquable, et des intuitions sur la physique, révolutionnaires pour l'époque. Au point de vue mathématique son importance revient à avoir traité la géométrie avec les méthodes de l'analyse mathématique et à avoir ainsi fondé, à la suite des résultats de Gauss sur la théorie des surfaces, la géométrie différentielle moderne.

"Recent Development in the Theory of Connections and Homology Groups", *Advances in Mathematics*, Vol. 1, Fasc. 1, 1961, pp. 2-3.

[4] Cette idée se trouve exprimée par le mathématicien et philosophe anglais W. K. Clifford (1845-1879) : "I hold in fact (1) that small analogous to little hills on a surface which is on the average flat ; namely, that the ordinary laws of geometry are not valid in them ; (2) that this property of being curved or distorced is continually being passed on from one portion of space to another after the manner of a wave ; (3) that this variation of the curvature of space is what really happens in that phenomenon which we call the motion of matter, whether ponderable or etheral ; (4) that in the physical world nothing else takes place but this variation, subject (possibly) to the laws of continuity." ; "On the space-theory of matter", in *Proceedings of Cambridge Philosophical Society*, Vol. 2 (1876), pp. 157-158.
Elle a été reprise et développée récemment par J. A. Wheeler : "Is the physical world in which we live a purely mathematical construct? Put the question in another way : Is space-time only an arena within which fields and particles move about as 'physical' and 'foreign' entities? Or is the four-dimensional continuum all there is? Is curved empty geometry a kind of magic building material out of which everything in the physical world is made : (1) slow curvature in one region of space describes a gravitational field ; (2) a rippled geometry with a different type of curvature somewhere else describes an electromagnetic field ; (3) a knotted-up region of high curvature describes a concentration of charge and mass-energy that moves like a particle? Are fields and particles foreign entities immersed in geometry, or are they nothing but geometry? It would be difficult to name any more central to the plan of physics that this : whether space-time is only an arena, or whether it is everything. Einstein from an early day was animated by the hope for a purely geometrical description of nature. He in turn derived inspiration from previous thinkers, among them Riemann and the Clifford to whom we owe the Clifford numbers of Dirac's theory of the electron. [...] The vision of Clifford and Einstein can be summarized in a single phrase, 'a geometrodynamical universe' : a world whose properties are describes by geometry, and a geometry whose curvature changes with time – a *dynamical* geometry." ; "Curved Empty Space-Time as the Building Material of the Physical World : an Assessment", in *Logic, Methodology and Philosophy of Science* , Proceedings of the 1960 International Congress, Stanford, Stanford University Press, 1962, pp. 361, 365, 368-69, 371-72.

Après avoir critiqué le caractère uniquement nominal des concepts qui ont servi à la construction de la géométrie, ainsi que le manque de clarté qui aurait caractérisé depuis Euclide jusqu'à Legendre les rapports entre ces mêmes notions et tout particulièrement le fait de n'être jamais parvenu à éclairer si ces rapports étaient de nature conventionnelle ou nécessaire, Riemann reconnaît qu'une telle situation est due au fait de n'avoir jamais pris en considération le concept général de "grandeur n-fois étendue" (*allgemeine Begriff einer n-fach ausgedehnten Grösse*). L'étude d'un tel concept appartient, selon Riemann, à une branche des mathématiques qu'il appelle "théorie de la grandeur" (*Grössenlehre*). Suivant que la définition d'un tel concept dépend de rapports de mesure, ou qu'elle en est indépendante, on pourra développer cette théorie selon les méthodes de la géométrie infinitésimale, ou bien, comme une discipline autonome qu'il nommera plus tard, à la suite de Leibniz et de Gauss, *Analysis situs*[5]. Mais sauf cette brève mention, un tel aspect de la question ne sera pas traité dans le mémoire en question.

Le concept de grandeur n-fois étendue ou, suivant le langage moderne, de variété, est susceptible de plusieurs rapports métriques (*Massverhältnisse*), et il s'ensuit que l'espace n'est qu'un cas particulier de variété à trois dimensions. Les hypothèses que l'on fait en vue de déterminer les relations métriques de l'espace doivent être des plus simples, et on ne peut arriver à distinguer celles qui caractérisent notre espace ordinaire des autres variétés étendues à 3-dimensions, qu'en faisant recours à l'expérience. Tant qu'on se place sur un plan abstrait et qu'on ne précise pas la nature de ses éléments, le concept de variété a pour Riemann la signification d'une n-uple de nombres ou, pour utiliser l'expression de Lie, celle d'une variété numérique (*Zahlenmannigfaltigkeit*)[6], que l'on peut étudier en se limitant à appliquer les méthodes de l'analyse. Le caractère essentiel d'une variété à n-dimensions consiste en ce que la détermination du lieu peut se ramener à n déterminations de grandeurs. Toutes les sortes de variétés, et en particulier celles que l'on peut reconnaître dans la réalité, ne possèdent pas un tel caractère, du moment que leur nature qualitative ne se laisse pas déterminer quantitativement, avec quelques exceptions comme celles des couleurs et des lieux des objets sensibles ("die Orte der Sinnengegenstände"), d'ailleurs citées par notre auteur. D'une façon générale, Riemann fait la distinction entre variétés discrètes et variétés continues : les premières sont constituées d'éléments, les secondes de points. Le moyen pour déterminer une variété discrète est le dénombrement ; pour une variété continue, c'est la mesure. Enfin, et c'est le point essentiel, Riemann pense que *l'espace forme une variété continue, tandis que les nombres forment une variété discrète*. D'une variété il exige, outre le fait que ses éléments soient déterminables par la donnée de n grandeurs - les "coordonnées", qui sont des fonctions

[5] Il remarque à ce propos : "Die Untersuchungen, welche sich, in diesem Falle über sie anstellen lassen, bilden einen allgemeinen von Massbestimmungen unabhängingen Teil der Grössenlehre, wo die Grössen nicht als unabhängig von der Lage existierend und nicht als durch eine Einhait ausdrückbar, *sondern als Gebiete in einer Mannigfaltigkeit betrachtet werden.*" ; "Ueber die Hypothesen..." (1854), in *Mathematische Werke*, B.G. Teubner, Leipzig, 1876, p. 256 (souligné par nous).

[6] "[...] Während Lobatchewski nur das Werke Euklids fortsetzen und gewissermassen eine zweiter Euklid sein wollte, schlug Riemann einen ganz andern Weg ein, indem er den Raum als Zahlenmannigfaltigkeit auffaste und auf diese Zahlenmannigfaltigkeit die ganze Analysis anwendete." ; *Theorie der Transformationsgruppen*, Dritter Abschnitt, B.G. Teubner, Leipzig, 1893, Vorrede, p. X.

continues à l'intérieur de la variété – que la variété soit continue, c'est-à-dire que l'on puisse passer de l'un de ses éléments à l'autre de façon continue.

Avouons cependant ici ne pas comprendre tout à fait la cohérence de l'affirmation de Riemann (que nous avons mise en relief), puisqu'il admet bien pourtant que l'espace est un ensemble de points, chacun étant déterminable par le moyen d'un système de coordonnées, donc par des nombres. Mais laissons cette apparente difficulté de côté. D'après notre interprétation, la conception riemannienne de l'espace présente trois aspects essentiels qui sont autant de modes d'existence du continu.

i) Il y a un premier niveau dans lequel la propriété de la continuité signifie la possibilité d'opérer une transition en un certain sens indéfinie d'un élément à l'autre d'une variété ; il s'agit du processus par lequel on peut engendrer une variété à $n-1$ dimensions d'une variété à n dimensions, et inversement. C'est le niveau topologique (intuitif) de la continuité, qui pourrait être désigné par celui de la *dimensionnalité*. Ce caractère amorphe du continu s'exprime également par la convention (l'hypothèse) que le monde où nous vivons est un *continu spatial* à 3 dimensions, c'est-à-dire une variété tridimensionnelle. C'est la première propriété perceptible de l'espace, et cela signifie que celui-ci peut être défini essentiellement par la donnée de trois nombres.

ii) Dans un deuxième sens, la propriété de la continuité est liée à la structure métrique de l'espace et signifie que chaque point, ainsi que ses variations infinitésimales, sont représentables par une fonction continue de ses différentielles. Mais Riemann exige davantage de ce type de continuité, à savoir que de telles fonctions ne soient pas simplement continues, mais en plus, qu'elles soient continûment différentiables ; en raison de cela, nous désignerons ce niveau par celui *différentiable* du continu. Ce dernier est fondamental au point de vue mathématique puisqu'il permet de connaître de nouvelles propriétés ainsi que des structures formelles plus riches, caractérisant la géométrie des variétés et donc de l'espace.

iii) Enfin, Riemann avance la possibilité qu'il existe un troisième niveau du continu, dont la nature n'est en rien assimilable aux autres que nous venons tout juste d'énoncer et où ses principes constitutifs ne feraient pas partie de la manière dont on se le représente (abstraitement), comme cela est le cas pour le discret. Bien qu'on ne trouve pas chez lui une réponse à la question, son raisonnement indique néanmoins une direction de recherche. Premièrement, il affirme que contrairement aux variétés discrètes, les variétés continues ont vraisemblablement une origine de nature dynamique, c'est-à-dire que la propriété de la continuité serait liée au contenu physique de l'espace. En deuxième lieu, tel qu'il ressort assez clairement de certains de ses travaux et de ses quelques réflexions philosophiques, Riemann a une vision essentiellement spatiale du continu. Par cela il faut entendre que les phénomènes physiques et le type d'espace dans lequel ceux-ci se déroulent sont indissociables ; l'espace imaginé par Riemann est "non-vide" (à la différence de celui pensé par Newton), et doué de continuité. Ces deux caractéristiques signifient : d'une part, que l'espace est affecté par des corps en état de mouvement constant ; et d'autre part, que les effets des phénomènes physiques de nature différente peuvent se transmettre de façon continue – c'est-à-dire de façon à ce que l'effet produit par eux demeure constant – d'une région à l'autre de l'espace. Il n'est pas exagéré d'affirmer que le mathématicien de Göttingen avait clairement conçu l'existence d'un "niveau" physico-mathématique du continu, qui lui paraissait déjà – en anticipant ainsi de cinquante ans la théorie de la relativité générale – non pas comme une simple hypothèse mathématique, mais comme le substrat réel des phénomènes naturels, et en particulier physiques. Ce point mériterait d'être approfondi davantage, ce que nous nous proposons de faire ailleurs.

Hermann Weyl, à la suite notamment de L. E. J. Brouwer, a essayé de montrer que formulé en termes plus modernes, le concept de variété continue au sens de Riemann se laisse expliquer par celui d'ensemble, une fois que ses éléments sont désignés comme étant des points, et qu'en particulier le concept du continu, devient plus intelligible si on l'explique par celui de *voisinage* :

> Die Mf (Mannigfaltigkeit) ist stetig, wenn die Punkte so miteinander verwachsen sind, dass es unmöglich ist, einen Punkt für sich herauszuheben, vielmehr immer nur zusammen mit einem vag begrenzten ihn umgebenden Hof, mit einer "Umgebung".[7]

---

[7] H. Weyl, *Riemanns geometrische Ideen, ihre Auswirkung und ihre Verknüpfung mit der Gruppentheorie*, Springer-Verlag, Berlin/Heidelberg, 1988, pp. 2-3. Weyl remarque qu'afin d'éliminer l'arbitraire que malgré tout le concept de voisinage (*Umgebung*) présente, il est utile de le remplacer par le concept fondamental (*Grundbegriff*) de domaine (*Gebiet*) ; on obtient ainsi le résultat : la coupure (*Durchschnitt*) de deux domaines et la réunion (*Vereinigung*) d'un ensemble quelconque de domaines est toujours un domaine ; en outre, soient P et Q deux points, il existe alors des domaines de points entre eux continuellement séparés (stets punktfremde Gebiete) dont l'un contient P, et l'autre Q, respectivement. Weyl écrit encore : "Der Prototyp der eindimensionalen stetigen Mannigfaltigkeit ist die Gesamtheit der reellen Zahlen x (Zahlgerade), der Prototyp der zweidimensionalen die Gesamtheit der reellen Zahlenpaare (x, y) (Zahlenebene), usw. Man kann sagen, dass die Zahlenebene, der Zahlenraum,... durch einen Prozess der Multiplikation, des Ineinandersteckens aus dem arithmetischen Fundamentalkontinuum der reellen Zahlen entstehen." (*Ibid*). La chose essentielle à mettre en évidence est que de cette manière on aboutit à traiter toutes les opérations sur une variété continue M par une application sur l'espace numérique Z_n, autrement dit, on assume que ce dernier constitue un modèle satisfaisant pour la variété. "An dem 'n-dimensionalen Zahlenraum' Z_n messen wir eine beliebige Mf $\mathcal{M}$, wenn wir die Definition aufstellen : $\mathcal{M}$ heisst n-dimensional, wenn sich jede Umgebung $U_\nu(P)(\nu = 1, 2, ...)$ eines Punktes P von $\mathcal{M}$ eindeutig und stetig auf Z_n abbilden lässt." (p. 4). Ce résultat a été obtenu et démontré par D. Hilbert dans "Über die Grundlagen der Geometrie", *Math. Ann.*, Bd. 56, 1903, pp. 281-380. Il faut mentionner ici que les résultats présentés par Weyl avaient été exposés et démontrés d'abord par Brouwer ("Über Abbildung von Mannigfaltigkeit", *Math. Ann.*, Bd. 7, 1911, pp. 97-115 ; "Beweis der Invarianz des n-dimensionalen Gebiets", *Math. Ann.*, Bd. 71, 1911, pp. 305-313), et exposés systématiquement par F. Hausdorff dans son traité fondamental, *Grundzüge der Mengenlehre* (1ère éd. : Berlin, 1914). L'approche de ces deux mathématiciens est complètement topologique. Dans cette nouvelle approche, en particulier chez Hausdorff, un concept qui joue un rôle très important est celui de connexité (*Zusammenhang*), qui lui servira à réinterpréter topologiquement le continu (cf. pp. 150-164 ; 3ème éd. : New York, 1965). Pour un exposé plus récent, voir G. Choquet, *Topology*, Academic Press, New York, 1966.
Rappellons qu'une tentative de définir le continu se trouve déjà Aristote [Livre V (E), 3, 227a] : "Le continu – écrit-il – est une détermination particulière du contigu, et j'affirme qu'il n'y a continuité que lorsque les limites de deux choses, par lesquelles l'une touche à l'autre, deviennent une seule et même limite, et comme le mot le dit, se tiennent ensemble."
Leibniz, de son côté, donne une définition de la propriété de la continuité qui, jusqu'à un certain point, fait penser à celle que plus tard développera Riemann : "Wenn in der Reihe der gegebenen Grössen zwei Fälle sich stetig einander nähern, sodass schliesslich der eine in den anderen übergeht, so muss notwendig in der entsprechenden Reihe der abgeleiteten oder abhängigen

Cette formulation intuitive on peut l'exprimer mathématiquement de la façon suivante :

> Jedem Punkte P entspricht eine Folge von Teilmengen der Mannigfaltigkeit $U_1(P)$, $U_2(P)$,... ; $U_n(P)$ heisst die n^{te} Umgebung von P. Man versteht also den Sinn der Relation zwischen zwei Punkten P, Q und der natürlichen Zahl n : "Der Punkte Q liegt in der n^{ten} Umgebung von P." Vorausgesetzet wird : P ist Element jedes $U_n(P)$; $U_{n+1}(P)$ ist in $U_n(P)$ ganz enthalten ; zu irgend zwei voneinander verschiedenen Punkten P, Q der Mf kann ein so hoher Index gefunden werden, dass die n^{te} Umgebung von P und die n^{te} Umgebung von Q punktfremd sind. Mit Hilfe der Umgebungen lassen sich alle Stetigkeitsbegriffe erklären, wie stetige Kurve, stetige Ortfunktion auf der Mf $\mathcal{M}$, abgeschlossene Menge, Gebiet. Eine Punktmenge $\mathcal{U}$ auf $\mathcal{M}$ heisst z. B. ein Gebiet, wenn zu jedem Punkt P von $\mathcal{U}$ ein Index n gehört, derart dass die ganze Umgebung $U_n(P)$ zu $\mathcal{U}$ gehört. Ein gebiet ist zusammenhängend, wenn sich seine Punkte nicht auf zwei Gebiete verteilen lassen - oder wenn irgend zwei seiner Punkte durch eine ganz im Gebiete verlaufende stetige Kurve verbunden werden können. Riemann setzt in seiner oben widergegebenen Erklärung voraus, dass die Mf selber in diesem Sinne ein zusammenhängendes Gebiet ist.[8]

On peut, cependant, remarquer avec Weyl que l'interprétation ensembliste de la propriété de la continuité d'une variété viole la nature du continu puisque ses éléments, à la différence de ceux appartenant à l'ensemble, ne peuvent pas être considérés comme étant isolés ou séparés. Ce ne sera pas la relation de l'élément à l'ensemble, mais plutôt celle de la partie au tout, que l'analyse du continu doit prendre en compte[9]. Riemann ne partage en rien une conception atomique du continu. Mais quant au problème de savoir comment une variété doit être engendrée, il n'est pas trop clair ; ce que l'on comprend de ses remarques est qu'une variété à n-dimensions doit pouvoir s'obtenir de la même

---

Grössen, die gesucht werden, dasselbe eintreten." ; "Über das Kontinuitätsprinzip", *Mathematische Schriften*, Bd. VI, 1687 (pp. 129-135) ; et in *Hauptschriften zur Grundlegung der Philosophie*, Erläuterungen herausgegeben von Ernst Cassirer, Band I [*Schriften zur Mathematik* , Hamburg, Verlag von Felix Meiner, ergänzte Auflage, 1966 (pp. 84-93)], pp. 84-85.
De même, Clifford, dans un texte où il donna un exposé élémentaire des nouvelles idées de Riemann sur l'espace, affirme que l'espace est un agrégat de points (a "point-aggregate") à trois dimensions – il faudrait plutôt parler de variété car les éléments d'un agrégat ne sont pas forcément homogènes –. La continuité d'une variété doit être expliquée, selon le géomètre anglais, par les propositions : i) qu'entre chaque deux points, il y a un nombre infini d'autres points, et ii) que deux parties adjacentes d'une variété ont la même limite. On pourrait de nos jours interpréter ces deux propriétés par celle topologique que la variété est *dense* (Pour une définition, cf. G. Choquet, *cit.*, pp. 17-20, 48-51, 67-69). Les variétés discrètes, par contre, on peut les caractériser de la manière suivante : i) il existe un nombre fini d'éléments entre deux éléments d'une série, et ii) les deux parties adjacentes qui divisent en deux parties les variétés discrètes, n'ont pas la même limite. Ces propriétés qui caractérisent les premières notions sur l'espace font que "the first postulate of the science of space is that space is a continuous aggregate of points, and not a discrete aggregate." ; "The Postulates of the Science of Space" (conférence, 1873), in *Lectures and Essays*, Vol. 1, Macmillan and Co., London, 1879, pp. 302-308.

[8] Weyl, *ibid.*

[9] *Ibid*, pp. 4-5.

manière qu'on obtient une variété à 3-(espace), 2-(surface) et 1-(ligne) dimensions. De ce que Riemann écrit à ce propos, on retiendra ceci :

> Geht man bei einem Begriffe, dessen Bestimmungsweisen eine stetige Mannigfaltigkeit bilden, von einer Bestimmungsweise auf eine bestimmte Art zu einer andern über, so bilden die durchlaufenen Bestimmungsweisen eine einfach ausgedehnte Mannigfaltigkeit, deren wesentliches Kennzeichen ist, dass in ihr von einem Punkte nur nach zwei Seiten, vorwärts oder rückwärts, ein stetiger Fortgang möglich ist. Denkt man sich nun, dass diese Mannigfaltigkeit wieder auf bestimmte Art, d. h. so, dass jeder Punkt in einen bestimmten Punkt der andern übergeht, so bilden sämmtliche so erhaltene Bestimmungsweisen eine zweifach ausgedehnte Mannigfaltigkeit. In ähnlicher Weise erhält man eine dreifach ausgedehnte Mannigfaltigkeit, wenn man sich vorstellt, dass eine zweifach ausgedehnte in eine völlig verschiedene auf bestimmte Art übergeht, und es ist leicht zu sehen, wie man diese Construction fortsetzen kann. Wenn man, anstatt den Begriff als bestimmbar, seinen Gegenstand als veränderlich betrachtet, so kann diese Construction bezeichnet werden als eine Zusammensetzung einer Veränderlichkeit von n+1 Dimensionen aus einer Veränderlichkeit von n Dimensionen und aus einer Veränderlichkeit von Einer Dimension.[10]

Ce qu'il faut en particulier souligner est la chose suivante : comme nous l'avons dit au début, le raisonnement de Riemann consiste dans un premier temps à appliquer les méthodes de l'analyse à la géométrie. Par conséquent, le concept de variété se lie tout naturellement dans sa conception à celui de fonction, et en fait, c'est ce dernier qui permet de mieux comprendre le premier. Afin de penser la variabilité dimensionnelle d'une variété, Riemann raisonne de la façon suivante : il imagine une partie (*Stück*) d'une variété, comptée à partir d'un point fixe, comme variable – ses valeurs doivent pouvoir être comparées entre elles –, et puis il suppose que cette partie prend une valeur déterminée en chaque point de la variété qui, en même temps que le point, change de façon continue. Cela correspond, selon Riemann, à se donner à l'intérieur de la variété une fonction continue du lieu. Il s'ensuit que tout système de points, pour lequel la fonction a une valeur constante, forme une variété continue d'un nombre moindre de dimensions.

---

[10] *Cit.*, p. 257. W. K. Clifford a été plus clair en ce qui concerne ce point. Il écrit : "We speak of the line as a point aggregate of one dimension, because, starting from one point, there are only two possible directions of change ; so that the line can be traced out in one motion. In the same way, a surface is a line-aggregate of one dimension, because it can be traced out by one motion of the line ; but it is a point-aggregate of two dimensions, because, in order to build it up of points, we have first to aggregate points into a line, and then, lines into a surface. It requires two motions of a point to trace it out. [...] The piece of solid space is an aggregate of surfaces, and an aggregate of the same kind as the line is of points ; it is a surface-aggregate of one dimension. But at the same time it is a line-aggregate of two dimensions, and a point-aggregate of three dimensions. [...] The piece of space is called a point-aggregate of three dimensions, because it takes three distinct motions to get it from a point. We must first aggregate points into a line, the line into a surface, then surfaces into a solid. [...] When we moved our line or our surface, the new line or surface contained no point whatever that was in the old one ; we started with one aggregate of points, and by moving it we got an entirely new aggregate, all the points of which were new. But this cannot be done with the solid ; so that the process is at the end. We arrive, then, at the result that *space is of three dimensions.*" ; "The Postulates of the Science of Space" (conférence, 1873), publiée dans *Lectures and Essays*, *cit.*, pp. 301-306.

Il suffit donc de faire varier une telle fonction pour qu'on puisse passer d'une variété à l'autre de façon continue[11].

Nous n'avons parlé jusqu'à maintenant que du caractère topologique (amorphe) du continu, qui représente le premier niveau de constitution mathématique de l'espace et dont le concept de variété en constitue un élément essentiel. Un tel concept, en plus de signifier une généralisation de celui d'espace, permet une classification nouvelle des formes spatiales. Il permet également de penser autrement les rapports des mathématiques à la physique. D'un point de vue épistémologique on peut dire que le résultat, peut-être le plus important qu'a produit le concept de variété, a été l'ébranlement du caractère absolu et immuable du concept d'espace. Ainsi qu'il a été remarqué :

> Riemann est passé de l'espace d'Euclide à l'espace et plus généralement aux espaces de Riemann par un double processus *d'abstraction* et *concrétisation*. Il passe de l'espace euclidien à la grandeur triplement étendue par abstraction, par *démétrisation*, et ensuite, par le processus inverse – concrétisation – il ajoute à la variété triplement étendue une métrique. L'avantage du double processus est évident : la métrique ajoutée est plus générale que la métrique enlevée. Riemann, en concevant la possibilité de concrétisation diverses, créait, aux côtés de l'espace d'Euclide, une gamme infinie d'espaces nouveaux.[12]

3. La Structure Métrique du Continu

Conférer une structure au continu consiste d'après Riemann dans la détermination métrique de celui-ci, une telle détermination se faisant de façon abstraite et analytique. Ce qui cependant l'intéresse, est de trouver l'expression géométrique d'une telle détermination, ou à partir de celle-ci, parvenir à des résultats ayant une signification géométrique. Une fois que le continu a été caractérisé comme une variété à n-dimensions, Riemann admet qu'on peut concevoir plusieurs déterminations métriques. Bien que le choix d'une d'entre elles revient dans un certain sens à être conventionnel, celui-ci, cependant, n'est dans aucun cas arbitraire. Dans cette partie de son travail Riemann s'inspire notamment du mémoire fondamental de Gauss, *Disquisitiones generales circa superficies curvas* (Göttingen, 1828). Mais il obtient une première généralisation par rapport au travail de ce dernier en considérant au lieu d'une surface, une variété, qu'il conçoit comme objet mathématique abstrait, indépendant du nombre de ses dimensions et de son plongement dans l'espace euclidien. Un point essentiel est que le choix du type de structure métrique pour le continu repose, dans la conception de Riemann, sur le principe à la fois mathématique et épistémologique général que l'Univers doit être compris d'après son comportement dans l'infiniment petit ; et d'ailleurs, c'est ce même principe qui éclaire tout le travail scientifique de Riemann[13]. En plus d'avoir servi de fondement pour la construction de sa nouvelle géométrie différentielle, la connaissance de l'infiniment petit est

[11] *Ibid.*

[12] P. Libois, "Quelques applications des idées de Riemann sur l'espace", in *Der Begriff des Raumes in der Geometrie*, Riemann-Tagung, Berlin, 1957, pp. 194-195.

[13] Cf. H. Weyl, *Vorworts* à la 3e éd. de "Ueber die Hypothesen, welche der Geometrie zu Grunde liegen", in *Das Kontinuum und andere Monographien*, Chelsea, New York, 1973. Voir aussi B. Kanitscheider, *Geometrie und Wirklichkeit*, Duncker & Humblot, Berlin, 1971, pp. 146-150.

pour Riemann la raison d'être de l'application des mathématiques à la physique. Weyl a mis en relief que cette nouvelle conception constitue un progrès dans la géométrie, comparable à celui que Faraday et Maxwell ont introduit dans la physique : ils ont en commun le fait d'avoir permis le passage de la théorie des "actions à distance" à celle des "actions de contact" (den Übergang von der Fernwirkungs- zur Nahewirkungstheorie), et les principes auxquels obéirait la structure métrique du continu sont en réalité analogues à ceux du champ gravitationnel et, jusqu'à un certain point, électromagnétique[14].

Or, pour que la géométrie différentielle de Riemann fût en mesure d'expliquer aussi les phénomènes électromagnétiques, il a fallu montrer que l'espace ne se laisse pas caractériser seulement par une connexion métrique, mais également par une connexion affine. Un développement mathématique fondamental dans cette direction a été possible grâce à la théorie du transport parallèle de Levi-Civita[15]. Une telle théorie permet justement de doter l'espace, outre d'une connexion métrique, d'une connexion affine[16]. Soient P et P' deux points quelconques liés par une courbe, on peut alors s'imaginer un vecteur donné en P qui est parallèle tout le long de la courbe à partir de P jusqu'à P'. En d'autres mots, nous dirons que le point P d'une variété est en connexion affine avec son voisinage, si l'on sait dans quel vecteur en P' un vecteur quelconque en P s'est transformé quand on l'a déplacé parallèlement à lui-même de P au point infinitésimal voisin P'. La notion de déplacement parallèle peut être aussi définie par le moyen du postulat suivant : il y a un système de coordonnées (pour le voisinage de P) dont les composantes d'un vecteur quelconque en P demeurent inaltérées lorsque ce même vecteur subit un déplacement parallèle infinitésimal. Cette condition caractérise le déplacement parallèle comme étant en fait une *propagation*. Un système de coordonnées jouissant de cette propriété est dit géodésique en P[17]. Le transport parallèle de vecteurs de P à P', qui induit un automorphisme des espaces de vecteurs au point initial P_0, n'est pas en général

[14]Précisons que dans la théorie de la gravitation d'Einstein, le concept fondamental était celui de métrique pseudo-riemannienne dans l'espace-temps. Weyl fut le premier à suggérer que le champ électromagnétique correspond au concept géométrique de connexion. Cf. "Gravitation und Elektrizität", *Sit. Ber. Kön. Preuss. Akad. Wiss.*, Bd. XXVI, 1918, pp. 465-469. Les développements ultérieurs sont dus notamment à J. A. Wheeler, *Geometrodynamics*, Academic Press, New York, 1962. Sur ce sujet, voir également Yu. I. Manin, *Mathematics and Physics*, Birkhäuser, Boston, 1981, pp. 22-24.

[15]Cf. "Nozione di parallelismo in una varietà qualunque", *Rend. Circ. Matem. di Palermo*, Vol. 42, pp. 173-205.

[16]A vrai dire, on en trouve déjà l'idée chez Riemann : "Endlich könnte man drittens, anstatt die Länge der Linien als unabhängig von Ort und Richtung anzunehmen, auch eine Unabhängigkeit ihrer Länge und Richtung vom Ort voraussetzen." ; "Ueber die Hypothesen...", *cit.*, p. 265.

[17]Donnons quelques développements techniques. Soient x_i^0 les coordonnées de P dans ce système, $x_i^0 + dx_i$ celles de P', ξ_i les composantes d'un vecteur en P, $\xi_i + d\xi_i$ les composantes du même vecteur quand on l'a déplacé parallèlement à lui-même de P en P'. Comme le déplacement parallèle de P en P' réalise une représentation affine ou linéaire de l'ensemble des vecteurs en P dans l'ensemble des vecteurs en P', il faut que les $d\xi_i$ s'expriment linéairement en les ξ_i :

$$d\xi_i = -d\gamma_r^i \xi^r \quad .$$

Ensuite les $d\gamma_r^i$ doivent être des formes différentielles linéaires des dx_i :

$$d\gamma_r^i = \Gamma_{rs}^i (dx)^s \ , \tag{1}$$

intégrable, c'est-à-dire que le vecteur en P', auquel on aboutit, dépend du chemin, le long duquel le déplacement a été réalisé. Cette difficulté peut être surmontée si l'on considère que la forme quadratique qui définit la métrique de l'espace, rend possible de comparer non seulement la longueur de deux vecteurs au même point, mais également en des points quelconques éloignés les uns des autres. La géométrie riemannienne ainsi généralisée pourrait expliquer les phénomènes gravitationnels et ceux du champ électromagnétique. Weyl fut le premier à développer cette nouvelle idée. A ce propos, il écrit :

> Eine *wahrhafte Nahe-Geometrie darf jedoch nur ein Prinzip der Übertragung einer Länge von einem Punkt zu einem unendlich benachbarten kennen*, und es ist dann von

---

et

$$\Gamma^i_{sr} = \Gamma^i_{sr} \quad . \tag{2}$$

Les deux dernières formules définissent la connexion affine, à la condition naturellement que les Γ^i_{rs}, qui sont des nombres quelconques, soient symétriques ; alors les formules de transformations

$$x_i - x_i^0 = \bar{x}_i - 1/2\Gamma^i_{rs}\bar{x}_r\bar{x}_s \tag{3}$$

définissent un système de coordonnées géodésiques $\bar{x}_i$ en P, donc on a :

$$\bar{x}_i = 0, \;\; d\bar{x}_i = dx_i(\alpha^i_k = \delta^i_k), \;\; \frac{\partial^2 f_i}{\partial\bar{x}_r\partial\bar{x}_s} = -\Gamma^i_{rs} \quad . \tag{4}$$

On remarquera que chaque modification virtuelle de la connexion affine $[\Gamma^i_{rs}]$ peut être un tenseur, car :

$$[d\xi^i] = [\Gamma^i_{rs}]\xi^r(dx)^s \tag{5}$$

est la différence de deux vecteurs qui résultent de deux propagations différemment effectuées par le vecteur ξ de P en P'. Le *déplacement parallèle d'un vecteur covariant* ξ_i du point P au point P' se définit univoquement par la condition que le déplacement simultané de ξ_i et d'un vecteur contravariant η^i n'altère pas le produit invariant $\xi_i\eta^i$:

$$d(\xi_i\eta^i) = (d\xi_i\eta^i) + (\xi_r d\eta^r) = (d\xi_i - d\gamma^r_i\xi)\eta^i = 0 \tag{6}$$

d'où

$$d\xi_i = \sum_r d\gamma^r_i\xi_r \quad . \tag{7}$$

Un *champ de vecteurs* ξ^i sera dit *stationnaire* en P, si les vecteurs au point voisin P', proviennent du vecteur en P par déplacement parallèle, c'est-à-dire si les équations aux différentielles totales

$$d\xi^i + d\gamma^i_r\xi^r = 0 \;\; (ou\frac{\partial\xi^i}{\partial x_s} + \Gamma^i_{rs}\xi^r = 0) \tag{8}$$

sont satisfaites.
Dans un système de coordonnées quelconque les Γ^i_{rs} sont des fonctions continues des coordonnées x_i du point P. Par un choix approprié du système, les Γ^i_{rs} peuvent être amenées à s'annuler toutes en un point arbitrairement choisi P, mais il est en général impossible de les annuler toutes, pour tous les points de la multiplicité. Il n'y a aucune différence entre les points de la multiplicité, relativement à la nature de leur connexion affine avec leur voisinage ; à cet égard la multiplicité est homogène...
Nous avons suivi ici de très près H. Weyl, *Raum-Zeit-Materie* [(1ère éd., 1918) ; Springer-Verlag, Berlin/Heidelberg, 1988, pp. 113-128], auquel je renvoie le lecteur pour un exposé complet.

vornherein ebensowenig anzunehmen, dass das Problem der Längenübertragung von einem Punkte zu einem endlich entefernten integrabel ist, wie sich das Problem der Richtungsübertragung als integrabel herausgestellt hat. Indem man die erwähnte Inkonsequenz beseitigt, kommt eine Geometrie zustande, die überraschenderweise, auf die Welt angewendet, *nicht nur die Gravitationserscheinungen, sondern auch die des elektromagnetischen Feldes erklärt.* Beide entspringen nach der so entstehenden Theorie aus derselben Quelle, *ja im allgemeinen kann man Gravitation und Elektrizität gar nicht in willkürloser Weise voneinander trennen.* In dieser Theorie haben *alle physikalischen Grössen eine weltgeometrische Bedeutung ; die Wirkungsgrösse insbesondere tritt in ihr von vornherein als reine Zahl auf. Sie führt zu einem im wesentlichen eindeutig bestimmten Weltgesetz ; ja sie gestattet sogar in einem gewissen Sinne zu begreifen, warum die Welt vierdimensional ist.* [18]

Revenons maintenant au problème de la métrique, dont le but est celui de donner une structure au continu et par ce moyen de le rendre intelligible. Doter l'espace d'une métrique revient à se donner une théorie de la mesure. Mesurer consiste dans une superposition de grandeurs à comparer, mais cela exige qu'on trouve une grandeur qui sert d'étalon de mesure pour les autres ("Das Messen besteht in einem Aufeinanderlegen der zu vergleichenden Grössen ; zum Messen wird also ein Mittel erfordet, die eine Grösse als Massstab für die andere fortzutragen." ; Riemann [1854], *cit.*, p. 256). Selon Riemann, toute détermination métrique doit exiger l'indépendance des grandeurs du lieu, ce qui conduit à faire l'hypothèse que la longueur des lignes est indépendante de sa position, et que par conséquent chacune d'entre elles est mesurable par le moyen des autres. L'hypothèse la plus simple qu'on peut faire en vue de caractériser les rapports métriques de l'espace consiste à montrer que le théorème de Pythagore, qui donne l'expression de la distance entre deux points, conserve sa validité en supposant que ces mêmes points sont infiniment voisins. Pour cela il faut exiger, comme on verra, quelques restrictions. Le deux "axiomes" à partir desquels Riemann construit sa géométrie infinitésimale sont les suivants :

(I) l'élément linéaire ds mesure la longueur d'arc d'une courbe entre deux points infiniment voisins, respectivement P et P' avec coordonnées (x_1, x_2, x_3) et $(x_1 + dx_1, x_2 + dx_2, x_3 + dx_3)$; de là, il suppose que la longueur du segment PP' est une fonction homogène du premier ordre des éléments différentiels dx ;

(II) puisque l'élément linéaire varie dans le même rapport que les éléments différentiels dx, il est possible d'admettre que ds exprime une forme quadratique définie positive qui a la forme suivante :

$$ds^2 = \sum_{ik} g_{ik} dx_i dx_k \ . \quad (g_{ki} = g_{ik}) \ .$$

Les coefficients $g_{ki} = g_{ik}$, qui représentent la *forme métrique fondamentale* invariante par rapport aux transformations des coordonnées, n'auront pas une valeur constante, bien qu'ils soient des fonctions continues des lieux. Ils décrivent en système de coordonnées x_i le *champ métrique*. Remarquons encore que l'espace est le cas le plus simple parmi les variétés métriques, et que son élément linéaire s'exprime (en coordonnées rectangulaires) par la forme $ds = \sqrt{\sum (dx)^2}$; elles sont appellées par

[18] Cf. "Gravitation und Elektrizität", *cit.*, pp. 466-467.

Riemann "variétés planes" (*Ebene Mannigfaltigkeiten*), et peuvent s'exprimer d'une manière tout à fait indépendante du choix des grandeurs variables.

L'hypothèse faite par Riemann en vue de caractériser la métrique, nous l'avons déjà dit, n'est pas le seule possible car, au lieu d'une forme quadratique définie positive,

$$f = (dx_1)^2 + (dx_2)^2 + ... + (dx_n)^2$$

– où f est une fonction homogène du premier ordre qui transforme linéairement les variables P (qui sont les points de la variété) –, on peut considérer une autre forme quadratique dans laquelle f_P est la racine quatrième d'une expression différentielle du quatrième degré[19]. Le choix de Riemann se laisse expliquer par le fait que les métriques satisfaisant au théorème de Pythagore dans l'infiniment petit, donnent lieu à une classe particulière d'espaces qui, du point de vue mathématique, apparaissent comme les plus simples et complets. Mais le résultat fondamental qui doit être mis en relief ici est le suivant : à chaque métrique choisie correspond une géométrie différente qui caractérise l'espace ou en général une variété, et en effet, une fois que le choix de la métrique (c'est-à-dire de son élément linéaire) a été fait, on peut en tirer toutes les propriétés mathématiques servant à caractériser la géométrie de l'espace. De cette manière, le concept général de variété se trouve non seulement spécifié, mais encore éclairé par celui de métrique : c'est-à-dire que la même forme spatiale (ou variété) peut recevoir différentes caractérisations métriques, donc différentes géométries. Par ailleurs, la même géométrie métrique peut servir à caractériser différents types de variétés. La métrique pluralise ainsi les concepts d'espace et de géométrie, et la variété en met en lumière les analogies profondes. Tandis que le concept de variété représente une généralisation de celui d'espace, le concept de métrique constitue, à son tour, une spécification de celui de variété. Ce sera grâce à cette idée que plus tard on pourra justifier et développer mathématiquement les géométries non-euclidiennes. Beltrami notamment, utilise la nouvelle géométrie infinitésimale de Gauss et Riemann pour montrer que la géométrie non-euclidienne de Lobatchewski est parfaitement intelligible comme théorie mathématique puisqu'elle est identique à la géométrie métrique d'une surface (variété à 2-dimensions) à courbure constante négative, à savoir la pseudosphère[20].

[19]Weyl remarque à ce propos : "Daher formuliert man das Raumproblem vielleicht besser folgendermassen : Alle Funktionen, welche aus einer, f, durch lineare Transformation der Variablen hervorgehen, rechne ich zu einer Klasse (f). Jeder solchen Klasse (f) von homogenen Funktionen erster Ordnung entspricht eine besondere Art von Geometrie : in einem metrischen Raum von der Art (f) gehört die Funktionen f_P, welche nach $(I)(ds = f_P(dx_1, dx_2, ..., dx_n))$ an jeder Stelle P des Raumes die Masszahlen der Linienelemente bestimmt, der Klasse (f) an. Diese Festsetzung ist unabhängig von der Wahl der Koordinaten x_i . Unter diesen Raumarten ist die Pythagoreisch-Riemannsche, die der Funktion $(II)(f = \sqrt{(dx_1)^2 + (dx_2)^2 + ... + (dx_n)^2})$ entspricht, eine einzige spezielle." ; in *Erläuterungen* à la 3ème éd. de "Die Hypothesen...", *cit.*, pp. 25-26.

[20]Beltrami a exposé ce résultat dans : "Saggio di interpretazione della geometria non-euclidea", *Giornale di Matematiche*, 6, 1868, pp. 284-312. Dans une lettre à Jules Hoüel du 1 avril 1868, il explique la signification de son résultat ainsi : "De la formule $ds = const. \sqrt{d\eta_1^2 + d\eta_2^2}$ que j'ai établie à la page 21 de mon dernier mémoire (il s'agit de "Teoria fondamentale degli spazi di curvatura costante", *Annali di Matematica Pura ed Applicata*, serie II, tomo II, 1869, pp. 232-255) on tire (ou plutôt on vérifie, car cela se trouve dans Lobatcheffsky) que la géométrie de la sphère-limite n'est pas autre chose que celle du plan euclidien. En disant que la courbure

L'importance de la nouvelle géométrie infinitésimale de Riemann ne réside pas seulement dans le fait d'avoir donné un fondement à la possibilité mathématique de géométries différentes. Sa signification est beaucoup plus profonde, elle a mené à un changement radical de notre conception de l'espace et de la géométrie, et le concept de variété a été l'élément essentiel en ce qui concerne la partie mathématique d'un tel changement. Cette nouvelle géométrie a notamment permis que s'affirme une conception intrinsèque de l'espace : on ne se contente plus d'épuiser le contenu idéal et objectif de celui-ci par la simple attribution de propriétés ayant uniquement une justification logique ; ce qu'on fait consiste plutôt à comprendre les propriétés qui peuvent caractériser la structure propre à l'espace, et à les conceptualiser mathématiquement. En ceci nous semble résider le sens du travail géométrique innovateur de Riemann. En un certain sens, sa vision est à l'origine de la définition moderne de géométrie, et en particulier de géométrie différentielle telle qu'on la retrouve, par exemple, chez O. Veblen et J. H. C. Whitehead :

> As the words are generally used at present a geometry is the theory of a space, and a space is a set of objects, usually called points, together with a set of relations in which these points are involved. A space, therefore, is not merely an abstract set of objects, but a set of objects with a definite system of properties. These properties will be referred to as the *structure* of the space.[21]

de cette surface est nulle je n'ai pas voulu dire autre chose. En d'autres termes j'ai voulu dire que toutes les propriétés métriques de cette surface sont les mêmes que celles du plan ordinaire, à cause de l'identité des éléments linéaires chez l'une et chez l'autre. Et cela a lieu indépendamment de l'axiome XI, seulement on ne doit pas se figurer la sphère-limite comme étant un plan euclidien véritable. Dans la géométrie ordinaire, aussi, on ne pourrait pas conclure, de ce que l'élément linéaire d'une surface serait réductible à la forme $\sqrt{dx^2 - dy^2}$, que cette surface est un plan : elle pourrait être aussi bien un cylindre, un cône, ou toute autre surface développable. Dans la géométrie abstraite on doit dire la même chose, c'est-à-dire que la surface de courbure nulle est identique, non pas sa forme mais ses propriétés métriques avec le plan euclidien, sur lequel on peut dire qu'elle est développable, ou constructible, sauf le mot barbare (je crois). Pour mon compte je dirais que la sphère-limite est une des formes sous lesquelles le plan euclidien existe dans l'espace non-euclidien, *en considérant le plan euclidien comme défini par la propriété d'avoir sa courbure nulle*. Quand, au contraire, on définit le plan d'après la propriété (beaucoup plus essentielle dans la géométrie élémentaire) de contenir toute entière chaque droite passant par deux de ses points, alors on trouve, comme surface analogue dans l'espace non-euclidien, celle que j'ai appelée surface de 1er ordre, c'est-à-dire une surface dont la courbure est négative, constante et (ce qui est très important) *égale à celle de l'espace lui même*. Par conséquent il y a bien, sous le rapport de la courbure, une infinité de surfaces (de courbure négative constante) intermédiaires entre le plan non-euclidien (ou surface de 1er ordre) et la sphère-limite, mais ces surfaces ne sont plus des plans non-euclidiens, elles sont des sphères non-euclidiennes, dont le rayon varie depuis la valeur constante jusqu'à l'infini [...]". Cette lettre fait partie d'un corpus de 62 lettres écrites par Beltrami à Hoüel entre 1868 et 1881, où l'auteur de la découverte de la géométrie pseudosphérique non-euclidienne explique les idées et les méthodes qui l'on conduit à une telle découverte ainsi que sa signification pour les mathématiques de l'époque. L'introduction, les commentaires à la fois mathématiques, historiques et épistémologiques de cette correspondance ainsi que sa transcription intégrale font l'objet actuellement d'un travail de publication d'un livre que nous sommes tout juste au point de compléter et qui paraîtra au cours de l'année 1992.

[21] *The Foundations of Differential Geometry*, University Press, Cambridge, 1932, pp. 17-18.

Ajoutons que la géométrie n'est plus conçue par Riemann comme la science des figures qui peuvent être construites dans l'espace, mais comme la science des différentes formes spatiales – le concept de variété exprime en fait, ce caractère qualitatif de forme spatiale, et une variété est un exemple parmi d'autres de "forme spatiale" – mathématiquement concevables, et en même temps, comme la science de l'espace lui-même, c'est-à-dire de l'espace physique. Ces deux aspects sont, au fond, solidaires dans la pensée de Riemann du fait qu'il concevait sa nouvelle géométrie infinitésimale comme nécessaire pour la connaissance des rapports de causalité des phénomènes dans l'infiniment petit ("Auf der Genauigkeit, mit welcher wir die Erscheinungen in's Unendlichkleine verfolgen, beruht wesentlich die Erkenntniss ihres Causalzusammenhangs").

4. Le Concept de Courbure et la Propriété de l'Homogénéité

Pour résumer cette dernière partie on dira qu'une variété ou un espace est riemannien s'il possède une métrique. Mais il existe une autre propriété essentielle qui caractérise la géométrie intrinsèque d'une variété, à savoir sa courbure. La courbure riemannienne est une généralisation pour les espaces à n-dimensions de celle donnée par Gauss pour les surfaces. Rappelons que ce dernier définissait la courbure d'une surface comme la mesure de la divergence de ses normales. Il considérait à ce propos une sphère de rayon unité centré en un point quelconque O, et il faisait correspondre, moyennement l'équipollence des normales, un élément dw de la sphère à un élément de surface do ; dw étant l'angle que les normales à do forment quand on leur donne une même origine. La limite du rapport dw/do est la courbure de Gauss. Il montre ensuite que cette courbure est une propriété intrinsèque de la surface elle-même, et qu'on peut connaître sa valeur si l'on connaît en même temps les coefficients de la forme différentielle quadratique qui définit la surface (appellée par Gauss "forme fondamentale") et leurs dérivées jusqu'au deuxième ordre. La courbure est ainsi un invariant différentiel de la géométrie de la surface ; c'est-à-dire que Gauss démontre analytiquement que si on étend et déchire la surface, la courbure ne change pas.

Riemann élargit la notion de courbure aux formes différentielles quadratiques de plusieurs variables et montre qu'elle fait intervenir la notion de tenseur au lieu de celle de scalaire. Autrement dit, il considère en chaque point et dans chaque direction de surface d'un espace donné, une courbure que ces éléments déterminent. D'après Weyl, on peut le montrer par le raisonnement suivant. Tendons sur une courbe fermée une surface qu'on décompose en parallélogrammes infiniments petits. L'altération subie par un vecteur qui se déplace sur la courbe frontière est le résultat des altérations qu'il subira en parcourant la frontière de chaque petit parallélogramme ; désignons par dx_i et δx_i les composantes des côtés d'un tel quadrilatère issus de P et calculons la variation sur le contour du parallélogramme. On constatera d'abord que l'accroissement $\Delta x = (\Delta\xi')$ que le vecteur $x = (\xi')$ subit, s'obtient à partir de x par une correspondance linéaire, qui détermine une matrice ΔF :

$$\Delta x = -\Delta F(x) \ ; \ \Delta\xi^\alpha = -\Delta F^\alpha_\beta \xi^\beta \quad . \tag{1}$$

Soit $\Delta F = 0$, la variété en P est *plane* dans l'élément de surface considéré ; si cela arrive pour tous les éléments d'une surface finie, chaque vecteur qui parcourt une courbe fermée située sur elle revient à son point de départ sans altération. $-\Delta F$ dépend aussi linéairement de l'élément de surface :

$$\Delta F = F_{ik} dx_i \delta x_k = 1/2 F_{ik} (\Delta x)^{ik} \quad (F_{ki} = -F_{ik}) \quad . \tag{2}$$

La forme différentielle que l'on vient d'écrire, caractérise la ***courbure***, c'est-à-dire l'écart entre la variété en P et une variété qui en P est plane dans toutes les directions superficielles ; les coefficients de cette forme ne sont pas des nombres, mais des matrices ; on peut dire que la courbure est définie par un ***tenseur-matrice linéaire du 2ème ordre*** (cf. H. Weyl [1918], *cit.*, pp. 117-121).

On peut affirmer en fait, que Riemann avait déjà une idée de la signification géométrique du tenseur de courbure (cf. [1854], II, 3 ; *Commentatio mathematica...*, [1861], 2ème partie) lorsqu'il considère les sous-variétés à 2-dimensions et un point $P_0 \in V$, et lorsqu'il suppose qu'il existe un sous-espace quelconque qui comprend (localement) tous les morceaux d'espaces de vecteurs en Po relatifs à une variété à n-dimensions ; dans ce cas, toutes les lignes géodésiques de la variété qui ont leur origine en Po, seront déterminées si elles passent par cet espace de vecteurs en Po[22].

Riemann définit de façon intuitive la courbure comme étant la quantité dont la variété s'écarte dans un point, suivant sa direction superficielle, de la planarité ("[...] und kann daher als das Mass der in diesem Punkte in dieser Flächenrichtung stattfindenden Abweichung der Mannigfaltigkeit von der Ebenheit angesehen werden"). Dans son analyse du concept de courbure, deux résultats paraissent particulièrement importants :

i) le premier affirme que les relations métriques d'une variété sont déterminées par la mesure de courbure en chacun de ses points (dans les mots de Riemann : "Zur Bestimmung der Massverhältnisse einer n-fach ausgedehnten in der vorausgesetzten Form darstellbaren Mannigfaltigkeit wurden vorhin $n(n-1)/2$ Funktionen des Orts nöthig gefunden : wenn also das Krümmungsmass in jedem Punkte in $n(n-1)/2$ Flächenrichtungen gegeben wird, so werden daraus die Massverhältnisse der Mannigfaltigkeit sich bestimmen lassen, wofern nur zwischen diesen Werthen keine identischen relationen stattfinden, was in der That, allgemeine zu reden, nicht der Fall ist. Die Massverhältnisse dieser Mannigfaltigkeiten, wo das Linienelement durch die Quadratwurzel aus einem Differentialausdruck zweiten Grades dargestellt wird, lassen sich so auf eine von der Wahl der veränderlichen Grössen völlig unabhängige Weise ausdrücken.")[23] ;

[22]Cf. J. Böhm et H. Reichardt, *Kommentierender Anhang*, in *Gaussche Flächentheorie, Riemannsche Räume und Minkowski-Welt*, Teubner-Archiv zur Mathematik, Band I, Leipzig, 1984, pp. 137-142.

[23]Pour un commentaire approfondi sur ce point, cf. R. Torretti, *Philosophy of Geometry from Riemann to Poincaré* (1978), D. Reidel, Dordrecht, 1984, pp. 98-102. Pour une traduction dans un langage mathématique moderne, très éclaircissante, de l'énoncé de Riemann, voir M. Spivak, *A Comprehensive Introduction to Differential Geometry* (second edition), Vol. II, Publish or Perish, Inc., Houston, 1979, pp. 175-178. Voici ce qu'il écrit : "In an n-dimensional vector space there are $n(n-1/2)$ 'independent' 2-dimensional subspaces : if $v_1, ..., v_n$ is a basis, we can choose the subspaces spanned by v_i and v_j , for $i < j$. Riemann claims that the metric $<,>$ is determined if $Q(W)$ is known for $n(n-1/2)$ independent 2-dimensional subspaces $W \subset M_q$ at each point q, for example, if Q is known for the subspaces spanned by each $\partial/\partial x^i|_q$ and $\partial/\partial x^j|_q$ $(i < j)$. A very special case of this general claim is the following, which we will henceforth call the *Test Case* :

(*) If M is n-dimensional and $Q = 0$ for $n(n-1/2)$ independent 2-dimensional subspaces of

ii) la courbure est un invariant de la géométrie de la variété et des surfaces qui s'y laissent représenter. Cette idée ouvre un horizon tout nouveau à la géométrie différentielle ainsi qu'à l'étude des structures qui caractérisent les variétés différentiables. Elle se développe suivant deux directions : d'un côté, elle conduit à la définition rigoureuse du tenseur de courbure et à la notion d'"espaces à torsion" dans le cadre de la théorie du transport parallèle due à Levi-Civita et de son développement grâce en particulier à Elie Cartan[24] et Hermann Weyl[25] ; et de l'autre, elle est à l'origine de la "théorie des espaces localement euclidiens" qui a surtout servi à montrer l'importance pour la géométrie du problème des rapports entre propriétés locales et globales dans une variété.

Le concept de courbure permet à Riemann une compréhension profonde de la structure de l'espace car il rend possible une nouvelle classification (typologie) des variétés. La classe la plus importante de variétés est celle dont la mesure de courbure est constante, qui comprend comme cas particulier celle dont la courbure est nulle. Les variétés à courbure constante sont caractérisées par les deux propriétés suivantes :

(i) elles admettent la libre mobilité des figures et des corps rigides dans l'espace, autrement dit, les figures peuvent s'y mouvoir sans subir d'extension ;

(ii) elles admettent, en outre, une homogénéité, bien que locale, de l'espace, c'est-à-dire que non seulement tous ses points, mais également toutes les directions de surface issues de ces mêmes points, sont équivalents. Ceci doit être interprété dans le sens que n'importe quel corps géométrique situé dans une position arbitraire, peut être amené à coïncider isométriquement avec n'importe quel autre point de l'espace et dans n'importe quelle direction. Réciproquement, une variété qui possède une telle propriété d'homogénéité, aura une courbure constante. Dans un système de coordonnées approprié, l'élément linéaire des variétés à courbure constante prendra nécessairement la forme :

$$ds^2 = \sum_i dx_i^2 + \frac{\alpha}{1-\alpha r^2}(\sum x_i dx_i)^2 \quad .$$

Il est important ici de remarquer que Riemann introduit la possibilité de l'homogénéité comme une hypothèse *a priori*, qu'il tire comme conséquence analytique de la propriété de l'espace d'avoir une courbure constante, sans pour autant en donner une justification théorique et encore moins, une fondation empirique. On reviendra sur ce point.

Un résultat important de cette partie du travail de Riemann (Section II, 5, de "Über die Hypothesen...") a été sans doute d'avoir montré qu'une surface (Σ) est développable,

each M_q, then M is flat, that is, M is locally isometric to $\mathbb{R}^n$ with its usual inner product. In connection with the *Test Case*, it should be pointed out that a local isometry with $\mathbb{R}^n$ is the best we can hope for, since there are Riemannian manifolds which are not homeomorphic to $\mathbb{R}^n$, but which are locally isometric to $\mathbb{R}^n$, and hence have $Q = 0$ everywhere. The simplest example of such manifold is the 'flat torus' [...]." (pp. 176-77).

[24]Cf. "Sur une généralisation de la notion de courbure de Riemann et les espaces à torsion", *Comptes-Rendus de l'Académie des Sciences*, tome 174, 1922, pp. 593-595 ou *Œuvres Complètes*, Partie III, Vol. 1, Paris, Gauthier-Villars, 1955, pp. 616-618 ; et surtout, "Les récentes généralisations de la notion d'espace", *Bull. Soc. Math.*, tome 48, pp. 294-320 ; *Œuvres Complètes*, *cit.*, pp. 863-889.

[25]Cf. *Raum-Zeit-Materie* (1ère éd., 1918), Springer-Verlag, Berlin/Heidelberg, 1988 ; *Matematische Analyse des Raumproblems*, Springer, Berlin, 1923.

selon que sa valeur de courbure K est $= O$, $> O$, ou $< O$, respectivement sur un cylindre (ou d'autres surfaces euclidiennes), sur une sphère et sur une surface de révolution (ou surface annulaire). Il ne faut pas pourtant oublier que l'équivalence entre les surfaces dont il est question ici, concerne uniquement les propriétés dites locales (donc, surtout métriques) de ces mêmes surfaces, mais non pas les propriétés globales (topologiques). Une telle distinction se trouve déjà implicitement chez Riemann.

Résumons maintenant les résultats de Riemann jusqu'ici exposés (dans le présent paragraphe), dans un langage mathématique moderne et sous la forme d'un théorème. Soit M une surface pseudo-riemannienne de dimensions $n \geq 2$ et soit K un nombre réel ; alors, les conditions suivantes sont équivalentes :

(i) M a courbure constante K ;

(ii) si $x \in M$, il y aura des coordonnées locales u^i d'un voisinage de x dans lequel la métrique est donnée par l'expression :

$$ds^2 = \frac{e_1 du^1 \oplus du^1 + \ldots + e_n du^n \oplus du^n}{\{1 + K/4 \sum e_i (u^i)^2\}^2}, \quad e_i = \pm 1 \quad ;$$

(iii) si $x \in M$, alors x possède un voisinage qui est isométrique à un ensemble ouvert, respectivement sur S^n_s si $K > O$, R^n_s si $K = O$ et H^n_s si $K < O$.

5. L'Approche Nouvelle de la Théorie des Groupes du Problème de l'Espace.

Au point de vue théorique l'assertion de Riemann (qui, rappelons-le, identifie l'homogénéité de l'espace au fait que la courbure soit constante) sera pour la première fois démontrée rigoureusement grâce à l'application du concept de groupe de transformations dû essentiellement à Sophus Lie, qui reformulera ce qu'il appelle le "problème de Riemann-Helmholtz" (*das Riemann-Helmholtzsche Problem*) dans les termes suivants : il s'agit de caractériser les groupes de mouvements des espaces de Riemann à courbure constante parmi les autres ensembles possibles de mouvements d'une variété numérique (*Zahlenmannigfaltigkeit*)[26]. La solution donnée par Lie, qui traduit dans une formulation mathématique rigoureuse une idée déjà mentionnée par Helmholtz, utilise la notion de transformation infinitésimale et s'applique à un espace à n-dimensions. Elle peut s'énoncer ainsi : l'espace est homogène s'il se laisse caractériser complètement par son groupe de transformations ; les figures que l'on obtient par chaque opération différente du groupe sont, du point de vue de la géométrie considérée, entre elles semblables. Sa

[26]Selon l'expression de Lie : "Fassen wir die bei Riemann und bei Herrn v. Helmholtz zu Grunde liegenden Gedanken zusammen, so können wir sagen, dass beide Forscher, wenn auch nur implicite, ein neues Problem aufgestellt haben, das wir als das Riemann-Helmholtzsche problem bezeichnen wollen, und das kurz so ausgesprochen werden kann : *Es sollen solche Eigenschaften gefunden werden, die sowohl der Schaar der Euklidischen als den beiden Schaaren von Nichteuklidischen Bewegungen zukommen, und durch die diese Schaaren vor allen andern möglichen Schaaren von Bewegungen einer Zahlenmannigfaltigkeit ausgezeichnet sind.*" ; in *Theorie der Transformationsgruppen*, Dritter Abschnitt, Abtheilung V : *Untersuchungen über die Grundlagen der Geometrie, cit.*, p. 397.

conception infinitésimale du problème exige uniquement que tous les points ainsi que les directions de surface à 1, 2,..., $(n-1)$-dimensions dans l'espace soient équivalentes (*Gleichberechtigen*) ; que par un mouvement quelconque on puisse transporter chaque point dans n'importe quel autre ; et enfin, que par le mouvement qui laisse fixe un point, on puisse également transporter n'importe quelle direction de ligne en ce point dans une autre direction quelconque. Si, par contre, un système incident de directions d'éléments de dimension 0 jusqu'à $n-1$ est donné, alors, en dehors de l'identité, aucun autre mouvement laisse fixe le même point. A partir de ces conditions on ne peut obtenir, selon Lie, que la géométrie euclidienne et les deux géométries non-euclidiennes, l'hyperbolique de Bolyai-Lobatchewski, et l'elliptique de Riemann[27].

Le résultat de Lie a la signification suivante : le problème de savoir quelles sont les variétés qui possèdent la propriété de l'homogénéité revient à reconnaître une structure de groupe à l'espace faisant en sorte que ses propriétés caractéristiques demeurent invariantes par n'importe quel ensemble d'opérations appartenant au groupe lui-même. En d'autres mots, le groupe assure l'invariance dans le changement ; il permet de retrouver l'identité globale de l'espace après chaque permutation de ses éléments. De plus, le groupe qu'on associe à une variété permet de caractériser complètement sa géométrie, de telle sorte que chaque groupe qui agit sur une variété différentiable induit une géométrie sur cette même variété. C'est cette idée, simple et profonde en même temps, qui a fourni un fondement mathématique décisif pour la reconnaissance sur le plan formel d'une pluralité de géométries, d'où la possibilité d'obtenir chacune d'entre elles comme modèle de l'autre. Ainsi, des théories géométriques différentes quant à leurs objets, sont équivalentes quant à leurs structures, et cela parce qu'elles sont constituées par des groupes entre eux isomorphes.

Une des idées qui a contribué le plus au progrès de la géométrie différentielle moderne fait référence à l'existence d'une classe générale d'espaces dits "espaces homogènes symétriques", dont la théorie a été développée par Elie Cartan[28]. L'idée est que toute

---

[27]Cf. S. Lie, *cit.*, pp. 393-543. Une analyse détaillée de la conception de Lie est donnée par Weyl dans *Mathematische Analyse des Raumproblems*, *cit.*, Ch. III : "Gruppentheoretische Analyse des Raumproblems", pp. 29-34 ; et dans *Riemanns geometrische Ideen...*, *cit.*, pp. 31-35. Dans son ouvrage de 1923 [*cit.*] Weyl montre de quelle manière on peut interpréter un espace de vecteurs à n dimensions en appliquant le concept de groupe. Il donne à ce propos le théorème suivant :

"Im n-dimensionalen Vektorraum mit den Koordinaten x_i liege eine Gruppe homogener linearer Tranformationen vor, welche dem Vektorkörper freie Beweglichkeit um sein Zentrum O verleiht.

D. h. es sei mit Hilfe einer zu der Gruppe gehörigen Transformation möglich, ein beliebiges System inzidenter Richtungselemente von der I^{ten} bis zur $(n-I)^{ten}$ Stufe in O in ein beliebiges anderes derartiges System überzuführen ; hingegen sei die Identität die einzige Operation der Gruppe, welche ein solches System von Richtungselementen festlässt. *Dann existiert notwendig eine positiv-definite quadratische Form*

$$\sum_{ik} g_{ik} x_i x_k ,$$

welche bei allen Transformationen der Gruppe ungeändert bleibt." (pp. 31-32).

[28]Voir "Sur une classe remarquable d'espaces de Riemann", Bull. Soc. Math. France, vol. 55 (1926), pp. 214-264 ; "Sur une classe remarquable d'espaces de Riemann", Bull. Soc. Math.

variété riemannienne M est homogène s'il existe un groupe G, qui lui est associé, lequel agit de façon différentiable et transitive sur M[29]. Un espace symétrique de Riemann peut être défini, soit comme un espace dont le parallélisme de Levi-Civita préserve la courbure sectionnelle de Riemann, soit comme un espace dont la symétrie autour d'un point est une isométrie. De cette deuxième définition, il découle que l'espace admet un groupe transitif d'isométries et que la composante connexe du sous-groupe d'isométries laissant fixe un point de l'espace est compact. Ce fait met en relation les espaces symétriques de Riemann avec les espaces homogènes, et il permet donc, de traiter les propriétés locales de ces espaces selon un point de vue global[30]. Le développement de la géométrie différentielle au XXème siècle, s'est réalisé à partir de l'idée fondamentale que le concept de variété différentiable (redéfini à la lumière de celui d'"espace topologique" introduit par Felix Hausdorff en 1914) et celui de groupe sont inséparables.

6. La Théorie de la Mesure, l'Espace, et la Géométrisation de la Physique

De même que Riemann, Helmholtz et Clifford reconnaîssaient dans la possibilité de mesurer l'espace, ainsi que les grandeurs et les distances des corps qui y sont situés, le fondement de la géométrie. Mais à la différence du mathématicien allemand, ils pensaient qu'une telle opération avait un caractère empirique et qu'elle constituait le véritable fondement de la géométrie physique. Pour que la géométrie puisse être appliquée à notre espace physique, il est nécessaire que l'on présuppose l'existence de corps rigides, qui sont en fait une idéalisation des corps existants dans la nature. Leur caractéristique consiste en ce qu'ils peuvent être transportés dans n'importe quelle partie de l'espace sans subir de déformation – ils gardent donc leur "identité" –. La possibilité de mesurer l'espace trouve son explication, selon Clifford, dans un postulat primitif de notre conception de l'espace ordinaire, à savoir que nous pouvons mouvoir un objet partout dans l'espace sans en altérer les grandeurs et les formes. On remarquera ici que Clifford ne fait, en réalité, qu'admettre de façon préalable l'homogénéité de l'espace. Mesurer la grandeur d'un corps correspond à mesurer la distance des différents points en lui. Mais afin d'effectuer cette opération on a besoin, selon Clifford, d'une sorte d'étalon de mesure. Il écrit à ce propos :

France, vol. 55 (1927), pp. 114-134 ; "Les espaces riemanniens symétriques", Verh. Int. Math. Kong. Zurich, vol. I, 1932, pp. 152-161.

[29] Pour une analyse détaillée de l'œuvre de Cartan et en particulier de cette théorie voir S.S. Chern, "Elie Cartan and his Mathematical Work", *Bull. Amer. Math. Soc.*, Vol. 58 (1952), pp. 217-250 ; ou *Selected Papers*, vol. II, Springer-Verlag, New-York, 1989, pp. 267-300. Chern commente cette nouvelle notion introduite par Cartan ainsi : "The most important work on Riemannian geometry is undoubtedly the theory of symmetric Riemann space. It is well known that the local properties of a Riemann metric are given by Riemann-Christoffel curvature tensor and its successive covariant derivative. Besides the locally Euclidean spaces the simplest Riemann spaces are therefore the ones for which the covariant derivative of the Riemann-Christoffel tensor is zero. These spaces, which include the Riemann spaces of constant curvature, are called symmetric by Cartan [...]". (pp. 294-295).

[30] Pour plus de détails, voir K. Nomizu, *cit.*, pp. 23-24, 39-40.

> The measurement of distance is only possible when we have something [...] which we can carry about and which does not alter its length while it is carried about. The measurement is then effected by holding this thing in the place of the distance to be measured, and observing what part of it coincides with this distance.[31]

Comment déterminer maintenant l'égalité des longueurs? La réponse de Clifford est la suivante :

> [...] we may say generally that two lengths or distance of any kind are equally, when, one of them being brought up close to the other, they can be made to fit without alteration.[32]

Ensuite, il remarque que puisque la longueur ou la distance a été définie "by means of a measure which can be carried about without changing its lenght" ("par le moyen d'une mesure qui peut être transportée sans en changer la longueur"), comment alors, se demande-t-il, cette opération de la mesure peut être expérimentée? Clifford semble en fait, douter de la possibilité de disposer de moyens permettant de vérifier si les longueurs demeurent inaltérées ; par conséquent, le critère consisterait plutôt en ce que "two things are always of the same length when they are in the same place, not that this length is unaltered." ("deux objets sont toujours de la même longueur lorsqu'ils occupent le même lieu, et non pas lorsque cette longueur reste inchangée.")[33].

Il nous semble indéniable du point de vue historique que les idées de Helmholtz ont influencé de façon décisive la conception générale d'Einstein sur la théorie de la relativité restreinte, en même temps que le nouveau calcul tensoriel créé au début de ce siècle par les mathématiciens italiens Ricci et Levi-Civita en a constitué son outil mathématique et conceptuel fondamental. L'influence de Clifford sur Einstein est plus difficile à prouver historiquement, et si elle a eu lieu, elle n'a pu être qu'indirecte. Par ailleurs, la pensée de Clifford a dû exercer un rôle majeur dans l'élaboration de la relativité générale, car l'idée essentielle de celle-ci se trouve déjà explicitement chez lui, lorsqu'il affirme que la géométrie et la matière se conditionnent réciproquement. L'idée d'expliquer les phénomènes physiques par des concepts géométriques – qui constitue le point de départ d'un programme plus général de géométrisation de la physique – n'était pas étrangère au géomètre anglais ; elle constitue en réalité le fil conducteur de toute sa pensée théorique et d'une part de sa production mathématique. Nous croyons en effet, que Riemann et encore plus explicitement Clifford, ont été les premiers à avoir théorisé un programme cohérent en vue d'une interprétation géométrique des phénomènes physiques. Les idées directrices d'un tel programme seraient les suivantes :

i) L'espace physique qui s'étend au delà de nos possibilités d'observation (macroscopique et microscopique) n'est pas forcément homogène, et par conséquent, ses rapports métriques ne sont pas les mêmes que ceux qui caractérisent notre espace ordinaire. On pourrait donc penser que la géométrie d'un tel espace varie de façon

---

[31] *The Common Sense of the Exact Science*, Dover, New York, 1955, Chap. II : "Space", pp. 48-49.

[32] *Ibid.*

[33] *Ibid.*

continue d'une région à l'autre. Cette idée est partagée aussi bien par Riemann que par Clifford.

ii) Il est tout à fait légitime de supposer que l'espace, au lieu d'être plat, soit courbe ; dans ce cas, les lois ordinaires de la géométrie ne pourraient plus lui être appliquées. Cette hypothèse, reconnue comme possible par Riemann[34], Clifford la considère comme un fait acquis[35].

iii) Les différentes formes de mouvement de la matière doivent être pensées non pas tellement comme existant dans l'espace, mais plutôt comme engendrées par sa structure géométrique. Sur ce point Clifford va plus loin que Riemann en avançant l'idée que la variation de la matière dépend plus précisément de la courbure de l'espace.

Bien que conscient du caractère incomplet de notre analyse, nous espérons néanmoins avoir apporté quelques éléments nouveaux et d'intérêt pour la compréhension du "problème de l'espace" et de ses rapports à la géométrie aussi bien qu'à la physique. Notre but était de mettre en relief les points suivants. Premièrement, que les différentes étapes par lesquelles s'articule la formalisation mathématique du concept *espace* visent une intelligibilité de plus en plus profonde (une sorte de "capture géométrique") du continu ; une telle formalisation s'organise en quatre niveaux fondamentaux, distincts et liés en même temps : le métrique, le différentiable, le "groupal", et le topologique. Ce dernier, dans un certain sens, les embrasse tous. Chacun d'eux, obéissant à une hiérarchie conceptuelle précise, est une réintérprétation de l'autre. En deuxième lieu, nous voulions montrer, en nous appuyant notamment sur la pensée de Riemann, Clifford et Weyl, que la formalisation mathématique n'épuise pas le contenu effectif de l'espace ; elle en est tout au plus une représentation abstraite. Elle est par ailleurs une source de modèles dont le mathématicien se sert pour approfondir sa connaissance de l'espace. Mais l'autre aspect essentiel est qu'à chaque concept géométrique important correspond un système physique. Mentionnons à ce propos quelques exemples pris des développements récents de la physique : un champ gravitationnel est une connexion dans un espace des degrés internes de liberté d'un giroscope, elle contrôle l'évolution du giroscope dans l'espace-temps ; un champ électromagnétique est une connection dans l'espace des degrés internes de liberté d'un électron quantique, la connection contrôle son évolution dans l'espace-temps, etc.[36] Une conclusion épistémologique que l'on pourrait tirer de ce fait est que la géométrie (la forme) tout autant que le contenu physique (la matière) de l'espace sont des éléments essentiels de sa constitution et, dans ce sens, ils sont indissociables.

---

[34] "Setzt man voraus, dass die Körper unabhängig vom Ort existiren, so ist das Krümmungsmass überall constant, und es folgt dann aus den astronomischen Messungen, dass es nicht von Null verschieden sein kann [...]. Wenn aber eine solche Unabhängigkeit der Körper vom Ort nicht stattfindet, so kann man aus den Massverhältnissen im Grossen nicht auf die Unendlichkleinen schliessen [...]. " ; "Ueber die hypothesen...", *cit.*, pp. 267-268.

[35] "On the space-theory of matter", *cit.*, pp. 157-158.

[36] Cf. Yu. I. Manin, *Mathematics and Physics*, Birkhäuser, Boston, 1981, pp. 22-24.

Bibliographie

d'Abro, A., *The Evolution of Scientific Thought from Newton to Einstein*, Dover Publications, New York, 1950 (revised and enlarged version of the work originally published in 1927).

Alhfors, L.V., Sario, L., *Riemann Surfaces*, Princeton University Press, Princeton, 1960.

Aristotele, *La Fisica*, Laterza, Bari, 1968.

Becker, O., *Beiträge zur phänomenologischen Begründung der Geometrie und ihre physikalischen Anwendung*, Max Niemeyer Verlag, Tübingen, 1973 [1923].

Beltrami, E., "Saggio di interpretazione della geometria non-euclidea", *Giornale di Matematiche*, **6**, 1968, 284-312.

Beltrami, E. "Teoria fondamentale degli spazi di curvatura costante", *Annali di Matematica Pura, ed Applicata*, serie **II**, tomo **II**, 232-255.

Bohm, J., Reichardt, H., *Gausssche Flächentheorie, Riemannsche Räume und Minkowski-Welt*, Teubner-Archiv zur Mathematik, Bd. **1**, Leipzig, 1984.

Boi, L., "Idéalisation et objectivation, ou des rapports entre géométrie et physique", *Fundamenta Scientiae*, Vol. **10**, No. **1**, 1989, 85-114.

Boi, L., "The 'Revolution' in the Geometrical Vision of Space in the Nineteenth Century and Hermeneutical Epistemology of Mathematics", in *Revolutions in Mathematics* (Edited by Donald Gillies), Oxford University Press, Oxford, 1992 [à paraître].

Brouwer, L. E. J., "Über Abbildung von Mannigfaltigkeiten", *Math. Ann.*, **71**, 97-115 ; ou *Collected Works* Vol. 2, North-Holland, Amsterdam, 1976 (pp. 454-474).

Brouwer, L. E. J., "Characterization of the Euclidean and non-Euclidean motion groups in $\mathbb{R}^n$", *Handelingen van het Nederlandsch Natuuren Geneeskundig Congres*, **12**, 189-199 (*Coll. Works :* pp. 185-192).

Cartan, E., *Œuvres Complètes*, Partie III, Vol. 1, Gauthier-Villars, Paris, 1955.

Cartan, E., *Leçons sur la géométrie des espaces de Riemann*, Gauthier-Villars, Paris, 1928.

Châtelet, G., "Sur une petite phrase de Riemann", *Analytiques*, **3**, 1979, 67-75.

Chern, S.-S., *Selected Papers*, Vol. II, Springer, New York, 1989.

Clifford, W. K., *Mathematical Papers*, Chelsea, New York, 1968 [1882].

Clifford, W. K., *The Common Sense of the Exact Sciences*, Dover , New York, 1955 [1885].

Clifford, W. K., *Lectures and Essays*, Vol. II, Macmillan and Co., London, 1879.

Clifford, W. K., "On the Space-Theory of Matter", *Proc. Camb. Phil. Soc.*, Vol. **II**, 1870, 157-158.

Doubrovine, B., Novikov, S., Omenko, A., *Géométrie Contemporaine*, 2 vols, Editions MIR, Moscou, 1982.

Einstein, A., "Geometrie und Erfahrung", *Sit. Ber. Preuss. Akad. Wiss.*, **I**, **1**, 1921, 122-130.

Freudenthal, H., "Lie Groups in the Foundations of Geometry", *Advances in Mathematics*, Vol. **1**, Fasc. **2**, 1964, 145-190.

Gauss, C. F., "Disquisitiones Generales Circa Superficies Curvas", *Commentationes societatis regiae scientiarum Gottingensis recentiores*, Vol. **VI**, Göttingen, 1828.

Gauss, C. F., *Werke*, Bd. VIII (*Grundlagen der Geometrie Nachlass :* pp. 159-452), B. G. Teubner, Göttingen, 1900.

Giorello, G., "Pratica geometrica e immagine della matematica in Thomas Hobbes", in *Hobbes Oggi. Filosofia e Scienza nel Cinquecento e nel Seicento*, Milano, Franco Angeli, 1990 (pp. 215-244).

Granger, G.-G., "Sur l'idée de concept mathématique 'naturel'", *Revue Internationale de Philosophie*, Vol. **42**, N° **167**, 1988, pp. 474-499.

Gray, J., "Non-Euclidean geometry, a reinterpretation", *Historia Mathematica*, Vol. **6**, 1979, pp. 236-258.

Gromov, M., "Manifolds of negative curvature", *Journal of Differential Geometry*, Vol. **13**, N° **2**, 1978, pp. 223-230.

Grünbaum, A., *Philosophical Problems of Space and Time* (Second, enlarged edition), D. Reidel, Dordrecht, 1973.

Hausdorff, F., *Grundzüge der Mengenlehre*, Veit & Co., Leipzig, 1914.

Helmholtz, H., *Wissenschaftliche Abhandlungen*, Bd. II, Johann Ambrosius Barth, Leipzig, 1883.

Herbart, J. F., *Sämtliche Werke*, Bd. VIII : *Allgemeine Metaphysik nebst den Anhängen der philosophischen Naturlehre*, Hermann Beyer & Söhne, Langensalza, 1893 [1829].

Hilbert, D., "Über die Grundlagen der Geometrie", *Math. Ann.*, **56**, 1903, 281-380.

Houzel, Ch., "Aperçu sur les mathématiques d'Henri Poincaré", *Préface*, n° **5**, décembre 1987-janvier 1988, pp. 120-126.

Husserl, E. [1889-90], "Geschichtlicher Überblick über die Entwicklung der Geometrie", in *Husserliana*, Bd. XXI : *Studien zur Arithmetik und Geometrie*, Martinus Nijhoff Publishers, The Hague, 1983 (pp. 312-347).

Husserl, E., *Idées directrices pour une phénoménologie et une philosophie phénoménologique pures* , Gallimard, Paris, 1950 [1913].

Jordan, P., *Schwerkraft und Weltall*, Friedr. Vieweg & Sohn, Braunschweig, 1955.

Kanitscheider, B., *Geometrie und Wirklichkeit*, Duncker & Humblot, Berlin, 1971.

Kant, I., *Kritik der reinen Vernunft*, tr. fr. in *Œuvres philosophiques*, vol. I, Gallimard, "Pléiade", Paris, 1980 [1781].

Killing, W., *Einführung in die Grundlagen der Geometrie*, Bd. II, F. Schöningh, Paderborn, 1898.

Klein, F., *Gesammelte Mathematische Abhandlungen*, Bd. I, Springer, Berlin, 1921.

Klein, F., *Vergleichende Betrachtungen über neuere geometrische Forschungen* (Erlanger Programm), A. Deichert, Erlangen, 1872.

Kulkarni, R. S., "Curvature and metric", *Annals of Mathematics*, Vol. **91**, 1970, pp. 311-331.

Lampariello, G., "B. Riemanns physikalisches Denken", in : *Der Begriff des Raumes in der Geometrie*, Riemann-Tagung, Akademie-Verlag, Berlin, 1957 (pp. 222-234).

Lanczos, C., *Space through the Ages*, Academic Press, London, 1970.

Leibniz, G., *Mathematischen Schriften*, herausgegeben von C. I. Gerhardt, Bd. VII (*Initia rerum mathematicarum metaphysica*, pp. 17-28), H. W. Schmidt, Halle, 1863.

Levi-Civita, T., "Nozione di parallelismo in una varietà qualunque", *Rend. Circ. Matem. di Palermo*, **42**, 173-205.

Libois, P., "Quelques applications des idées de Riemann sur l'espace", Riemann-Tagung, *cit.* (pp. 195-201).

Lie, S., *Theorie der Transformationsgruppen*, Bd. III, B.G. Teubner Leipzig, 1893

Lie, S., *Gesammelte Abhandlungen*, Bd. II, Teil 1, B. G. Teubner, Leipzig, 1935.

Manin Yu. I., *Mathematics and Physics*, Birkhäuser, Boston, 1981.

Maxwell, J. C., *A Treatise on Electricity and Magnetism*, 2 Vols, Dover, New York, 1954 [1881].

Nikulin, V. V., Shafarevich, I. R., *Geometries and Groups*, Springer, Berlin/Heidelberg, 1987.

Nomizu, K., "Recent development in the Theory of Connections and Holonomy Groups", *Advances in Mathematics*, Vol. **1**, Fasc. **1**, 1961, pp. 1-47.

Paty, M., "Interprétation et construction dans le rapport des mathématiques à la physique", *Fundamenta Scientiae*, Vol. **10**, N° **1**, 1989, pp. 35-55.

Petitot, J., "Idéalités mathématiques et réalité objective", in *Hommage à Jean Toussaint Desanti*, Editions TER, Paris, 1991 (pp. 213-282).

Petitot, J., "Refaire le 'Timée'. Introduction à la philosophie mathématique d'Albert Lautman", *Revue d'Histoire des Sciences*, Vol. **40**, No **1**, 1987, pp. 79-115.

Poincaré, H., *Œuvres*, tome II, Gauthier-Villars, Paris, 1916.

Poincaré, H., "Analysis Situs", *Journal de l'Ecole Polytechnique*, tome 1, 1895, pp. 1-121 ; ou, *Œuvres*, tome VI, Gauthier-Villars, Paris, 1953 (pp. 193-288).

Poincaré H. [1898], *Des fondements de la géométrie*, Chiron, Paris, 1921.

Poincaré, H., "Sur les hypothèses fondamentales de la géométrie", *Bull. Soc. Math. France*, tome **15**, 1887, pp. 203-216.

Pontrjagin, L., *Topological Groups*, Princeton University Press, Princeton, 1939.

Riemann, B., *Mathematische Werke*, B. G. Teubner, Leipzig, 1876.

Riemann, B., "Über die Hypothesen, welche der Geometrie zu Grunde liegen" [Habil.-Vortrag, 1854], *Abh. Königl. Gesell. Wiss. Gött.*, **XIII**, 1867 ; (*Math. Werke :* pp. 254-269).

Riemann, B., *Fragmente philosophischen Inhalts* [notes manuscrites non datées] ; (*Math. Werke :* pp. 475-506).

Riemann, B., *Commentatio mathematica, qua respondere tentatur quaestioni ab Illma Academia Parisiensi propositiae :* "Trouver quel doit être l'état calorifique d'un corps solide homogène indéfini pour qu'un système de courbes isothermes, à un instant donné, restent isothermes après un temps quelconque, de telle sorte que la température d'un point puisse s'exprimer en fonction du temps et de deux autres variables indépendantes." [manuscrit de 1861] ; (*Math. Werke* : pp. 370-399).

Salanskis, J.-M., *L'herméneutique formelle*, Editions du CNRS, Paris, 1991.

Scholz, E., *Geschichte des Mannigfaltigkeitsbegriffs von Riemann bis Poincaré*, Birkhäuser, Boston, 1980.

Spivak, M., *A Comprehensive Introduction to Differential Geometry*, Vol. II, Publish or Perish, Inc., Houston, 1979.

Thom, R., "L'aporia fondatrice delle matematiche", in *Enciclopedia*, Vol. **15**, Einaudi, Torino (pp. 1133-1146).

Tits, J., "Transitivité des groupes de mouvements", Riemann-Tagung, *cit.* (pp. 98-111).

Tits, J., "Sur certaines classes d'espaces homogènes de groupes de Lie", *Mémoires de l'Académie Royale de Belgique*, Classe des Sciences, Série **II**, Tome **XXIX**, fasc. **3**, 1955, pp. 1-268.

Torretti, R., *Philosophy of Geometry from Riemann to Poincaré* , D. Reidel , Dordrecht, 1984 [1978].

Varga, O., "Über Riemannsche Räume, die freie Beweglichkeit besitzen", Riemann-Tagung, *cit.* (pp. 124-130).

Veblen, O., Whitehead, J. H. C., *The Foundations of Differential Geometry*, Cambridge University Press, Cambridge, 1932.

Vuillemin, J., *La Philosophie de l'Algèbre*, Presses Universitaires de France, Paris, 1962.

Weyl, H., "Gravitation und Elektrizität", *Sit. Ber. Königl. Preuss. Akad. Wiss.*, **XXVI**, 1918, 465-480.

Weyl, H., *Raum-Zeit-Materie* , Springer, Berlin/Heidelberg, 1988 [1918].

Weyl, H., *Mathematische Analyse des Raumproblems*, Springer, Berlin, 1923.

Weyl, H. [1925], *Riemanns geometrische Ideen, ihre Auswirkung und ihre Verknüpfung mit der Gruppentheorie* , Springer, Berlin, 1988.

Wheeler, J. A., *Geometrodynamics*, Academic Press, New York, 1962.

Wolf, J. A., *Spaces of Constant Curvature*, McGraw-Hill, Inc., San Francisco, 1967.

Geometrie und Erfahrung

F. Balibar

UFR de Physique. Université Paris 7(France)

Although I have chosen to use the title Einstein gave to one of his most famous non scientific papers, I am not going to speak of the way geometry is to be considered by the physicist after the emergence of the theory of general relativity, but rather of the change induced by Faraday's achievements in the physicist's conception of force, space and geometry. I derive an encouragement to introduce Faraday in a conference held by mathematicians on geometry from the following statement uttered by James Clerk Maxwell himself :

> "The high place which we assign to Faraday in electro-magnetic science may appear to some inconsistent with the fact that the electro-magnetic science is an exact science, and that in some of its branches it had already assumed a mathematical form before the time of Faraday, whereas Faraday was not a professed mathematician, and in his writings we find none of those integrations of differential equations which are supposed to be of the very essence of an exact science. Open Poisson and Ampère, who went before him, or Weber and Neumann, who came after him, and you will find their pages full of symbols, not one of which Faraday would have understood [...] It is true that no one can essentially cultivate any exact science without understanding the mathematics of that science. But we are not to suppose that the calculations and equations which mathematicians find so useful constitute the whole of mathematics. The calculus is but a part of mathematics. The geometry of position is an example of a mathematical science established without the aid of a single calculation. Now Faraday's lines of force occupy the same position in electro-magnetic science that pencils of lines do in the geometry of position ".[1]

This paper will be devoted to understanding in detail what Maxwell means when qualifying Faraday as a mathematician and how this is related to the invention, by Faraday, of the lines of force. It will be shown that, up to a certain extent, Faraday's work can be viewed as a "return to geometry" motivated by the deficiences of the physics of his time, which deprived space of all possible physical features and for which the way to answer physical questions was to transform them in mathematical (analytical) problems. It will be shown that this return to geometry has its origin in experiment, thus justifying the title chosen for this paper (although "Geometrie und Experiment" would have been a more suitable title, since what is at stake here is not the experience of everyday life, but an ensemble of specifically planned and designed experiments, which Faraday has resumed under the general title *Experimental Researches*).

[1] J.C. Maxwell : *Faraday*, Nature, vol. VIII, reprinted in *J.C. Maxwell's Scientific Papers*, (2 vols.), Cambridge, 1890, p.355.

As already stated, my argument will be centered on the concept of lines of force. These are introduced by Faraday, in his *Experimental Researches*[2] , in three steps :

1) "A line of force may be defined as that line which is described by a very small magnetic needle, when it is so moved in either direction correspondent to its strength, that the needle is constantly a tangent to the line of motion". This definition would lack completeness if it were not augmented by the two following addenda, which can be considered as the second and third part of the definition proper :
2) "These lines have not merely a determinate direction recognizable as above, but because they are related to a polar or antithetical power, have opposite qualities or conditions in opposite directions ; these qualities, which have to be distin- guished and identified, are made manifest to us, either by the position of the ends of the magnetic needle, or by the direction of the current induced in the moving wire."
3) "A point equally important to the definition of these lines is that they represent a determinate and unchanging amount of force. Though their forms, as they exist between two or more centers or sources of magnetic power, may vary very greatly, and also the space through which they may be traced, yet the sum of power contained in any section of a given portion of the lines is exactly equal to the sum of the power in any other section of the same lines, however altered in form, or however convergent or divergent they may be at the second place. The experimental proof of this will be given ."

We are thus left with a definition in three parts and I shall comment on each of them separately.

The first part of the definition points to the essentially *geometric character of the new concept* which is specified in terms of definite elements of geometry, namely *lines*. It should be stressed that the geometry involved is by no means of an axiomatic type, since it is completely determined by experiment, as is obvious from the prompt reference to a measuring instrument (a magnetized needle). Faraday aims at replacing – or at least complementing – the "cartesian" description of space by coordinates, i.e. numbers associated to a rigid body of reference, by a description of a new type, where the coordinates are to be supplemented and even replaced, by the specification of a *direction* in space, that along which the magnetic needle points when placed at the considered place. The modern physicist will have recognized in this procedure the introduction of "generalized coordinates" in place of "ordinary coordinates", a current procedure in quantum electrodynamics. When one is willing, as is Faraday here, to replace a commonly used description by another one, one needs to exhibit the necessity for such a change. This is done by Faraday in the following terms :

> "It is evident that our physical views are very doubtful ; and I think good would result from an endeavour to shake ourselves loose from such preconceptions as are contained in them, that we may contemplate for a time the force as much as possible in its purity"[3].

[2] M. Faraday : *Experimental Researches in Electricity,* (3 vols.), Londres 1839-55, paragraphe 3071.

[3] *Experimental Researches* ..., op. cit., paragraphe 3304.

Contemplate the force in its purity ; this is what Faraday is aiming at, an objective which is obviously not fullfilled by the common way of looking at forces, namely the analytical description inherited from Newton and developped by those whom Maxwell calls "the french philosophers", Ampère, Fourier, Cauchy, Poisson and their likes. Faraday's endeavour therefore appears as being essentially motivated by a *critique of the classical concept of force.* One may speculate that Faraday was feeling more easily led to such a critique than the contempory physicists by the fact that he had not been taught in "Oxbridge" and therefore was insensible to the power of conviction conveyed by symbols which meant nothing to him. Faraday, it would seem, was perfectly aware of his superiority in this respect :

> "It is better to be aware, or even to suspect, we are wrong, than to be unconsciously or easily led to accept an error as right".[4]

It is easy, even at this stage of the analysis, to tell on which point does Faraday's critique of the classical concept of force bear. The lines of force are *continuous* ; they connect a point in space to another point in space without any break of continuity. As such they clearly stand in opposition to the concept of *forces acting at a distance*, on which the whole of the newtonian and post-newtonian physics relies. Given the importance of the concept he is going to attack, it is no surprise that Faraday would look in Newton himself for an argument against an idea, that of forces acting at a distance, which he suspects cannot have been rightly attributed to Newton. Invoking the famous passage of Newton's third letter to Bentley, Faraday writes :

> "I am encouraged to make this ebdeavour by [...] the strong conviction expressed by Sir Isaac Newton that even gravity cannot be carried on to produce a distant effect except by some interposed agent fullfilling the conditions of a physical line of force".[5]

Faraday is obviously aiming at a physics where things act locally (at the place where they lie) and where interactions are propagated from place to place :

> "If we could tell the *disposition* of the force of a magnet, first at the place of its origin, and next in the space around, we should then have attained to a very important position in the pursuit of our subject ; and if we could do that, assuming little or nothing, then we would be in the very best conditions for carrying the pursuit further".[6]

Faraday's physics, since it is concerned with such a continuous propagation from place to place, does not "jump" over space and cannot ignore the local structure of space. This is of course in complete contrast to the way the "french philosophers", and their followers, had handled the problem up to then : for them, two points were sufficiently specified through their coordinates, or positions, and what lies in the gap between those two points was by no means a physical object to be treated as such. While the "french philosophers" had considered space as "l'ordre des choses entre elles", and nothing more,

[4] *Experimental Researches* ..., op. cit., paragraphe 3323.
[5] *Experimental Researches* ..., op. cit., paragraphe 3304.
[6] *Experimental Researches* ..., op. cit., paragraphe 3304.

Faraday who had taught himself by performing experiments, was willing to explore, with the aid of a suitable physical apparatus, what lies in the gap between two points, therefore imparting an experimental reality to what had up to then been considered as a non existing and non intervening object : space itself.

Such an endeavour would be meaningless if the gap which lies between two interacting points did not actually exhibit physical qualities ; in other words, if the geometrical lines drawn by Faraday were only geometrical, and were not litterally lines of *force*. As already stated, this is precisely the case : the magnetic needle feels something, is acted by something in the gap. But an adept of the theory of forces acting at a distance could object that the rotation of the needle in the gap between the two points proves nothing since it can be easily explained in the frame of his theory. A needle, he would claim, can be viewed as a dipole, and the torque formed by the two opposite magnetic forces acting (at a distance) on the north and south poles is responsible for the observed rotation. This would be a serious objection, were it no for the second part of Faraday's definition which says that the lines of force are oriented (they have "opposite qualities or conditions in opposite directions"). It is not obvious at first sight that this brings about a radical change in the conception of space and force : after all, forces, in the traditionnal conception, are vectors, and therefore also oriented ; moreover, it is well known that particles of the same polarity repell each other while particles of different polarities attract each other. Actually the change introduced by Faraday is very subtle, and at the same time very drastic : while, according to the usual conception of force, the polar character of the world is concentrated in matter (in the particles themselves which have definite polarity), in Faraday's view of Nature the polarity lies in the *lines* of force, therefore in *space* itself. One is led to a geometrical, rather than analytical, conception of space and force :

> "For just as either geometry or analysis may be employed to solve correctly a particular problem, though one has far more power and capability, generally speaking, than the other, [...] so may the idea of the attraction and repulsion of centres, or that [...] of the lines of force, be applied on the consideration of magnetic phaenomena. It is the occasionnal and more frequent use of the latter which I at present wish to advocate".[7]

The difference between the two conceptions can equally be described as follows. In the traditionnal picture, a force can only exist if there are *two* particles : a single particle has no physical effect of its own whatsoever and the force does not exist until it is raised, so to say, by the introduction of another particle. Faraday who had studied the effect of paramagnetic substances on a light beam (a "non material" object) had become convinced that a magnetic force is present even when there is *only one "material" object in space* ; he had thus reached the conclusion that a single magnetic object modifies space around itself – a modification which can be best represented through the concept of lines of force, precisely. This was a very new idea, and nobody was better aware of its importance than Faraday's follower, James Clerk Maxwell who, having a probably better feeling of what was revolutionnary in Faraday's work than the author himself, wrote to Faraday :

[7] *Experimental Researches* ..., op. cit., paragraphe 3374.

"Now, as far as I know, you are the first person in whom the idea of bodies acting at a distance by throwing the surrounding medium into a state of constraint has arisen, as a principle to be actually believed in. We have had streams of hooks and eyes flying around magnets, and even pictures of them to beset, but nothing is clearer than your descriptions of all sources of force keeping up a state of energy in all that surrounds them, which state by its increase or diminution measures the work done by any change in the system".[8]

The geometrical implications of such a change in the way of looking at things cannot be overestimated. Up to then, a point in physical space was a true geometrical point and the interaction of two of these points was represented by what we now call a vector, lying along the straight line joining these two points. Now, in the post-faradayean area, points in physical space should be seen as *sources* from which a certain "power" emanates, a power which is sensible by the change in the condition of space that it brings about. The points in physical space can no longer be represented by geometrical points ; this picture has to be replaced by that of a source out of which (or towards which) lines of force (or "power") diverge (or converge).

Moreover the interaction can no more be visualized as a vector ; these must be replaced by a network of lines of force which stretch from one point to the other along more or less curved paths. As Mxwell writes to Faraday (in the letter just cited) :

"Your lines of force can weave a web across the sky...You seem to see the lines of force curving round obstacles and driving plump at conductors and swerving towards certain directions in crystals, and carrying with them everywhere the same amount of attractive power spread wider or denser as the lines widen or contract".

And Maxwell adds :

"Now the mode of looking at Nature which belongs to those who can see the lines of force deals very little with 'resultant forces' but with a network of lines of action of which these are the final results".

In other words, the lines of force are more fundamental than the forces themselves which are just a consequence of the spatial disposition of the lines of force. These who at first sight could have been considered as a mere mode of visualization are actually of a *conceptual* nature; they are *the* new concept on which any explanation of a phenomenon has to be based. This is so much the case in Faraday's thought that he even goes to stating a new fundamental principle, nothing less than an alternative to the newtonian laws of motion :

"All the phaenomena resolve themselves into this, that a portion of such matter, when under magnetic action, tends to move from stronger to weaker places or points of force. When the substance is surrounded by lines of magnetic force of equal power on all sides, it does not tend to move [9]".

[8] Maxwell to Faraday, november 9, 1857.

[9] *Experimental Researches* ..., op. cit., paragraphe 2418.

If material objects are to move "from stronger to weaker places of force", the question is : "How does Faraday quantify the force at a given place?" To this, the third part of the definition of the lines of force (as reproduced above) gives an answer. The lines of force, it is there stated, "represent a determinate and unchanging amount of force". Once translated in modern terms this says that the flux of the force (one should say, in appropiate modern words : "of the field") is kept constant along a given tube of force. This conception is completely alien to the traditionnal view of forces acting at a distance : something is kept constant *in the space which lies in between the interacting "points"*, precisely where in the traditionnal view one is left with a mere passive void. It would seem at first sight that the third part of the definition only stresses what has already been stated in the second part. A closer examination will show that what is involved here is nothing less than a radical change in the definition of the nature of forces. An object which has been neglected up to that point must now come to the front, namely the "moving wire" where some current is to be induced, according to the definition of the lines of force. This measuring instrument of a very specific type, which is absent in the first part of the definition, appears, on the same footing as the magnetic needle, in its second part, and is the only device which is mentionned in the third part. It would seem that, as the definition of the lines of force becomes more and more explicit, the magnetic needle is progressively replaced by the moving loop of current, which thus appears as the measuring device which is adequate to the new concept and appropriate to the determination of the changes induced in the condition of space around a source of power.

This shift from a *straight* magnetic needle to a *loop* of current brings along with it more about the "true" nature of space that it would seem at first sight. As already said, forces were, up to then, represented by vectors, along a straight line, and the rotation of an object in space could only be explained by the existence of *two* forces, acting as a torque. In this respect, Cavendish's torsion apparatus for measuring the gravitationnal constant is paradigmatic of classical physics – as is the magnetic needle whose rotation is interpreted as due to the torque resulting from the two opposite forces acting on its two ends. The point now is that where classicaly trained physicists would see a rotation as due to a torque Faraday "with the eyes of his mind" (this is Maxwell's expression) saw the action of a *single* force. This force, of course, could not be of a newtonian type. Nevertheless it could be detected if one was willing to use an appropiate instrument, whose shape and structure had been specifically designed to evidence rotations, namely a *loop*. Contrary to the magnetic needle which can serve both purposes, observe a straight force and a rotationnal one, and is therefore ambiguous, the loop of current is an instrument dedicated to rotations, and thus to (electro -)magnetic effects. "A moving wire may be accepted as a correct philosophical indication of the presence of magnetic force[10]". The value of the moving wire as an investigator is greater than that of the needle in that although "the needle gives true experimental indications", it is not so sure

> "that we always interpret them correctly. To assume that pointing is always the direct effect of attractive and repulsive forces acting in couples is to shut out ideas, in relation to magnetism, which are already applied in the theory of the nature of light [an allusion to the polarisation of light] ; and the shutting out of such ideas *may be*

[10] *Experimental Researches* ..., op. cit., paragraphe 3083.

an obstruction to the advancement of truth and the defence of wrong assumptions and error[11]".

If, as a conclusion, one is willing to resume in a brief statement what Faraday has changed, one might say that Faraday, by looking at space on a local scale, has put an end to the idea that straight lines are adequate elements for the description of both space and forces. As he himself puts it, one has been induced in the error of vewing straight lines as characterizing space and forces by the consideration of a specific type of force : gravitation ; but gravitation is just a special case and one should not infer from its structure to that of all types of physical forces :

" In gravity, the line of force is a *straight* line ; it cannot be deflected or even affected in its course. Gravity presents us with the simplest case of attraction ; and appearing to have no relation to any physical process by which the power of the particles is carried on between them, seems to be a pure case of attraction (or action at a distance)[12]".

Returning to Maxwell's eloge of Faraday given as an entry to this paper, one would be inclined to say that Faraday, independantly of Riemann, has open a new field for what Einstein used to call a "practical geometry", a geometry in which not only the *position* of an object but also the *local structure* of space around this position, is to be considered. In this regard, Faraday's *Experimental Researches* and Riemann's dissertation *Über die Hypothesen welche der Geometrie zu grunde liegen*, are not only contemporary, but also very similar in mind – as has already been pointed out by Hermann Weyl in *Raum, Zeit und Materie* :

"Geometry has become a doctrine of space itself, and not only as was the case with euclidean geometry, and with all what is usually called 'geometry', a doctrine of possible forms in space"[13].

[11] *Experimental Researches* ..., op. cit., paragraphe 3122.

[12] M. Faraday, *On the Physical Character of the Lines of Force*, Philosophical Magazine, june 1852.

[13] H. Weyl, *Raum, Zeit und Materie*, Berlin, 1918.

The Geometric Challenge of Riemann and Clifford

Ruth Farwell and Christopher Knee

Respectively, Department of Mathematical Sciences, Brighton Polytechnic, Moulsecoomb, Brighton BN2 4GJ, UK and Department of Sociology, St. Mary's College, Strawberry Hill, Twickenham TW1 4SX, UK.

1 Introduction

This presentation will consist of an amalgam of ideas that are appearing as several different published articles [1]. Those articles include the details of the research which we do not propose to explore here. What we shall present here, however, is a more general argument in the history of ideas concerning the fate of a particular view of space.

We start with an examination of Riemann's metric geometry and develop the contention that his geometry, in its mathematical content, presented a challenge both to mathematics and to the orthodox theory of knowledge of time.

We continue with the work of W. K. Clifford who translated and presented Riemann's ideas in more popular form as well as drawing out some of the philosophical implications of Riemann's geometrical ideas. Clifford, however, added something new to Riemann's geometry. He attempted to geometrise physics by suggesting that matter is an aspect of space, not separate from it.

We then address the question of what happened to these ideas. Indeed, the geometrical ideas of Riemann and Clifford did not surface again generally until after Einstein's theory of general relativity. Metric geometry was not developed in the intervening period. We argue that Riemann's and Clifford's geometrical thinking was too challenging to be developed. Only after a "safer" framework had been elaborated, could the geometry re-emerge. In the meantime, another aspect of Riemann's work was developed by, amongst others, Christoffel, Lipschitz, Ricci and Levi-Civita. This became known as tensor analysis.

Tensor analysis was crucial to the development of Einstein's theory of general relativity. Furthermore, tensor analysis was the mathematics which provided a less challenging framework within which Riemann's and Clifford's geometrical ideas could re-emerge. Paradoxically, general relativity was a return to a more conventional epistemology than that suggested by Riemann and explored by Clifford.

We want to emphasise several characteristics of our argument. The first is that both the challenge and the resolution of that challenge are to be found in the mathematics. Both Clifford and Einstein were philosophers as well as mathematicians and theoretical physicists. Second, we introduce W. K. Clifford's original and often overlooked contribution to geometry. Third, we trace the history of tensor analysis back to a little read paper of Riemann on heat conduction. Lastly, we present general relativity both in its mathematics and in its philosophy as a "safe" aproach. It was less revolutionary than is sometimes supposed.

2 Riemann's metric challenge

Riemann delivered his it Habilitationsvortrag [2] or qualifying lecture, to the faculty of the University of Goettingen in 1854. This lecture marks the beginning of metric geometry. Although not presented mathematically, the mathematical details implicit in the lecture and their philosophical implications were quite revolutionary. The concomitant of his ideas was the possibility of the existence of many different geometries unimagined until then.

By 1854 the debate over the non-Euclidean geometries of Lobachevski and Bolyai had been carrying on for half a century. Nevertheless Euclidean geometry still had a unique status. Physical space it was Euclidean geometry and conversely. By physical space we mean our perception of the space in which we live. It has a collection of properties attributed to it as a result of our experience or our pre-conceived ideas about its nature. Physical space was synonymous with Euclidean geometry. Although, geometries other than that of Euclid existed, there was an absolute certainty about the geometry of physical space : it was Euclidean.

Riemann's innovative approach to geometry involved the introduction of a "multiply extended quantity". This can be considered as a collection of points, or a space. The modern terminology for his multiply extended quantity is a manifold. The means by which he was able to conceive of geometries other than those of Euclid, Lobachechevski and Bolyai was via the introduction of "measure relations", or metric relations, on the multiply extended quantity. In Euclidean geometry, the shortest distance between two points is a measure relation. To define multiply extended quantities, Riemann chose to generalise the Euclidean concept of shortest distance.

The formula for the distance between two points (x_1, x_2) and $(x_1 + dx_1, x_2 + dx_2)$ in the Euclidean plane is given by the square root of

$$(dx_1)^2 + (dx_2)^2. \tag{1}$$

In (1) the length is given by the Pythagorean formula which can alternatively be written as

$$g_{ij} dx_i dx_j \tag{2}$$

where summation of i and j from 1 to 2 is implied and $g_{11} = g_{22} = 1, g_{12} = g_{21} = 0$. The form (2) was generalised by Riemann to

$$ds^2 = g_{ij}(x) dx_i dx_j \tag{3}$$

where the summation of i and j is now from 1 to n, so the multiply extended quantity has n dimensions. Most importantly the metric coefficients, that is the g_{ij}, are generalised to be continuous functions of x. The expression ds^2 is known as the generalised line element.

Thus Riemann generalised the concept of length in Euclidean geometry to a line element, which involved "abstract notions of quantity", that is, they could no longer be interpreted intuitively as "length". The measure relations were given by abstract mathematical expressions. Once the measure relations have been assigned to a space, the geometry of the space is automatically determined. In defining the line element with general metric coefficients g_{ij}, Riemann defined a multiplicity of metrics and consequently a multiplicity of different geometries.

Hence although it has the appearance of an abstract mathematical expression, the general line element is precisely what lies at the heart of the metric challenge of Euclidean geometry. The general metric enables us to conceive of geometries other than tho se of Euclid, Lobachevski and Bolyai.

Now the appearance of the line element depends on the coordinates used to describe the space. For example, expression (1) gives the line element in the Euclidean plane in Cartesian coordinates (x_1, x_2). By writing it in plane polar coordinates (r, ϕ), we obtain a different form :

$$dr^2 + r^2 d\phi^2 \tag{4}$$

We have indicated that the geometry of the space is determined once the line element has been defined. However the appearance of the line element depends on the coordinates employed. The coordinates are merely an aid to the description of the space ; they do not themselves define the space. Thus our conclusion about the geometry of the space is that it should not be dependent on the description employed, that is, it should be independent of the 'mode of representation'.

Riemann investigated the independent measure relations that can be defined on spaces, that is, measure relations which cannot be transformed into one another by a change of coordinates alone. Essentially this was an investigation of the different classes of geometry which can be associated with the spaces. Within each class the spaces have line elements which may be transformed into one another by a change of coordinates.

The classification depended upon the introduction of the concept of curvature. We have noted that what is special about Euclidean space is that the line element or distance function can be written as a sum of squares of the quantities dx_i with constant coefficients. In general this is not possible.

However, the generalised line element (3) can be expanded as a series of terms the first one of which has the Euclidean form :

$$ds^2 = \{(dx_1)^2 + (dx_2)^2\} + \{\text{higher order terms}\}. \tag{5}$$

The next order term in the series contains what Riemann called the sectional curvatures. The sectional curvatures for Euclidean space are therefore zero and Riemann chose to call such spaces flat. Spaces having the same curvature are geometrically equivalent.

The consequence of Riemann's work is that, unlike before, there is not one preferred geometry which must automatically be consistent with the proper ties of physical space. The physics of Riemann's time indicated that physical space must be a triply extended quantity, but there remained the question which measure relations, or which geometry, should be associated with it. Riemann concluded that "the properties which distinguish space from other conceivable triply extended magnitudes are only to be deduced from experience". But our experiences could be consistent with many different geometries and not just one particular geometry. Gone was the absolute certainty about the nature of physical space. It was precisely the mathematical expression giving the generalised line element in Riemann's geometries that provided the cause for the rejection of certainty.

3 Clifford's extension of Riemann's ideas and his introduction of an original element

W. K. Clifford is not now as famous as Reimann. Clifford was a British mathematician (1845-1879) whose name crops up in connection with biquaternions, Reimann surfaces, associative algebras and, sometimes, the history of geometry. It was Clifford who was responsible for the first translation into English of Riemann's it Habilitationsvortrag of 1854 [3]. He was indeed influenced by the work of Riemann ;but significantly also by that of the physiologist and geometer Hermann von Helmholtz. Clifford was active as a populariser of mathematical and physical ideas of his age. Clifford also made a significant original contribution to geometry in the form of a revolutionary conception of space which goes beyond that of Riemann.

In 1870, Clifford addressed the Cambridge Philosophical Society. This was published as a short note it On the space-theory of Matter [4]. In it, Clifford said the following :

> "Riemann has shown that as there are different kinds of lines and surfaces, so there are different kinds of space of three dimensions ;and that we can only find out by experience to which of these kinds the space in which we live belongs...
>
> I wish here to indicate a manner in which these speculations may be applied to the investigation of physical phenomena. I hold in fact
>
> (1) That small portions of space it are in fact analogous to little hills on a surface which is on average flat ;namely that the ordinary laws of geometry are not valid in them.
> (2) That this property of being curved or distorted is continually being passed on from one portion of space to another after the manner of a wave.
> (3) That this variation of the curvature of space is what really happens in that phenomenon that we call the it motion of matter, whether ponderable or ethereal.
> (4) That in the physical world nothing else takes place but this varition, subject (possibly) to the laws of continuity".

Some popular accounts of the history of geometry mention this extract as a significant contribution to a new conception of space which was developed much later with the general theory of relativity. Some mathematicians and theoretical physicists also acknowledge Clifford's contribution (Levi-Civita [5] and Eddington [6], for example). Nevertheless, Clifford's work is not generally regarded as important as that of Riemann in the field of geometry. We argue that Clifford saw clearly the philosophical and geometrical implications of Riemann's ideas. Furthermore, we argue that the consequences of both Riemann's and Clifford's ideas were too challenging to orthodox epistemology to be pursued. It was not until the development of tensor analysis that a safer framework became available within which these geometrical ideas could be developed further.

Because Clifford spent so much time and energy delivering popular lectures on philosophical and scientific ideas, his perspective should be sought both in his mathematical papers and his popular lectures. He used these not only to present some of the current ideas of his time but also to develop some of his own contributions. The principal characteristic of Clifford's perspective stemmed from his concern with how physical space may be defined, the role of perception in that process of definition, and the implications for geometry.

The orthodox theory of knowledge at Clifford's time was represented by Kant's *Critique of Pure Reason* [7] according to which "all empirical laws are only special determinations of the pure laws of the understanding, under which, and according to the norms of which, they first become possible". For Kant ""space is not an empirical concept which has been derived from outer experiences". Rather, in order that we may talk of experiences at all, "the representation of space must be presupposed". Kant then privileges the concept of physical space and holds that Euclidean geometry is the only one fully to describe physical space.

The essence of the challenge of non-Euclidean geometry to this perspective is first that there exist more than one geometry, and second that experience is the only way to decide which of those geometries may be appropriate as a description of physical space. Thus the extent of human knowledge can only be limited. For Clifford, though we may be capable of conceiving the meaning of absolute certainty, we cannot claim to possess it. At least, in his words, "this knowledge would be absolutely different in kind from any knowledge that we possess at present". The orthodox theory of knowledge held that exactness of a theoretical kind existed in fact, and that it was represented in the form of Euclidean geometry. For Clifford, the work of Lobachevski and Gauss, developed by Reimann and Helmholtz disputed this contention.

For Clifford, synthetic apriori statements are no longer possible. Instead, experience yields contingent truth which could correspond to a number of different consistent, logical mathematical truths. It is interesting to note that Clifford, like T.S. Kuhn [8], uses Copernicus as an example of change equivalent to the one he is attempting to describe and that he also talks of the change of perspective as a "revolution". What Clifford was presenting was not a technical advance in a limited area of geometrical thinking but a radical shift of paradigm involving a different perception of the world.

For Clifford, geometry and philosophy were not separate disciplines. It is clear that, in the realm of ideas, Clifford's discussion was an intervention in a philosophical debate around idealism and empiricism as well as a challenge to accepted views on the nature of space. That conception of space included for Clifford, elements of physics which did not belong to the orthodox Euclidean model. This is a major departure from the work of Riemann.

Indeed, as a mathematician, Riemann's prime concern was to define space. Clifford's contribution was to move beyond the mathematicians definition of space and to hint at a perspective which includes also physical phenomena. In other words, Clifford replaced the definition of space as an arena in which physical events take place and explored the possibility of defining physical phenomena themselves as characteristics of space.

This is a significant difference of perspective between Riemann and Clifford. As Riemann said at the end of his it Habilitationsvortrag, "this leads us into the domain of another science, of physic, into which the object of this work does not allow us to go today". Clifford, on the other hand, described objects as characteristics of physical space and thus, so to speak, geometrised physics which was precisely what general relativity did some forty or fifty years later.

4 What happened to the ideas?

Thus, so far it can be said that Clifford took up the philosophical implications of Riemann's work. In the course of presenting and extending those ideas, Clifford proposed a different definition of space from Riemann's. Both agreed on the possibility of many different geometries. Both also agreed that the decision as to which of those geometries best describes physical space can only be made on the basis of our experiences of the nature of that space. Clifford went further and explored the ideas that the motion of matter is no more than the variation of the curvature of space and that in the physical world "nothing else takes place but this variation".

The nature of physical space is at the heart of general relativity. General relativity rests on a different conception of space from the Euclidean one. Furthermore, general relativity cannot be described using only the tools of Euclidean geometry. Riemann's work provides the mathematical techniques needed for the description of general relativity.

More interestingly, Clifford's work provides the conceptual basis of general relativity by suggesting that there is an inseparable link between physical matter and space. The connection is made through the curvature of space which is a metric, therefore a geometrical, property of space. His formulation is remarkably close to Einsteins. "According to the general theory of relativity, the geometrical properties of space are not independent, but they are determined by matter". [9]

The reason for which Riemann's and Clifford's work was not developed was that it suggested that there no longer existed any certainty about which of many possible geometries described physical space. This uncertainty constituted too great a challenge to conventional thinking to be developed at the time. Geometrical developments after Clifford's death tended to be within what is known as projective geometry.

The reason for the appeal of projective geometry was its apparent ability to project the unpalatable spaces associated with non-Euclidean geometries into Euclidean spaces. The uncertainty over which geometry corresponded to physical space was thus overcome by subsuming non-Euclidean geometries under Euclidean geometry. However, there is a whole class of Riemannian geometries, those with variable curvature, later used in the theory of general relativity, which could not be unified within projective geometry.

Although Riemann's geometrical ideas themselves remained dormant, some of the technical ideas embodied in the Habilitationsvortrag were developed. They stimulated work in the area now known as tensor analysis. A paper submitted by Riemann for a prize to the Academie des Sciences in 1861 on heat conduction is relevant. It was written in Latin and entitled it Commentatio Mathematica.... [10]. We hold, in contrast to certain other commentators, that this paper does not contain the conscious workings of the geometrical ideas contained in this Habilitationsvortrag of 1854. Rather, it contains the derivation of a condition necessary for one quadratic form to be transformed into another with constant coefficients, that is, the condition necessary for the form $\Sigma\beta_{i,i'}ds_i ds'_i$ to be transformed into the form Σdx_i^2. We note that the forms considered here resemble the line elements (3) and (1) above.

Thus, we hold that Riemann's revolutionary geometrical ideas did not initially stimulate further interest in manifolds, but rather in the theory of quadratic forms, the precursor of tensor analysis. Riemann's metric geometry remained formant because it was challenging. It triggered developments in a different non-geometrical branch of mathematics which is now referred to as tensor analysis.

5 How did general relativity allow them to re-emerge

The mathematical description of Einsteins general relativity crucially depends on tensor analysis. A tensor quantity defined on a space is expressed in terms of some coordinate system. What distinguishes a tensor quantity from any other quantity is that it has a prescribed behaviour under coordinate transformations. If an equation is written in terms of tensors then even though the individual expressions appearing in the equation might change under the coordinate transformation, the overall form of the tensor equation does not.

This is the feature that is important in the mathematical description of physical laws and thus tensors are useful in physical theories. The mathematical description is in terms of equations involving coordinate systems but coordinate systems are features of the model, not of the physics in other words, the equations must be form invariant under a coordinate transformation.

Einstein was seeking in the theory of general relativity an extension to the special theory which allowed for the inclusion of gravity. The significant step taken by Einstein to incorporate the gravitation law into the theory was the acknowlegment that all observers, even those accelerating relative to others, are equivalent as regards the formulation of physical laws – the special the ory only allows for equivalence of privileged classes of observers. The law of gravitation must have the same interpretation for all observers – or, put mathematically, when the equation representing the law is represented in any coordinate system. Thus Einstein was looking for a tensor equation to describe gravity. The tensor that Einstein ended up using to represent gravity was the metric tensor. It consists of components which are the metric coefficients of Riemann's generalised line element (3).

Thus Riemann's geometry reappeared in the theory of relativity. What was characteristic of that theory was its applicability to all observers – we know exactly how to express Einsteins equation for gravity for any observer. This certainty was achieved by employing tensors. As Eddington puts it :

> "The nearest we can get to a non-subjective, but nevertheless observational view is to have before us the reports of all possible dummy observers, and pass in our minds so rapidly from one to other that we identify ourselves, as it were, with all the dummy observers at once... [the mathematician] has invented a transformation process which enables us to pass very quickly from one dummy observers account to anothers. The knowledge is expressed in terms of tensors... A tensor may be said to symbolise absolute knowledge ;but that is because it stands for the subjective knowledge of all possible subjects at once". [11]

Now it is well-known that Einstein was significantly influenced by the theories of Ernst Mach and owed him considerable debt in relation to his early formulation of special relativity. It is also clear, however, that as Einstein moved closer to including gravitation in his general theory of relativity, he moved further away from Mach's empiricism and closer to what he himself described as rationalism. In a letter to his friend Cornelius Lanczos in 1938, Einstein looked back on his trajectory and said the following :

> **"Coming from sceptical empiricism of somewhat the kind of Machs, I was made, by the problem of gravitation, into a believing rationalist, that is, one who seeks the only trustworthy source of truth in mathematical simplicity. The logically simple does not, of course, have to be physically true ;but the physically true is logically simple, that is, it has unity at the foundation". [12]**

The framework of general relativity was therefore different and less challenging in its philosophical implications than the framework of uncertainty characterising the geometrical ideas of Riemann and Clifford. We hold that once tensors had been used in the theory of general relativity, the framework existed in which a geometrical perspective in physics could be developed and allowed the challenging geometrical conceptions of Riemann and Clifford to be rediscovered. What made it possible for these ideas to re-emerge and, importantly, find acceptance, was that in the intervening period, a particular aspect of Riemann's non-geometrical work had been developed. What was to be known as tensor analysis after Einstein but was described as quadratic forms or covariant systems before that, provided a powerful antidote to the uncertainty which had previously characterised metric geometry.

Tensor analysis provided first a mathematical tool to solve certain specific problems in theoretical physics and thus united geometry and physics on an empirical basis reminiscent of Riemann's and Clifford's aspirations. Second, tensor analysis also provided a way of incorporating Riemannian geometry into a philosophical paradigm that did not suffer from the challenging uncertainties of its earlier formulation. The general relativity of Einstein re-establishes, through the use of tensors, a particular framework. This framework does not include a concept of absolute like that of Newtonian mechanics, but nevertheless includes the certainty provided by tensors. This certainty was lacking in the work of Riemann and Clifford and was the reason for the rejection of their geometrical ideas.

Thus it was in the mathematics of metric geometry that can be found the challenge which caused its rejection ;it was also in the mathematics of general relativity that can be found the safe framework which allowed that geometry to re-emerge.

References

1 R. Farwell and C. Knee, 'The End of the Absolute : a Nineteenth-Century Contribution to General Relativity', *Stud. Hist. Phil. Sci.*, **21** (1990) 91-121 ; The Missing Link : Riemann's 'Commentatio', Differential Geometry and Tensor Analysis', *Historia Mathematica* , (September 1990) ; 'The Recovery of Certainity, the Role of Tensors in Changing Perceptions of Physical Space' (submitted for publication).

2 G.F.B. Riemann, 'Ueber die Hypothesen, welche der Geometrie zu Gründe liegen', Bernhard Riemann's *Gesammelte mathematische Werke und wissenschaftlicher Nachlass*, H. Weber and R. Dedekind (eds.) (orig : Leipzig : 1876 ; reprint of 2nd edition of 1882, including supplement prepared by M. Noether and W. Witinger in 1902, New York : Dover, 1953).

3 W.K. Clifford, 'On the Hypotheses which lie at the bases of Geometry', *Mathematical Papers* (London : Macmillan, 1882), pp. 55-69.
4 *Op. cit.*, note 3, pp. 21, 22.
5 T. Levi-Civita, *The Absolute Differential Calculus (Calculus of Tensors)* (New York : Dover, 1977), p. 372.
6 A.S. Eddington, *Space, Time and Gravitation* (Cambridge : University Press, 1929), p. 192.
7 I. Kant, *Critique of Pure Reason* (tr. by N. Kemp Smith) (London : Macmillan, 1933) p. 503.
8 T.S. Kuhn, *The Structure of Scientific Revolutions* (Chicago : University Press, 1970).
9 A. Einstein, *Relativity : The Special and the General Theory* (London : Methuen, orig. 1920, reprint 1968), p. 113.
10 *Op. cit.* note 2.
11 A.S. Eddington, *The Philosophy of Physical Sciences* (Cambridge : University Press, 1949), pp. 86-7.
12 Quoted in G. Holton, 'Mach, Einstein and the Search for Reality', *Darwin to Einstein* , C. Chant and J. Fauvel (eds.) (Harlow :) p. 253.

Poincaré et Enriques : deux points de vue différents sur les relations entre géométrie, mécanique et physique

Giorgio Israel

Dipartimento di Matematica, Università degli Studi "La Sapienza", Piazzale A. Moro, 2 00122 - ROMA (Italy)

Le but de cet article n'est pas d'établir une comparaison entre la conception scientifique de Poincaré et celle d'Enriques : il me semble que le point de vue des "comparaisons" entre savants donne, en général, des résultats assez pauvres en histoire des sciences. Son but est plutôt celui d'étudier deux points de vue différents, au tournant du siècle passé, sur le thème des relations entre les mathématiques d'un côté et la mécanique et la physique de l'autre. Notre analyse peut être considérée comme une contribution à une *histoire de l'analyse qualitative* qui reste en grande partie à écrire. De plus, les points de vue de Poincaré et de Enriques représentent un intéressant "case study" sur le thème de la "tension essentielle", c'est-à-dire de la dialectique entre *tradition* et *innovation*.[1] En effet on se trouve en présence ici d'une inversion paradoxale qu'on pourrait ainsi synthétiser : ce qui caractérise aussi bien la démarche de Enriques que celle de Poincaré dans le sens de la *tradition* (et qui est à l'origine de l'oubli contingent de certains aspects de leur contribution scientifique) est le fondement même de leur *modernité.*

La thèse principale de ce travail est qu'il n'est pas possible d'expliquer le caractère fondamentalement *novateur* de l'analyse qualitative de Poincaré autrement que sur la base du caractère *conservateur* de son programme de recherche (surtout en ce qui concerne les rapports entre mathématiques, mécanique et physique). Tandis que l'attitude plus ouverte d'Enriques à l'égard de la physique moderne est (toujours d'une façon assez paradoxale) à l'origine de la caducité de ses points de vue sur le thème des rapports entre analyse mathématique, géométrie et physique. D'autre part, certaines de ses élaborations oubliées pendant une longue période, comme la critique du mécanisme, la réévaluation du rôle de la métaphysique et des modèles, apparaissent aujourd'hui d'une actualité inattendue.

En étudiant l'œuvre scientifique de Poincaré, il est aisé de constater combien la tension entre tradition et innovation qui est présente en elle, a été mal comprise. On rencontre souvent des avis opposés. Selon certains, Poincaré serait un novateur, voire même un révolutionnaire : son analyse qualitative représenterait une rupture du cadre classique qui aurait produit une crise du déterminisme classique du type laplacien.[2]Si l'on accepte ce point de vue, il faudrait bien expliquer les raisons de l'incontestable conservatisme de

---

[1] "The essential tension", dans le sens introduit par Kuhn dans [Kuhn 1977].

[2] C'est, à peu près, le point de vue de I. Ekeland dans [Ekeland 1984].

Poincaré dans le domaine de la physique-mathématique et de la physique. En général, si l'on insiste exagérément sur le côté novateur de l'œuvre de Poincaré, il est toujours possible de produire de exemples opposés qui témoignent de son conservatisme.

Parmi ces exemples, on pourrait choisir l'attitude de Poincaré sur le rôle du calcul des probabilité (et de la prétendue objectivité du "probable") et sur le rôle du déterminisme. Il est bien évident que le point de vue de Poincaré sur ces thèmes est bien moins rigide que celui de Painlevé, qui est un tenant strict du déterminisme laplacien. Dans son introduction au *Calcul des probabilités*,[3] Poincaré remarque que "les anciens distinguaient les phénomènes qui semblaient obéir à des lois harmonieuses" et "ceux qu'il attribuaient au hasard". La hasard avait, pour eux, un sens objectif. Et, il ajoute, en subtile et évidente polémique avec le déterminisme strict "à la Laplace" :

> cette conception n'est plus la nôtre ; nous sommes devenus des déterministes absolus, et ceux mêmes qui veulent réserver les droits du libre arbitre humain laissent du moins le déterminisme régner sans partage dans le monde inorganique. Tout phénomène, si minime qu'il soit, a une cause, et un esprit infiniment puissant, infiniment bien informé des lois de la nature, aurait pu le prévoir dès le commencement des siècles. [...] Pour lui, en effet, le mot de hasard n'aurait pas de sens, ou plutôt il n'y aurait pas de hasard. [...] Le hasard n'est que la mesure de notre ignorance. [...] Mais cette définition est-elle bien satisfaisante? [...] Il faut bien que le hasard soit autre chose que le nom que nous donnons à notre ignorance, que parmi les phénomènes dont nous ignorons les causes, nous devions distinguer les phénomènes fortuits sur lesquels le calcul des probabilités nous renseignera provisoirement, et ceux qui ne sont pas fortuits et sur lesquels nous ne pouvons rien dire, tant que nous n'aurons pas déterminé les lois qui les régissent. Et pour les phénomènes fortuits eux-mêmes, il est clair que les renseignements que nous fournit le calcul des probabilités ne cesseront pas d'être vrais le jour où ces phénomènes seront mieux connus."[4]

Dans ce texte Poincaré introduit aussi une intéressante discussion du phénomène (aujourd'hui très à la mode) de la "sensibilité aux conditions initiales". Dans cette discussion il identifie les situations dans lesquelles la prédiction est impossible avec les situations dans les lesquelles le phénomène doit être considéré comme fortuit.

> "Une cause très petite, qui nous échappe, détermine un effet considérable que nous ne pouvons pas ne pas voir, et alors nous disons que cet effet est dû au hasard. Si nous connaissions exactement les lois de la nature et la situation de l'univers à l'instant initial, nous pourrions prédire exactement la situation de ce même univers à un instant ultérieur. Mais lors même que les lois naturelles n'auraient plus de secret pour nous, nous ne pourrions connaître la situation qu'*approximativement*.. Si cela nous permet de prévoir la situation ultérieure *avec la même approximation*, c'est tout ce qu'il nous faut, nous disons que le phénomène a été prévu, qu'il est régi par des lois ; mais il n'en est pas toujours ainsi, il peut arriver que des petites différences dans les conditions initiales en engendrent de très grandes dans les phénomènes finaux ; une petite erreur sur les premières produirait une erreur énorme sur les derniers. La prédiction devient impossible et nous avons le phénomène fortuit."[5]

---

[3] [Poincaré 1912a].

[4] [Poincaré 1912a], p. 2-3.

[5] [Poincaré 1912a], p. 4-5.

Et Poincaré continue en proposant un exemple puisé de la météorologie qui anticipe d'une façon surprenante les thèmes récemment suscités par l'introduction des modèles de Lorenz.[6]

Dans l'analyse de Poincaré on décèle une évidente confusion entre le niveau ontologique et le niveau prédictif de la question, qui semble être l'expression d'un point de vue neo-positiviste.[7] Mais cette attitude curieusement "moderne" disparaît quand Poincaré commence à discuter le problème de l'objectivité du hasard. Il se demande :

> "Le hasard [...] a-t-il un caractère objectif?"[8]

Ici Poincaré, pour éviter d'adhérer complètement à la thèse selon laquelle le hasard aurait un caractère objectif, introduit une forme assez vague de "relativisme de l'objectivité". On a parlé – il observe – de "très petit" ou de "très complexe". Un intervalle est très petit quand dans les limites de cet intervalle, la probabilité reste sensiblement constante. Mais tout cela est relatif. L'univers tend à l'uniformité et donc à rendre "lisse" la courbe des probabilités : ce qui aujourd'hui n'est pas très petit sera très petit dans quelque milliard de siècles. Donc le concept de "très petit" est relatif et les probabilités ont un caractère objectif dans un sens "relatif".

Comment peut-on concilier ce Poincaré avec le Poincaré qui défend la centralité des équations différentielles et l'hypothèse de continuité contre la physique quantique de Planck? Et à propos de laquelle il observe :

> " On ne se demande plus seulement si les équations différentielles de la dynamique doivent être modifiées, mais si les lois du mouvement pourront encore être exprimées par des équations différentielles. Et ce serait là la révolution la plus profonde que la philosophie naturelle ait subi depuis Newton."[9]

Et encore

> "La discontinuité va-t-elle régner sur l'univers physique et son triomphe est-il définitif? ou bien reconnaîtra- t-on que cette discontinuité n'est qu'apparente et dissimule une série de processus continus? Le premier qui a vu un choc a cru observer un phénomène discontinu, et nous savons aujourd'hui qu'il n'a vu que l'effet de changements de vitesse très rapides, mais continus. Chercher dès aujourd'hui à donner un avis sur ces questions, ce serait perdre son encre."[10]

C'est le même Poincaré qui se réfère très explicitement au paradigme de Fourier en disant :

> " La théorie de la chaleur de Fourier est un des premiers exemples d'application de l'analyse à la physique ; en partant d'hypothèses simples qui ne sont autre chose que

[6] [Poincaré 1912a], p. 5-6. Cf. [Lorenz 1961] et [Lorenz 1964].

[7] Sur ces thèmes voir [Israel 1991a].

[8] Cf. [Poincaré 1912a] note 3, p. 16.

[9] Cf. [Poincaré 1912b], p. 225.

[10] [Poincaré 1912b], p. 232.

des faits expérimentaux généralisés, Fourier en a déduit une série de conséquences dont l'ensemble constitue une théorie complète et cohérente. Les résultats qu'il a obtenus sont certes intéressants par eux-mêmes, mais ce qui l'est plus encore c'est la méthode qu'il a employée pour y parvenir et qui servira toujours de modèle à tous ceux qui voudront cultiver une branche quelconque de la physique mathématique."[11]

On peut trouver une preuve indirecte mais très intéressante de ce caractère du programme scientifique de Poincaré – c'est-à- dire de la centralité des équations différentielles dans la physique mathématique et, plus en général, dans la physique – dans la motivation avec laquelle Volterra (et donc un savant qui considérait Poincaré comme un "modèle" scientifique) proposa d'attribuer le Prix Nobel pour la physique à Poincaré dans les années 1909, 1910 et 1911.[12]

Comment peut-on concilier ces aspects apparemment contradictoires et qui, à défaut d'une explication cohérente, divisent les historiens de l'œuvre de Poincaré en deux partis : ceux qui le considèrent comme un novateur et ceux qui le présentent comme un conservateur?

Il me semble qu'on ne peut pas trouver une meilleure description du paradigme de Poincaré que celle que Poincaré lui-même a suggérée dans son article de 1912 sur "Les rapports entre la matière et l'éther" :

"Les théories anciennes reposent sur un grand nombre de coïncidences numériques qui ne peuvent pas être attribuées au hasard ; nous ne pouvons donc disjoindre ce qu'elles ont réuni ; nous ne pouvons plus briser les cadres, nous devons chercher à les plier ; et ils ne s'y prêtent pas toujours.[...] Dans l'état actuel de la science nous ne pouvons que constater ces difficultés sans les résoudre."[13]

Le commentateur de Poincaré qui a saisi le mieux cette caractéristique fondamentale de son paradigme scientifique est probablement Arthur Miller. Par exemple, quand il observe :

"... selon Poincaré, il était admissible de multiplier les hypothèses dans le but de sauver une théorie qui expliquait d'une façon adéquate une large classe de phénomènes, et dont la structure violait pourtant une partie de la mécanique classique. Cependant cette procédure n'était pas valable si elle était utilisée pour expliquer une partie isolée de donnée expérimentales ; sur ce point Poincaré était adamantin, par exemple [...] il critiqua d'une façon âpre Lorentz pour l'hypothèse de la contraction – "les hypothèses sont ce qui nous manque le moins"."[14]

---

[11] Cf. [Poincaré 1895], p. 1.

[12] Cf. [Israel 1984a].

[13] Cf. [Poincaré 1912c], p. 360.

[14] "... in Poincaré's view, it was permissable to multiply hypotheses in order to save a theory that explained adequately a wide range of phenomena, yet whose structure violated a part of classical mechanics. However, this procedure was invalid if it was a matter of explaining a single piece of experimental data ; on this point Poincaré was adamant, for, [...] he harshly criticized Lorentz for the hypothesis of contraction – "hypotheses are what we lack least."" ([Miller 1981], p. 44).

A notre avis, l'introduction de l'analyse qualitative dans la théorie des équations différentielles ordinaires peut être considérée du point de vue de cette définition du paradigme de Poincaré. Il ne sera pas possible ici d'entrer dans les détails des célèbres mémoires de Poincaré sur les équations différentielles ordinaires.[15] Nous nous bornerons à mettre en évidence la structure du raisonnement sur lequel elles sont bâties. Le point de départ de Poincaré est une critique des limites du point de vue local dans la recherche des solutions d'un système de deux équations différentielles ordinaires : il souligne l'importance voire même la nécessité d'un point de vue global qui permette de suivre les intégrales sur tout le plan. Le point de vue global peut être développé dans deux directions : la recherche de nouvelles transcendantes et l'analyse qualitative. Ces deux points de vue sont interprétés et justifiés par Poincaré comme des extensions du point de vue classique. Cela est évident dans les cas des fonctions fuchsiennes et dans le cas de l'intégration à l'aide de fonctions algébriques et abeliennes. La justification que Poincaré donne de l'introduction de la méthode qualitative est particulièrement intéressante. Cette méthode n'est pas vue comme un nouveau point de vue "révolutionnaire", mais est au contraire justifiée à partir du paradigme classique qui – en accord avec le point de vue de Fourier – a comme but final la solution numérique. Poincaré observe :

> "L'étude complète d'une fonction comprend deux parties : I° partie qualitative (pour ainsi dire), ou étude géométrique de la courbe définie par la fonction ; 2° partie quantitative, ou calcul numérique des valeurs de la fonction.
>
> Ainsi, par exemple, pour étudier une équation algébrique, on commence par rechercher, à l'aide du théorème de Sturm, quel est le nombre des racines réelles : c'est la partie qualitative ; puis on calcule la valeur numérique de ces racines, ce qui constitue l'étude quantitative de l'équation. De même, pour étudier une courbe algébrique, on commence par *construire* cette courbe, comme on dit dans les cours de Mathématiques spéciales, c'est-á-dire qu'on cherche quelles sont les branches de courbe fermées, les branches infinies, etc. Après cette étude qualitative de la courbe, on peut en déterminer exactement un certain nombre de points.
>
> C'est naturellement par la partie qualitative qu'on doit aborder la théorie de toute fonction et c'est pourquoi le problème qui se présente en premier lieu est le suivant : *Construire les courbes définies par des équations différentielles*.[16]

Poincaré insiste sur l'utilité de l'analyse qualitative par rapport au but primaire : le calcul numérique de la fonction. Il ne serait donc pas fondé de parler d'une prise de distance du paradigme classique. Le schéma de Poincaré est bien différent et très clair. Le point de départ est la tentative d'élargir le point de vue classique. Dans le contexte de cette tentative on rencontre des obstacles, des difficultés qui imposent un élargissement des "cadres" : dans ce processus les rapports avec la structure analytique précédente doivent être clairement définis.

Nous nous trouvons donc en présence d'une véritable "dynamique de déformation" du paradigme classique soutenue par la ferme intention d'en conserver le noyau central. Ainsi, dans le passage au point de vue "géométrique-qualitatif", Poincaré sent immédiatement la nécessité de définir les rapports avec l'"ancien" point de vue : son système de référence

---

[15]Cf. [Poincaré 1881-86]. Pour une analyse plus détaillée, voir [Dell'Aglio 1987].

[16]Cf. [Poincaré 1951], p. xxii.

est toujours le modèle de la physique-mathématique classique tel qu'il a été codifié par Fourier. La triade "Analyse 'générale' – analyse 'spéciale' – calcul numérique" est pour Poincaré, comme pour Fourier, une étoile polaire de la recherche.[17] Mais le point de vue de Poincaré est dynamique et souple et le processus de déformation continue se développe dans le cours de sa production scientifique. Par exemple, presque vingt ans plus tard, dans ses *Méthodes nouvelles de la Mécanique Céleste*,[18] en parlant des solutions homoclines il fait une remarque qui est un argument pour tous ceux qui prétendent que la démarche qualitative de Poincaré a un caractère consciemment révolutionnaire :

> "On sera frappé par la complexité de cette figure, que je ne cherche même pas à tracer. Rien n'est plus propre à nous donner une idée de la complication du problème des trois corps et en général de tous les problèmes de la Dynamique où il n'y a pas d'intégrale uniforme et où les séries de Bohlin sont divergentes."[19]

Mais une remarque successive montre clairement que le cadre conceptuel de Poincaré n'est pas du tout changé par rapport à celui que nous avons décrit auparavant, même s'il ne se cache pas les difficultés croissantes posées par la découverte de l'étrange phénoménologie des solutions du problème des trois corps :

> "Cette remarque est encore de nature à nous faire comprendre toute la complication du problème des trois corps et combien les transcendantes qu'il faudrait imaginer pour le résoudre diffèrent de toutes celles que nous connaissons."[20]

Donc, rien ne permet de conclure, à partir de phrases comme celles qui précèdent, que Poincaré soit un anti-déterministe *ante-litteram*. Au contraire, son programme doit être considéré comme un programme réductionniste, même s'il s'agit d'un réductionnisme plutôt *sui generis* et que nous pourrions définir, en nous référant à la métaphore de la "déformation des cadres" comme un programme de *réductionnisme souple*.

Si Poincaré est aujourd'hui à la mode ce n'est pas parce-que les théories du chaos sont à la mode, mais c'est plutôt à cause du fait qu'après une phase de l'histoire des mathématiques (entre les années vingt et cinquante) inspirée par un paradigme "anti-continuiste" et "anti-différentiel", et qui a vu un certain déclin de intérêt pour la mécanique classique, nous assistons aujourd'hui à une renouveau de l'intérêt pour le point de vue continuiste- différentiel et pour les problèmes de la mécanique classique. Nous sommes en train de traverser un période de renaissance des "mathématiques du temps", pour utiliser la définition donnée par Smale il y a quelques années[21].

Il est intéressant de souligner une conséquence du point de vue qualitatif de Poincaré en ce qui concerne la conception de l'espace et du rapport entre géométrie et mécanique.

On pourrait résumer le point de vue qui dominait avant Poincaré, en disant qu'il était fondé sur l'idée d'une correspondance entre les équations différentielles et l'espace des configurations. Non seulement l'attention était concentrée sur la recherche des so-

---

[17] Sur ces thèmes, voir [Israel 1981].
[18][Poincaré 1892-99].
[19]Cf. [Poincaré 1899], p. 389.
[20] Cf. [Poincaré 1899], p. 391.
[21]Cf. [Smale 1980].

lutions particulières, mais cette correspondance liait d'une façon directe l'interprétation des solutions avec l'espace "réel", "physique", avec l'espace dans lequel les phénomènes mécaniques se manifestent. Au contraire, dans le point de vue introduit par Poincaré, la relation centrale est celle qui existe entre équations différentielles et espace des phases. Donc, non seulement l'attention est dirigée vers la recherche des propriétés globales de ce qu'aujourd'hui on appelle le "flux" du champ vectoriel, mais les propriétés des solutions sont interprétés dans un espace *abstrait* (l'espace des phases) qui représente les *états* mécaniques et les trajectoires d'évolution du système.

L'ancienne relation était apparemment plus concrète parce qu'elle liait les solutions des équations avec la trajectoire *réelle* dans l'espace physique *réel*. Mais elle était beaucoup moins intuitive, bien que plus directe. Avec Poincaré, moyennant un processus d'abstraction, on aboutit à un niveau des rapports entre mathématiques et réalité qui est beaucoup plus subtil et sophistiqué mais beaucoup plus "naturel". On voit ici à l'œuvre un style de raisonnement qui a ses origines dans le conceptions de Galilée, et selon lequel, pour mieux comprendre le réalité concrète il est nécessaire, dans un certain sens, de s'en éloigner par un processus d'abstraction[22]. Dans ce tournant de la pensée mathématique et physico-mathématique, la géométrie et l'intuition géométrique ont un rôle crucial.

Ici une une question se pose d'une façon naturelle : peut-on-dire que dans la conception de Poincaré soit présent un renversement hiérarchique entre géométrie et analyse?

La réponse et sans doute négative. Au contraire, c'est l'analyse qui conserve encore son rôle central : l'introduction de la méthode géométrique-qualitative représente seulement une extension du pouvoir d'explication de l'analyse et non le fait qu'elle soit en train de céder le pas à la géométrie. De ce point de vue, Poincaré se place plutôt dans le sillon de Fourier que dans celui de Lagrange.

La distinction entre ces deux traditions est importante pour comprendre la dynamique tout-à-fait différente qui caractérise le développement des mathématiques italiennes, dans lesquelles la tradition lagrangienne est plus forte et l'idée d'une hiérarchie géométrie-mécanique est plus présente, même si dans un sens un peu différent de celui de Lagrange (dans son cas la hiérarchie est plutôt dans le sens de la triade algèbre-géométrie-mécanique tandis que le premier terme est presque absent dans la tradition italienne). C'est ce genre de rapport entre géométrie et analyse qui caractérise les mathématiques italiennes au tournant du siècle : et il s'agit d'un rapport qui, en un certain sens, renverse le point de vue de la physique-mathématique classique et rétablit la suprématie de la géométrie.

On peut mieux saisir ce que nous voulons dire en examinant la manière avec laquelle les milieux des mathématiques italiennes accueillirent la "nouvelle physique" (c'est-à-dire la relativité). Volterra eut une attitude assez ouverte, bien que sa pensée ait été très influencée par les idées de Poincaré[23]. On remarque une attitude encore plus ouverte chez Levi-Civita, qui était pourtant un "conservateur" (selon la définition donnée par lui même) inspiré par une tradition de style lagrangien qui subordonnait le point de vue mécanique-physique à une démarche géométrique-formelle[24]. C'est ici que réside la raison de sa disponibilité à concevoir l'espace d'une façon abstraite et mathématique.

[22] On peut voir sur ce thème le fameux passage du *Dialogo dei Massimi Sistemi*, et l'analyse de ce passage dans [Israel 1986].

[23] Sur ces thèmes voir [Israel 1984b].

[24] Sur ces thèmes voir [Del'Aglio-Israel 1989] et [Israel1991b].

On peut même dire que Levi-Civita est intéressé à la relativité parce-qu'il en saisit le côté plus classique et traditionnel, celui qui renvoie à la physique-mathématique du dix-septième siècle et à l'idéologie sous-jacente à la mécanique analytique (inspirée par la conception de D'Alembert selon laquelle une science et d'autant plus vraie qu'elle est plus abstraite)[25]. On doit aussi mentionner l'attitude de Guido Castelnuovo qui fut très vite séduit par la relativité et contribua beaucoup à sa diffusion en Italie, et finalement celle de Federigo Enriques, qui voyait dans la "nouvelle physique" le triomphe d'une conception géométrique-synthétique opposée à la conception analytique-réductionniste qui caractérise le courant physico-mathématique qui va de Laplace à Fourier.

Nous allons résumer très rapidement les caractéristiques fondamentales de la pensée philosophique et scientifique de Enriques[26].

Le premier aspect est la polémique contre ce qu'il appelle le "conventionnalisme" de Poincaré, qui est considéré comme l'extrême produit négatif du criticisme kantien que Enriques accuse d'avoir produit une sorte de "pusillanimité" gnoséologique, à cause de son idée selon laquelle il existe un niveau inconnaissable de la réalité. C'est la raison pourquoi, selon Enriques, le conventionnalisme aboutit à une vraie forme de *nominalisme*.

Le deuxième aspect est la polémique contre la tradition de la physique-mathématique classique. Enriques est opposé à ce qu'il appelle l'idéal "anti-historique" du réductionnisme physique-mathématique classique, et qu'il considère comme une forme d'objectivisme exaspéré et dogmatique niant tout rôle à l'"activité de l'esprit" et qui trouve son expression la plus claire dans l'idéal mathématique de Laplace :

> "l'idéal mathématique de Laplace, [...] aspire à représenter toute la réalité *sub specie aeternitatis* dans les équations de l'Univers, à partir desquelles, en surmontant les difficultés d'intégration, on obtiendrait la prévision de tout événement particulier."[27]

Dans la fameuse opposition entre Fourier et Jacobi,[28] Enriques ne se place pas complètement du côté de Jacobi parce qu'il croit à la valeur de la science dans les applications, mais il est encore moins du côté de Fourier.

Le troisième aspect est la réévaluation, même si partielle, du rôle de la métaphysique comme système d'"images" de la réalité et comme une première forme d'approche qualitative à une vision scientifique de la réalité.

Le dernier aspect est représenté par le rôle central qu'Enriques attribue à la pensée géométrique par rapport à l'Analyse. Dans les intentions annoncées, ce rapport devrait être paritaire mais en réalité l'intention profonde est d'établir une hégémonie de la géométrie. La tendance fondamentale de la science moderne est, selon Enriques, la "substitution des calculs avec la pensée" et dans ce développement synthétique, qualitatif et

---

[25] Voir encore [Israel1991b].

[26] Pour une analyse plus détaillée, voir [Israel 1989].

[27] "l'ideale matematico di Laplace, [...] aspira a vedere rappresentata l'intera realté sub specie aeternitatis nelle equazioni dell'Universo, da cui, superando le difficolté d'integrazione si ricaverebbe la previsione di ogni evento particolare" ([Enriques 1938], p. 130).

[28] Nous nous référons ici au contenu de la fameuse lettre écrite par Jacobi à Legendre en 1830 : "Il est bien vrai que Fourier pense que l'objet principal des mathématiques est l'utilité publique et l'explication des phénomènes naturels ; mais un savant comme lui devrait savoir que le seul objet de la science est l'honneur de l'esprit humain et que, sur cette base, une question de la théorie des nombres est aussi importante qu'une question concernant le système planétaire."

conceptuel de la "nouvelle science" (qui est opposé au point de vue fondé sur la centralité du calcul) c'est la géométrie qui doit jouer un rôle central : la relativité, avec sa vision géométrique de l'espace, est un excellent exemple de cette tendance.

Quelles sont les origines de ces caractéristiques assez singulières de la pensée d'Enriques? Pour les déceler, il faut tenir compte de deux aspects.

Le premier aspect est la propension d'Enriques à une considération philosophique des problèmes de la science (Enriques lui-même disait que son intérêt pour les mathématiques avait son origine dans une "infection philosophique contractée au lycée"). Sous cet aspect, il est hors de doute qu'Enriques ait été un complète autodidacte : cela doit amener à examiner sa pensée sans essayer de l'encadrer dans des catégories traditionnelles.

Et, en deuxième lieu, il faut tenir compte de sa formation de savant dans le contexte de l'école géométrique italienne qu'Enriques lui-même contribua puissamment à transformer. Il est bien évident que les méthodes et l'image de la science qu'il tira de cette expérience de recherche influencèrent beaucoup ses conceptions de la philosophie de la science. Pour mieux comprendre cet aspect, il faut tenir compte des caractéristiques de l'école géométrique italienne et de la tradition géométrique dite "puriste".

L'intérêt pour les recherches géométriques se développa, en Italie, au cours du dix-huitième siècle, dans une direction qui synthétisait la démarche intuitive riemanienne avec la tendance à un point de vue affranchi des calculs de l'algèbre : il s'agissait d'une géométrie non plus bornée à l'étude des propriétés analytiques des courbes élémentaires, limitation qui avait pesé d'une façon si grave sur les mathématiques italiennes de la période pré-unitaire. Cette géométrie se référait à la *méthode synthétique* des Grecs et avait ses représentants en Chasles et de Jonquières en France, Mœbius, Plücker et Steiner en Allemagne, Salmon, Cayley et Sylvester en Angleterre. Le protagoniste italien de cette tendance fut Luigi Cremona. Dans ce cadre aucun intérêt ne se manifestait pour des tendances que nous pourrions appeler "pré-axiomatiques" et qui se développaient surtout en Allemagne autour des premières recherches dans le domaine de l'*algèbre abstraite*.

La redécouverte de la méthode synthétique et du rôle de l'intuition géométrique fut due, au moins en partie, à certains excès des partisans de l'analyse weierstrassienne. Dans un article publié en 1920 en espagnol[29], Enriques décrivait ainsi ces excès :

> « [...] les analystes, avec les yeux fixés sur l'étoile weierstrassienne, se complaisaient de dire que le temps était venu de libérer l'Analyse des intuitions spatiales *fallacieuses* ou au moins *étrangères* ; la finalité de l'Analyse pour quiconque était lié à l'idée de la dignité logique de la science était de transformer tout l'édifice des Mathématiques modernes en une théorie rigidement formaliste, séparée du monde extérieur et de supprimer tout dynamisme des concepts, en substituant les pseudo-passages à l'infini avec des chaînes d'inégalités.»[30]

---

[29]Cf. [Enriques 1920].

[30] « [...] los analistas, con la vista fija en la estrella weierstrassiana, se complacían en manifestar que era ya llegado el tiempo de librar el Análisis de la *falaces*, o al menos *extrañas*, intuiciones espaciales ; transformar todo el edificio de la Matemática moderna en una teoría rigidamente formalista, ajena al mundo externo, y suprimir todo dinamismo de conceptos, sustituyendo los pseudopasos al infinito con cadenas de desigualdades, parecía ser la finalidad del Análisis para quienquiera que tuviese idea de la dignidad lógica de la ciencia.» [Enriques 1920], p. 2.

La critique de l'analyse inspirée au point de vue de la"rigueur" se transformait en la critique de toute démarche formaliste, qui était accusée de pencher du côté de l'abstraction. Les géomètres synthétiques semblaient unis par un sentiment d'hostilité à l'égard de la domination des analystes. L'idéal du "purisme", bien que n'étant pas un produit exclusivement italien, atteignit en Italie des sommets vraiment extrêmes :

> " Perfectionner la méthode synthétique jusqu'au point de rendre non nécessaire toute aide de l'Algèbre, était l'aspiration, plus ou moins explicite, de tous les géomètres ; il y en avaient qui arrivaient au point de déclarer rigidement que la Géométrie finit quand on parle de nombre."[31]

Il est bien vrai que l'hostilité était surtout à l'gard de l'analyse : l'instrument des géomètres synthétiques était quand même l'algèbre classique,

> "mais si heureusement guidée par l'intuition qu'elle semble presque transfigurée, [...] l'algèbre dans laquelle on ne voit pas de développement algorithmique, mais le contenu qualitatif [...], qui habilement interprété conduit d'une façon simple et surprenante à des résultats fondamentaux et cachés."[32]

Il ne faut pas sous-estimer la fertilité de la nouvelle méthode qui semblait ouvrir des perspectives d'une ampleur insoupçonnée.

> "... la vision géométrique des entités algébriques attirait en raison de sa nouveauté, étant donné que les objets de recherche se présentaient chaque fois plus abondants et faciles. C'était comme si un monde nouveau s'ouvrait au géomètre, un monde dans lequel il était suffisant d'ouvrir la main pour faire une récolte abondante de découvertes et dans lequel l'imagination, en marche triomphale, ouvrait toujours de nouvelles portes enchantées, comme dans un palais bâti par les fées."[33]

Enriques décrit le succès triomphal de la méthode géométrique puriste dans le monde mathématique italien :

> " Quand les géomètres virent ce monde enchanté, l'annonce de la terre promise attira rapidement les hommes émerveillés. Les géomètres se multiplièrent de tous les côtés. [...]

---

[31] " Perfeccionar el método sintético hasta el punto de hacer innecesaria toda ayuda del Álgebra, era, más o menos explícita, la aspiración de todo geómetra ; había quien llegaba a proclamar, rígidamente, que la Geometría acaba cuando se habla de numero." [Enriques 1920].

[32] "ma cosìfelicemente guidata dall'intuizione geometrica da sembrar quasi trasfigurata, [...] l'algebra di cui non apparisce lo sviluppo algoritmico, bensìil contenuto qualitativo [...], che interpretato abilmente conduce in modo semplice e sorprendente a risultati fondamentali e riposti", [Castelnuovo 1928], p. 192.

[33] "... la visión geométrica de los entes algébricos atraía por su novedad, ya que los objetos de estudio se presentaban cada vez más abundantes y fáciles. Parecía come si al géometra se abriese un mundo nuevo, en el que bastaba abrir la mano para recoger abundante cosecha de descubrimientos, y donde la imaginación, en triunfal carrera, abría siempre nuevas puertas encantandas, come en un palacio construído por hadas." [Enriques 1920], p. 3.

c'était la période dans laquelle, selon l'amusante expression d'un ami et maître, il suffisait de semer un haricot pour voir naître un géomètre."[34]

Les conséquences de cette inondation "puriste" ne manquèrent cependant pas de se manifester. D'une part elles furent le résultat de l'hostilité assez fanatique à l'égard de l'analyse, qui produisit une division des mathématiques italiennes en deux champs ennemis. D'autre part, le charme des nouvelles méthodes engendra une passion immodérée pour les techniques de démonstration considérées en elles-mêmes et, en définitive, une subordination des problèmes aux méthodes. La recherche n'était plus centrée autour de noyaux de grands problèmes, dont l'importance objective était reconnue par la communauté scientifique, mais elle était guidée par un choix individuel. Corrado Segre, dans un article fondamental qui ouvrit la voie à une nouvelle phase de le géométrie italienne,[35] observa qu'on avait créé une géométrie artificieuse dans laquelle la fantaisie n'avait plus de limites et le problème était souvent inventé au seul but de mettre en œuvre des techniques spécifiques.

L'œuvre de Corrado Segre eut le rôle de rétablir la centralité des problèmes par rapport aux méthodes.[36] La voie à suivre était encore une fois celle indiquée par l'école de Riemann et de Klein et la recherche d'une interrelation entre problèmes analytiques et démarche géométrique intuitive.

Deux périodes furent décisives dans la formation de Federigo Enriques (qui était né à Livorno en 1871, avait fréquenté à Pise l'Université et la Scuola Normale Superiore et avait obtenu son diplôme en mathématiques en 1891) : une année de perfectionnement à Rome à partir de la fin de 1892 et une court séjour à Turin auprès de Corrado Segre. Dans cette deuxième période il fut certainement influencé par le point de vue de Segre, tandis qu'à Rome il suivit les cours de Cremona. Mais il en fut très déçu et la rencontre avec Guido Castelnuovo fut d'une importance fondamentale pour lui. Castelnuovo, décrivit dans une commémoration les traits caractéristiques de la manière de penser de Enriques :

" J'avais l'intention de lui suggérer la lecture de livres et mémoires, mais je me rendis compte immédiatement qu'il ne s'agissait pas de la démarche la plus convenable. Federigo Enriques était un lecteur médiocre. Dans la page qu'il avait sous les yeux il ne voyait pas ce qu'il y était écrit, mais ce que son esprit y projetait. Je décidai alors de choisir une autre méthode : la conversation. Non pas une conversation devant une table avec une feuille et une plume, mais une conversation péripatétique. Nous commençames alors ces interminables promenades dans les rues de Rome, pendant lesquelles la géométrie algébrique était le thème préféré des nos discours. Après avoir assimilé très rapidement les conquêtes de l'école italienne dans le domaine des courbes algébriques, Enriques commença hardiment à traiter la géométrie sur une surface algébrique. Il me tenait tous les jours au courant des progrès de ses recherches, que je soumettais à une critique sévère. Il

[34] " Apena los geómetras vislumbraron este mundo encantado, el anuncio de la tierra prometida atrajo rápidamente a los hombres maravillados. Por todas partes se multiplicaron los geómetras. [...] aquella fué la época en que, según decía graciosamente un compañero y maestro mío, bastaba sembrar una alubia para ver nacer un geómetra." ([Enriques 1920]).

[35] Cf. [Segre 1891].

[36] Sur ces thèmes voir [Menghini 1986].

n'est pas exagéré d'affirmer que la théorie des surfaces algébriques selon le point de vue italien fut bâtie dans ces conversations."[37]

C'est une description qui met en évidence des thèmes centraux. En premier lieu la tendance de Enriques à aborder le problèmes d'une façon intuitive et souvent même approximative, son antipathie pour les études méthodiques et pédants. En deuxième lieu, le rôle de la pensée géométrique dans sa formation scientifique et dans la formation de ses conceptions épistémologiques. L'intérêt de Enriques pour les problèmes de la science (et de la géométrie en particulier) ne fut jamais seulement un intérêt purement technique, mais il fut toujours stimulé par des problèmes de "culture générale", par des réflexions sur le thème du rôle de la pensée scientifique dans la culture et les activités humaines. Ses recherches mathématiques sont noués avec la tentative de donner une réponse aux grands thèmes philosophiques concernant les fondements de la science. Dans la personnalité de Enriques, il est impossible de séparer le philosophe de la science du mathématicien sans courir le risque de ne rien comprendre ni de l'un ni de l'autre.

Il est vraiment difficile de reconstruire un panorama cohérent de la pensée de Enriques. Nous ne nous trouvons pas en présence d'un système de pensée organisé : il s'agit plutôt de la persistance de certaines idées fondamentales dans un contexte de considérations, d'opinions et d'analyses dans lesquelles il est assez facile de déceler des contradictions et des changements d'attitude. Il ne faut pas surtout prendre à la lettre les définitions et les catégories philosophiques dont Enriques fait usage. Ainsi, la définition qu'il donne de son système quand il l'appelle "positivisme critique", ne doit pas suggérer une liaison quelconque avec le positivisme, sinon dans un sens négatif : en effet, il s'agit plutôt d'une critique du positivisme. L'attitude "libre" et en quelque sorte "ingénue" avec laquelle Enriques aborde les questions philosophique a des avantages évidents : elle lui permet d'avancer ses thèses sans préjugés ni réticences. Ainsi, il aborde le problème du rôle de la métaphysique sans se laisser intimider par l'atmosphère de "honte" que la pensée positiviste avait répandu autour de ce thème.

La question du rôle de la métaphysique nous ramène au thème de la polémique de Enriques contre le kantisme et ses "épigones", parmi lesquels le "nominaliste" Poincaré.

Un des reproches principales que Enriques fait à la science du dix- huitième siècle c'est d'avoir complètement mis de côté les problèmes posés par la *métaphysique*, ce qui a conduit à une vraie forme de lâcheté de la pensée. Selon Enriques, les problèmes insolubles n'existent pas : nous nous trouvons seulement en face de problèmes qui n'ont pas encore été formulés d'une façon correcte. Il n'existe pas une "réalité inconnaissable", mais plutôt une série d'objets tous accessibles à la pensée scientifique : cette série est infinie et donc

[37] " Stavo per suggerirgli la lettura di libri e memorie, ma mi accorsi subito che non sarebbe stata questa la via più conveniente. Federigo Enriques era un mediocre lettore. Nella pagina che aveva sotto gli occhi egli non vedeva ciò che era scritto, ma quel che la sua mente vi proiettava. Adottai quindi un altro metodo : la conversazione. Non gié la conversazione davanti a un tavolo col foglio e la penna, ma la conversazione peripatetica. Cominciarono allora quelle interminabili passeggiate per le vie di Roma, durante le quali la geometria algebrica fu il tema preferito dei nostri discorsi. Assimilate in breve tempo le conquiste della scuola italiana nel campo delle curve algebriche, l'Enriques si accinse arditamente a trattare la geometria sopra una superficie algebrica. Egli mi teneva quotidianamente al corrente dei progressi delle sue ricerche, che io sottoponevo ad una critica severa. Non è esagerato affermare che in quelle conversazioni fu costruita la teoria delle superficie algebriche secondo l'indirizzo italiano." [Castelnuovo 1947].

l'impossibilité d'épuiser la connaissance en un temps fini est à l'origine de l'idée fausse selon laquelle la réalité est inconnaissable.

Enriques se demande alors si la métaphysique a encore quelque chose à offrir. Sa réponse est que la positivisme a essayé de condamner et détruire la métaphysique en l'accusant de s'être posée comme objet de connaissance l'absolu qui n'est pas connaissable. Le fait est que l'absolu, selon Enriques, n'est pas inconnaissable : il s'agit tout simplement d'un symbole dépourvu de sens. Dans ce sens la métaphysique s'est posée devant un objet inexistant, mais le positivisme a tort de prétendre que le objets de la métaphysique ne soient pas représentés avec des images qui ont une valeur concrète. Les systèmes ontologiques bâtissent des entités qui sont des images de choses réelles, même si elles sont "différentes des objets concrets". Et Enriques observe :

> "En dernière analyse une ontologie est une représentation subjective de la réalité, un modèle façonné par l'esprit humain, dont les éléments, qui sont extraits des objets réels, sont combinés de façon telle qu'ils rendent compte d'un certain ordre des connaissances, selon un certain point de vue que l'on assume arbitrairement comme universel."[38]

En conclusion, les systèmes métaphysiques contiennent un système d'images, un modèle qui peut engendrer des associations utiles au progrès de la science. Ces systèmes doivent être refusés quand ils prétendent de fournir des connaissances objectives ; "mais la thèse que ces théories (disons métaphysiques) n'aient aucune valeur en tant que représentations psychologiques, dans le progrès génétique de la science, ne peut pas être acceptée sans une étude critique plus poussée..."[39]. Nous ne traiterons pas ici la question du rôle de la métaphysique par rapport à la critique du mécanisme et du réductionnisme développée par Enriques, et surtout par rapport aux tentatives d'exporter le mécanisme dans les domaines des sciences non physiques (biologie, physiologie, psychologie)[40]. Au contraire, la question du rôle des constructions métaphysiques en tant que représentations psychologiques de la réalité est directement liée au thème du rapport géométrie-réalité, qui nous intéresse ici. Selon Enriques, la connaissance scientifique doit faire recours aussi à des formes de modélisation caractéristiques de la pensée "métaphysique" : il est donc nécessaire d'accepter une idée plus large de la "connaissance" et pour mieux comprendre les aspects de cette idée élargie il peut être utile sinon nécessaire d'analyser la genèse psychologique des concepts scientifiques. La nouvelle "gnoséologie positive" doit fournir à l'épistémologie scientifique les instruments pour connaître cette genèse psychologique. La conception "psychologiste" de Enriques intervient d'une façon évidente dans sa méthodologie de recherche mathématique. Nous ne pourrons pas entrer dans les détails de cette question qui ont été étudiés dans [Israel 1989].

Nous nous bornerons à observer que, selon Enriques, la connaissance des formes de l'acquisition psychologique des concepts scientifiques est non moins importante de leur

[38] "In ultima analisi una ontologia è una rappresentazione subiettiva della realtè, un modello foggiato dallo spirito umano, i cui elementi, tratti da oggetti reali, vengono combinati per modo da render conto di un certo ordine di conoscenze, secondo un certo punto di vista, che si prende arbitrariamente come universale." ([Enriques 1906]).

[39] "ma, che codeste teorie (diciamo pur metafisiche) non abbiano proprio alcun valore, come rappresentazioni psicologiche, nel progresso genetico della Scienza, è una tesi che non può essere accolta senza uno studio critico più approfondito..." ([Enriques 1906]).

[40] Sur ces thèmes voir [Israel 1989] et [Cannizzaro].

vérification sur le terrain de la logique formelle, car la structure des concepts scientifiques est *déterminée* par la voie psychologique de leur acquisition. Voilà pourquoi l'application de la méthode de la logique formelle est un aspect seulement du procès de formation d'une théorie mathématique ; et souvent il s'agit de l'aspect le moins important, car il est restreint au seul aspect de la vérification, qui est en plus totalement facultative, car elle dépend d'une manière tout-à-fait particulière de considérer un résultat mathématique.

On peut comprendre maintenant pourquoi l'école de géométrie algébrique italienne, inspirée par les conceptions de Enriques, ne considérait pas l'acquisition des résultats mathématiques obtenus par la voie de l'intuition comme une procédure défectueuse, illégitime ou qui peut être acceptée provisoirement, pourvu qu'elle soit remplacée le plus tôt possible par l'intervention définitive de la logique formelle : cette procédure caractérise, au contraire, la pensée géométrique et, plus généralement, la pensée mathématique.

Il est inutile d'insister sur l'originalité et les caractéristiques tout-à-fait particulières de la pensée scientifique de Enriques, sur son autonomie des systèmes philosophiques de l'époque, ce qui fait qu'il apparait comme une sorte d'anomalie et comme un étrange mélange de points de vue "anciens" et "modernes". Ainsi sa distance du point de vue axiomatique et de tout point de vue abstrait a des caractéristiques "anciennes" : en partie il ne s'agit pas seulement d'une divergence mais aussi d'une incompréhension de certaines acquisitions techniques des mathématiques de l'époque. D'autre part, la sensibilité de Enriques au thèmes de la nouvelle physique théorique est très "moderne" : on a déjà mentionné le fait que Enriques a été un des plus importants partisans de la théorie de la relativité en Italie, tandis que la plupart des physiciens et des physico-mathématiciens étaient plutôt méfiants sinon hostiles à cette théorie.

Il est possible de comprendre ce mélange étrange de modernité et de points de vue "anciens", qui quelques fois frisent le vrai retard culturel, seulement si l'on détermine d'une façon précise les caractéristiques fondamentales de la pensée de Enriques. C'est ce que nous avons fait en bonne partie. Nous soulignons encore un autre aspect. Il s'agit du fait qu'une des plus importantes de ces caractéristiques est la référence systématique à une démarche synthétique et unitaire et le refus de toute forme de dualisme. Donc Enriques essaye toujours de "concilier" et de "synthétiser" tout ce qui apparait comme un couple de termes opposés. Ainsi il refuse l'opposition entre positivisme et kantisme en philosophie des sciences et plus en général toute opposition entre la tendance subjectiviste et la tendance objectiviste. Le dualisme positivisme-kantisme se représente dans l'opposition entre empirisme et ce qu'il appelle "nativisme" : d'où sa polémique contre Poincaré qui est considéré comme un partisan de la deuxième tendance. Et pourtant le point de vue synthétique de Enriques ne se situe pas dans une position parfaitement équilibrée entre le pôle objectiviste et le pôle subjectiviste : il est au contraire évident qu'entre les deux il penche nettement vers le pôle subjectiviste car, à son avis, l'objectivisme a eu une place tout-à-fait exagérée dans la pensée scientifique dont il est nécessaire de délimiter nettement l'extension.

Cela nous ramène à la question du rôle du réductionnisme et à l'opposition entre le point de vue de Fourier et celui de Jacobi : nous avons déjà observé que, dans cette opposition, Enriques se place plutôt du côté de Jacobi. Et non seulement pour les raisons que nous avons déjà exposé, mais aussi à cause de l'hostilité que Enriques nourrit à l'égard de l'idéal "anti-historique" du réductionnisme physico-mathématique classique, et de ce qu'il considère son objectivisme dogmatique et exaspéré niant tout rôle à l'"activité de l'esprit". La critique du réductionnisme et la démarche strictement objectiviste et quantitative de la physique-mathématique classique est certainement une des racines de son opposition à la pensée scientifique de Poincaré.

En définitive, la caractéristique fondamentale de la pensée de Enriques est l'adhésion à un point de vue *synthétique* et *qualitatif*, accompagnée par une polémique contre le "dogmatisme" de la démarche analytique, quantitative et objectiviste du paradigme physico-mathématique classique et la proposition de l'analyse psychologique de la genèse des concepts scientifiques comme une voie pour revitaliser l'empirisme provisoirement battu par le criticisme kantien.

Donc, si de point de vue *qualitatif* il s'agit, c'est dans un sens tout-à-fait différent de celui de Poincaré. Car le couple "synthétique-qualitatif" est inséparable dans la pensée de Enriques, tandis que, dans le cas de Poincaré, on pourrait plutôt parler d'un couple "analytique-qualitatif". Chez Enriques l'adoption d'un point de vue qualitatif implique un abandon de la démarche de l'analyse quantitative, donc une rupture avec le paradigme de la physique-mathématique classique à la Laplace ou à la Fourier ; tandis que chez Poincaré l'adoption du point de vue qualitatif a le rôle d'une d'une "déformation" de ce paradigme dans un but exactement opposé à celui de Enriques, c'est-é -dire d'en sauvegarder le noyau.

Nous allons voir les conséquences de cette vision de l'analyse qualitative de Enriques, si différente de celle de Poincaré, dans la façon de concevoir le rôle de la géométrie et de ses rapports avec l'analyse, la mécanique et la physique, jusqu'á la théorie des équations différentielles.

On a déjà vu qu'Enriques opposait au "purisme" l'exigence d'une synthèse entre géométrie et analyse . Mais il s'agissait d'une synthèse, d'un "accord" tout-à-fait inique dans lequel la géométrie se taillait la part du lion. Le "purisme" en tant que séparation de la géométrie de l'analyse disparaissait, mais le noyau du programme, c'est-à-dire la méthode synthétique et qualitative était préservée : au contraire, on essayait de faire devenir de moyen la méthode de toutes les mathématiques. Selon Enriques, l'analyse historique des tendances modernes des mathématiques et la pensée de Riemann en particulier, montrent qu'un grand développement est en train de se produire dans la direction d'une vision qualitative centrée autour de l'idée d'une "substitution des calculs à la pensée". C'est une tendance qui, selon Enriques, est confirmée par les développements de la nouvelle physique qui rendent inutile et surpassé le vieux réductionnisme quantitatif à la Fourier et à la Poincaré. La physique théorique, – et la relativité d'Einstein en particulier, avec l'attention qu'elle dé veloppe autour des concepts "globaux" et "synthétiques" – montre la nécessité d'une nouvelle mathématique non plus fondée sur la centralité des équations différentielles (et donc sur la centralité du point de vue quantitatif). Selon Enriques les développements de l'analyse mathématique montrent une tendance (liée aux recherches de Lie) qui ferait *tomber toute la théorie de l'intégration des équations différentielles dans le domaine de l'activité géométrique*.

En conclusion, Enriques propose une forme nouvelle et plus sophistiquée d'hégémonie de la géométrie sur l'analyse. Et ce projet aura des conséquences assez concrètes (et négatives) dans le développement des mathématiques italiennes.

Nous soulignons encore une fois que dans la pensée de Enriques le thème de l'*analyse qualitative* a une signification tout-à-fait différente de celle de Poincaré. Ainsi, le point de vue qualitatif amène plutôt à une réduction de la théorie des équations algébriques à la théorie des groupes de transformations et la théorie des équations algébriques à plusieurs variables à une vision directement géométrique. Enriques observe :

> "Le géomètre de nos jours identifie, en fait, la théorie des entités dont nous avons parlé avec la théorie des fonctions algébriques qui forme, comme on sait, la branche la théorie des fonctions analytiques dans laquelle la détermination qualitative atteint sa plus grande signification. C'est pour cette raison que les géomètres modernes tendent chaque fois à identifier l'orientation de leurs études avec le point de vue de Riemann, c'est-à-dire du penseur qui semble avoir exercé l'influence la plus profonde dans les Mathématiques du dix- neuvième siècle[41]."

Pour ce qui concerne les équations différentielles, le point de vue de Enriques est plutôt radical et montre quelle est la vraie nature du rapport qu'il prétend établir entre analyse et géométrie. C'est le domaine de recherche de la première qui est en train de se laisser absorber par celui de la seconde :

> " [...] toute la théorie de l'intégration des équations différentielles est tombée dans le domaine de l'activité géométrique, à partir du moment où Sophus Lie l'a fondée sur le concept de groupe de transformations. Klein a observé plusieurs fois, que Lie conçut sa théorie en forme géométrique et seulement plus tard il se décida à lui donner un développement plus analytique dans le but de la rendre plus populaire parmi les mathématiciens[42]."

La référence qu'Enriques fait aux travaux de Poincaré (et le rapprochement qu'il en fait avec les travaux de Picard et Painlevé) montre qu'ilse réfère à l'analyse qualitative dans un sens tout-à-fait différent :

> "Si à ce que nous avons dit nous ajoutons les magnifiques travaux de Poincaré, Picard et Painlevé sur les équations algébriques-différentielles (en connexion avec les courbes et les surfaces algébriques), on aura une idée plus adéquate de ce que représente l'esprit géométrique pour le développement actuel des problèmes du calcul intégral. En un autre sens, l'intégration des équations aux dérivées partielles a reçu un nouvel élan des méthodes de la Géométrie différentielle, dont le plus grand maître en Italie est Luigi Bianchi[43]."

41 "El geómetra de nuestros días identifica, en efecto, la teoría de los entes arriba nombrados con la teoria de la funciones algébricas, que constituye, come es sabido, la rama de la teoría de funciones analíticas en que la determinación cualitativa adquiere el máximo significado. Por esto, los geómetras modernos se inclinan cada vez más a identificar la propia orientación de sus estudios con la fundada por Riemann, el pensador que parece haber ejercido influencia más honda en la Matemática del siglo XIX." ([Enriques 1920], p. 9).

42 " [...] toda la teoria de la integración de las ecuaciones diferenciales ha venido a caer en el campo de la actividad geométrica, desde el momento en que Sophus Lie las ha basado sobre el concepto de los grupos de transformaciones. Ya Klein observó, repetidamente, que la mente de Lie concibió su propia doctrina en forma geométrica, y solo más tarde decidióse a darle un desarrollo más analítico por el deseo de hacerla más popular entre los matemáticos." ([Enriques 1920], p.11-12).

43 "Si a lo dicho se añaden los magníficos trabajos de Poincaré, Picard y Painlevé sobre las ecuaciones algébrico-diferenciales (en conexión con las curvas y las superficies algébricas), se tendrá una idea más adecuada de lo que significa el espíritu geométrico para el desarrollo

Donc le domaine de recherche du géomètre et de l'analyste sont en fait identiques. Ce qui les sépare est seulement la *mentalité* et, comme la démarche qualitative met en évidence l'utilité croissante du point de vue géométrique, la suprématie du géomètre apparaît comme une sorte de nécessité :

> "Les considérations et les exemples précédents me semblent suffisants pour démontrer ce que j'ai affirmé au début : c'est-à-dire, que l'activité du géomètre peut se développer aujourd'hui, et en fait se développe, dans tous les domaines de l'Analyse mathématique ; et que, en un mot, il n'existe pas une *diversité d'objets* séparant l'Analyse de la Géométrie, sinon une différence d'esprit, qui est due non seulement à la tradition historique mais aussi à la *différente mentalité* des deux types de mathématiciens[44]."

Selon Enriques les questions d'analyse qualitative qui intéressent les esprits géométriques sont les plus indiquées au contexte des applications physiques, "dont naît l'importance des mathématiques".[45] Ceux qui opposent le point de vue analytique (quantitatif) au point de vue géométrique (qualitatif) "invoquent souvent un lointain intérêt physique comme prétexte pour justifier le développement de théories purement analytiques"[46]. Au contraire, pour Enriques, "la mentalité strictement analytique ne s'approche pas plus que la mentalité géométrique d'une compréhension féconde des problèmes physiques."[47].

L'idée centrale de la démarche analytique est que le but de l'Analyse soit l'intégration des équations de la physique-mathématique. Mais "il suffit de connaître un peu l'histoire de la physique théorique pour admettre que les grands progrès de la science n'ont jamais été en accord avec ce programme"[48] :

> " [...] tout-à-fait différent me semble le vrai intérêt de la physique théorique, qui est en définitive – comme la Géométrie – une construction synthétique de la pensée, avec laquelle on veut simultanément comprendre, en une vision unifiée, un monde de relations imaginées, et qui apparait donc stérilement vide et abstraite si, – en laissant de côté toute explication intuitive des phénomènes – elle se réduit à une description pure de ses relations quantitatives. [...] Le mathématicien ne doit pas apporter à la Physique seulement des

actual de los problemas del cálculo integral. En otro sentido, la integración de las ecuaciones entre derivadas parciales ha recibido buen impulso con los métodos de la Geometría diferencial, de la cual tiene hoy Italia el mayor maestro en Luis Bianchi." ([Enriques 1920], p. 12).

[44] "Las consideracions y ejemplos que preceden me parecen suficientes para demostrar cuanto he afirmado el principio : que la actividad del geómetra puede hoy desarrolarse y, efectivamente, se desarrolla, en cualquier campo del Análisis matemático ; que, en una palabra, non existe *diversidad de objetos* que separe el Análisis y la Geometría, sino una diferencia de *espiritu*, debida no sólo a la tradición histórica, sino más bien a la *diversa mentalidad* de dos tipos de matematicos." ([Enriques 1920], p. 14-15).

[45] "... de las cuales nace la importancia de las Matemáticas." [Enriques 1920], p. 15.

[46] "... los vemos invocar un interés fisíco remoto, como un simple pretexto para justificar el desarrollo de doctrinas puramente analíticas." [Enriques 1920], p. 16.

[47] "... la mentalidad estrictamente analítica no se aproxima más a una fecunda comprensión de los problemas físicos que la mentalidad geométrica." [Enriques 1920].

[48] "Pero basta un pequeño conocimiento de la historia de la Física teórica para reconocer que los grandes progresos de la ciencia jamás han estado en consonancia con este programa." [Enriques 1920].

méthodes de calcul pour des déterminations quantitatives nécessaires, mais aussi son esprit constructif, qui doit se développer en un travail de coordination des images et de l'élaboration des concepts. De ce point de vue, la mentalité du géomètre se révèle apte à la compréhension et à la critique des théories physiques, bien plus que la mentalité strictement analytique."[49]

Ce qui précède est probablement suffisant pour démontrer que le programme de Enriques est radicalement différent de celui de Poincaré.

Mais la situation est renversée aussi dans l'évaluation historique des deux contributions scientifiques.

Le point de vue de Enriques apparaît peut-être plus moderne en relation à la période historique dans laquelle il fut avancé, et cela en raison de la vision synthétique qu'il propose à la recherche mathématique dans ses rapports avec la nouvelle physique ; mais son programme concret de recherche, dans le domaine analytique, a été totalement stérile. Ce programme fut suivi surtout par Enrico Bompiani qui poursuivit, pendant toute sa vie, la tentative de réduire la théorie des équations aux dérivées partielles à une méthodologie de recherche géométrique[50]. Cette tentative aboutit une impasse totale. Au contraire, il est presque superflu d'observer, à la lumière des développements les plus récents, combien le programme de Poincaré ait montré sa fertilité. D'autre part, il ne faut pas cacher que le problèmes philosophiques posés par Enriques, surtout pour ce qui concerne les limites du ***réductionnisme*** et la nécessité de rendre un rôle à la ***métaphysique*** dans le processus de la découverte scientifique sont très actuels et vivants. Et cela pour deux raisons : en premier lieu, à cause des limites de plus en plus évidents de tout programme réductionniste-mécaniste intégral ; et en deuxième lieu, à cause du rôle croissant des modèles mathématiques en tant qu'instrument herméneutique bien plus que comme un instrument de description directe de la réalité.

Bibliographie

Cannizzaro L. – Federigo Enriques e il riduzionismo nell'ambito delle scienze della vita, Dipartimento di Matematica, Université di Roma "La Sapienza", *preprint.*

Castelnuovo G. [1928] – La Geometria algebrica e la Scuola Italiana, in *Atti del Congresso Internazionale dei Matematici, Bologna, 3-10 Settembre 1928*, Bologna, Zanichelli, 1929.

Castelnuovo G. [1947] – Commemorazione di Federigo Enriques, *Periodico di Matematiche*, **25**, n.4, pp. 81-2.

[49] " [...] me parece muy otro el verdadero interés de la Física teórica, que es al fin – como la Geometría – una construcción sintética del pensamiento, con la cual se quiere también abrazar, en una visión unificada, un mundo de relaciones imaginadas, y que, por lo tanto, aparece estérilmente vacía y abstracta, si – dejando apárte toda explicación intuitiva de los fenómenos – se reduce a una pura descripción de sus relaciones cuantitativas. [...] El matemático no debe aportar a la Física solamente métodos de cálculo para determinaciones cuantitativas necesarias, sino también su espíritu constructivo, que ha de desenvolverse en un trabajo de coordinación de las imágenes y de la elaboración de los conceptos. Desde este punto de vista, la mentalidad del geómetra se revela apta para la comprensión y crítica de las teorías físicas, bastante más aún que la mentalidad estrictamente analítica. " ([Enriques 1920], p. 16-17).

[50] Cf. [Israel 1988].

Dell'Aglio L. [1987] – Sui concetti originari della teoria qualitativa delle equazioni differenziali ordinarie, *Rivista di Storia della Scienza*, **4**, n.3, 1987, pp. 377-390.

Dell'Aglio L. & Israel G. [1989] – La théorie de la stabilité et l'analyse qualitative des équations différentielles ordinaires dans les mathématiques italiennes : le point de vue de Tullio Levi-Civita, *Cahiers du Séminaire d'Histoire des Mathématiques*, Université Pierre et Marie Curie, Laboratoire de Mathématiques Fondamentales, Paris, Cahier n. 10, pp. 283-322.

Ekeland I. [1984] – *Le Calcul, l'Imprévu*, Paris, Editions du Seuil.

Enriques F. [1906] – *Problemi della Scienza*, Bologna, Zanichelli, p. 27-9 (1909^2).

Enriques F. [1920] – "La evolución del concepto de la Geometría y la Escuela Italiana durante los últimos cincuenta años", *Revista Matemática Hispano- Americana*, T.II, N.1-2, pp. 1-17.

Enriques F. [1938] – "Importanza della storia del pensiero scientifico nella cultura nazionale", *Scientia*, Vol. LXIII, pp. 125-134.

Israel G. [1981] – 'Rigor' and 'Axiomatics' in Modern Mathematics, *Fundamenta Scientiae*, **2**, n.2, pp. 205-219.

Israel G. [1984a] – Sulle proposte di Vito Volterra per il conferimento del premio Nobel per la fisica a Henri Poincaré, *Atti del Vá1 Congresso Nazionale di Storia della Fisica, Roma, 29-31 Ottobre 1984, Rendiconti della Accademia Nazionale delle Scienze detta dei XL, Memorie di Scienze Fisiche e Naturali*,**103**, Serie V, Vol. IX, p. II, 1985, pp. 227-229.

Israel G. [1984b] – Vito Volterra : un fisico matematico di fronte ai problemi della fisica del Novecento, *Rivista di Storia della Scienza*, **1**, n. 1, Marzo 1984, pp. 39-72.

Israel G. [1986] – *Modelli Matematici*, Roma, Editori Riuniti,pp. 71-80.

Israel G. [1988] – Bompiani Enrico, *Dizionario Biografico degli Italiani*, Primo Supplemento, Vol. XXXIV, Dicembre 1988, Istituto della Enciclopedia Italiana, pp. 471-473.

Israel G. [1989] – Federigo Enriques : a Psychologistic Approach for the Working Mathematician, in *Perspectives on Psychologism*, (M. A. Notturno, ed.), Leiden, Brill pp. 426-457.

Israel G. [1991a] – Il determinismo e la teoria delle equazioni differenziali, *Physis*,Vol. XXIX, 1991 (Nuova Serie), Fasc. 2, sous presse.

Israel G. [1991b] – Volterra's Analytical Mechanics of Biological Associations, à paraître dans *Archives Internationales d'Histoire des Sciences*.

Kuhn T. S. [1977] – *The Essential Tension*, Chicago, The University of Chicago Press.

Lorenz E.N. [1961] – Deterministic non-periodic flow, *J. Atmos. Sci.*, **20**, p. 130-141 .

Lorenz E.N. [1964] – The problem of deducing the climate from the governing equations, *Tellus*, **16**, p. 1-11.

Menghini M. [1986] – Sul ruolo di Corrado Segre nella geometria algebrica italiana, *Rivista di Storia della Scienza*, **3**, n.3, pp. 303-322.

Miller A. [1981] – *A. Einstein's Special Theory of Relativity*, Addison-Wesley, Reading, Mass..

Poincaré H. [1881-86] – Mémoire sur les courbes définies par une équation différentielle, *Journal de Mathématiques Pures et Appliquées*, s. 3, **7**, 1881, pp. 375-422 ; s. 3, **8**, 1882, pp. 251-296 ; s.4, **1**, 1885, pp. 167-244 ; s.4, **2**, 1886, pp. 151-217.

Poincaré H. [1892-1899] – *Les Méthodes nouvelles de la Mécanique Céleste*, 3 tomes, Paris, Gauthier-Villars, 1892, 1893, 1899.

Poincaré H. [1895], *Théorie Analytique de la propagation de la chaleur*, Paris.

Poincaré H. [1899] – *Les Méthodes nouvelles de la Mécanique Céleste*, T. III.

Poincaré H. [1912a] – *Calcul des Probabilités*, Paris, Gauthier-Villars, 1912^2.

Poincaré H. [1912b] – L'hypothèse des quanta, *Revue Scientifique, Revue Rose*, 50° année, 1re semaine, 24 février 1912, p. 225- 232.

Poincaré H. [1912c] – Les rapports de la matière et de l'éther,*Journal de Physique Théorique et appliquée*, 5e série, **2** , 1912, pp. 347-360.

Poincaré H. [1951] – "Analyse des travaux scientifiques de Henri Poincaré faite par lui même", *Œuvres de Henri Poincaré*, **I**, Paris, Gauthier-Villars.

Segre C. [1891] – Su alcuni indirizzi nelle investigazioni geometriche, *Rivista di Matematica*, I, pp. 42-66.
Smale S. [1980] – *The Mathematics of Time*, New York, Springer.

Physical Geometry and Special Relativity. Einstein et Poincaré

Michel Paty

Equipe REHSEIS (UPR 318 du CNRS), Université de Paris-7, 2 Place Jussieu, 75251 PARIS-Cedex 05, France. [1]

1. Introduction

The advent of General Relativity has been the occasion of a renewal of the debate among mathematicians, theoretical physicists and philosophers of science, about the relations between Geometry and Physics, with particular emphasis to the problem of the possibility of deciding which Geometry is appropriate to the representation of physical space. Taking aside the conceptions of neo-criticists as expound by Cassirer[2], the two most significant positions facing each other were that of logical positivism and empiricism as notably represented by Carnap and Reichenbach[3], and that of critical realism and rationalism as advocated by Einstein[4]. In essence, the arguments were partly borrowed from those which arose when non-euclidean geometries came to the forefront in the field of mathematics, being adapted and somewhat modified to take into account the kind of evidence for the physical concern of non-euclidean geometries which originated from General Relativity.

According to thc theory of General Relativity, the dynamics of the gravitation field is brought by the geometrical structure of physical space. General Relativity thus appeared as that theory which made possible to decide, from its experimental tests, what is the geometry of space. In the opinion of many, this entailed the strongest refutation of kantian synthetic a priori; logical positivists and empiricists concluded from it to empiricism as the only possible philosophy henceafter, which they identifed as precisely that philosophy compulsorily required by the theory of Relativity, and more generally by contemporary Physics (and science)[5] .

On this philosophical background, Einstein's position sounds somewhat different. He also concluded to the possibility of deciding experimentally about the Geometry appropriate to physical space, but denied the statements of empiricism, invoking Poincaré's

---

[1] This paper was prepared when the author was on leave of CNRS, at the Departamento de Filosofia, Letras e Ciências Humanas, Universidade de São Paulo, Brasil.

[2] Cassirer 1921.

[3] See for instance Carnap 1921, 1924, 1925, Reichenbach 1920, 1922, 1928,1949.

[4] Einstein 1921 b, 1949. See below, and also Paty 1989, and in press.

[5] See the works of Schlick (1917, 1921 a, 1922), and those already quoted of Reichenbach and Carnap.

philosophy of Geometry as providing decisive arguments in favour of some kind of conventionalism, that helped him to advocate, for the interpretation of the relations between Physics and Geometry, a philosophical view which we may characterize as critical rationalism and realism.

Here arises a first historical and epistemological problem, related to the more general one of the philosophy that was underlying Einstein's scientific achievements in, respectively, Special and General Relativity. The problem to which we shall restrict ourselves in this respect through the present paper is that of his conceptions of the relation between Geometry and Physics at these two stages of his scientific work. He stated explicitely his philosophy of these relations after having elaborated the General Theory of Relativity; and, as it is well known, he got interested in non-euclidean Geometries when he needed them to formulate this theory. The case of Special Relativity is not so clear, for he did not use indeed those Geometries, nor did he mention that there was eventually a problem of the relationship of Geometry and Physics when he was re-defining the concept of space (and time). But, indeed, in such a re-casting from newtonian concepts, an implicit conception of this relationship was at work. This leads us directly to his separation of kinematics from dynamics which is at the heart of Special Relativity. An important feature of Poincaré's own approach to the electrodynamics of moving bodies and to Relativity is, on the opposite, the strong linking of relative physical space and time (considered separately one from each other) with dynamics. We are thus led again to Poincaré's conceptions about space and about Geometry, this time "à propos" of Special Relativity.

Our first problem can then be formulated as follows: did Einstein's thought about space (and time) and about Geometry evoluate from an absence of concern for Geometry and a separation between space and dynamics, which would be typical of his path to Special Relativity, towards an almost complete alinement on Poincaré's conception of Geometry and of dynamical space (and time), occasioned by his work on General Relativity?

A second epistemological and historical problem thus comes on the forefront, and it in fact is twofold: did Poincaré actually thought that Geometry and Physics are indissociable in the way that has been generally considered after the establishment of General Relativity, by logical positivists and empiricists and by Einstein as well ? i.e., in essence, on the same ground as his indissociability of space and dynamics, the first one being but a reflexion of the last one ? And, as a kind of a corollary : did Poincaré's philosophy of Geometry have an effect on his own approach to Relativity, by which, although he developped the right behaviour of relativistic space and time (in the sense of Special Relativity), he insisted on maintaining the classical and absolute ones as well ?

We shall begin by recalling briefly Einstein's conception of physical Geometry and his call to Poincaré's views in his claim for the non empiricist character of the decision for a Geometry from experiment. Then we shall turn to Poincaré's true position about the so-called "indissociability of Geometry and Physics". Next, we shall inquire his approach to relativistic Mechanics through the point of view of his philosophy of Geometry. Finally we shall come back to Einstein and examine his exact conception of the use of Geometry in Physics at the time of the Special Theory of Relativity.

2. Einstein's Physical Geometry and his Reference to Poincaré.

Almost immediately after having fully developed his theory of General Relativity, Einstein was led to expound his views on the relationship of Geometry and Physics, in order to make understand that fundamental property of the theory which relates the space-time metrics and the gravitation field. We shall come back in the last section of this paper to some aspects of the circumstances of his own commitment with the question of Geometry, occasioned by his approach of the generalization of the relativity principle from inertial to accelerated motions, which he had related through the principle of equivalence (of inertial and gravitational masses) with the properties of gravitation. At that stage, we shall also inquire further about the deep roots of his conception of what he called "physical Geometry", or "practical Geometry". Let us, for the moment, summarize the essential of this conception, as it is when he stated it that Einstein called for Poincaré's philosophy of Geometry, on which, so he claimed, his own view was partially based[6].

"Physical Geometry", or equivalently "practical Geometry", is defined by him from pure, mathematical, or "axiomatic Geometry", by "adding to it", or "completing it with", relations of coordination that relate geometrical quantities (such as, for example, distance) to corresponding quantities considered for "practically rigid" bodies. The practical Geometry thus constructed, through the "interpretation" of mathematical (geometrical) quantities applied to these abstract and idealized objects is, for Einstein, a kind of a physical theory (and indeed, he said, it has been the oldest branch of physics). It is the theory of an idealized physical space, which he liked to call "space of reference", constructed from an abstraction and simplification of physical bodies, these being reduced to the consideration of their spatial properties only and extended by thought to build a space.

The concept of "body of reference", or of "space of reference" (abstracted from the latter), is so to speak a kind of an intermediate between the purely abstract space of mathematical Geometry (which is devoid of any connexion with the material world) and material bodies which are the objects of our experience. It is this "space of reference" that determines practical or physical Geometry as the *theory* of *that space*. Let us observe that this object and this theory are (abstract) constructions of the mind aimed at the description of some aspects of physical reality (namely, the purely spatial properties of bodies); they are not different, in this respect, from any other physical object or theory. From this, Einstein's answer to the question of the experimental decidability of the Geometry of the physical world does not differ from his answer to the question of the relations between theory and experiment in Physics. Experiment helps in choosing among various theories, but is in no way the only element of our decision. His critical rationalism and realism was at variance with empiricism, even in its more sophisticated logical version[7] .

When advocating this conception, Einstein quoted Poincaré's statements about the impossibility of any compulsory decision from experiment about the nature of the Geometry. Whatever be the results of our experiments with physical bodies, so Poincaré said, we shall always be free to choose euclidean or any non euclidean Geometry (and,

[6] Einstein 1921 a, d, 1926, 1949, 1954, etc.

[7] Paty 1989, and in press.

if we want so, to keep euclidean Geometry). Einstein rightly recalled the first consideration which led Poincaré to this conclusion: Geometry in itself, as a mathematical theory, says nothing about physical space or bodies. Then, at the same time he expounded his conception of "practical" or "physical" Geometry, Einstein mentioned a second consideration referred to Poincaré, which he described as the "indissociability of Geometry and Physics"[8], from which it will never be possible to decide from physical experiments for a Geometry appropriate to the physical world, as we can always choose a modification of Physics rather than of Geometry.

Such has been, since Einstein's 1921 paper "Geometry and experience", the interpretation of Poincaré's conception commonly taken for granted in the debate which opposed the various currents in philosophy of science (neo-criticists, logical empiricists, critical realists and rationalists). For his part, Einstein used this conventionalist argument as an evidence in favour of the rational aspect of the problem, opposed to the idea of a purely empirical conclusion. But he departed from Poincaré's strictly conventionalist position, by stating that we finally conclude as to the nature of the appropriate Geometry, through the choice of an approximation which makes us dissociate in pratice those elements that were indissociable in principle.

It is useful to consider in more details Einstein's use of the alleged Poincaré's argument. Poincaré stated, recalled Einstein, that there is no such a thing, in nature, as perfect rigid bodies, bodies being always affected by physical properties such as temperature, electric and magnetic quantities, etc., which modify their geometrical behaviour. Thus it is not Geometry alone that provides statements on the behaviour of real objects, but Geometry (G) combined with the whole of physical laws (P): "It is the sum ($G+P$) alone which is submitted to the control of experiment. One can consequently choose (G) arbitrarily, and parts of (P) as well: all these laws are conventions. (...) With this conception, axiomatical Geometry and those laws of nature to which the character of conventions is attributed appear, from the epistemological point of view, as being of an equal value". Concluding his evocation of this alleged point of view, Einstein gives the following appreciation: "*Sub specie aeterni* Poincaré's conception is in my opinion correct", emphasizing that actually, there does not exist, in the real world, objects corresponding exactly to the ideal standard objects of Geometry[9].

Another formulation of the conception of the indissociability of Geometry and Physics as attributed to Poincaré can be found in Einstein's 1949 "Reply to criticism", where he argues against Reichenbach, invoking Poincaré's argument that "verification (...) refers (...) not merely to Geometry but to the entire system of physical laws which constitute its foundation. An examination of geometry by itself is consequently not thinkable". As a conclusion we can choose that Geometry which is most convenient to us (i.e., euclidean Geometry) and fit the remaining physical laws in such a way as to obtain agreement with experiment[10].

Actually this is not exactly Poincaré's point of view, but a translation of it made by Einstein in his own perspective, that is according to his conception of physical Geometry. For, in Poincaré 's conception, Geometry enters in the considerations of Physics only through definitions and is not on an equal footing with it. The argument about physical

---

[8] Einstein 1921 a.

[9] Einstein 1921 d.

[10] Einstein 1949, p. 677.

properties of rigid bodies was used by Poincaré only with respect to the question of relativity of space, and of the possibility to obtain evidence for it by measuring bodies. When Poincaré made use of the concept of standard object, idealized as it may be, he was not considering Geometry but Physics (to him, the relativity of space is a physical property which bodies ought to observe through their positions and directions). As for Einstein, when he is referring to standard objects, he is considering "practical" or "physical" Geometry itself, and not any more the axiomatical one. But such an idea was alien to Poincaré. We shall see from his texts that Poincaré never related the consideration of the physical properties of bodies to a combination of Geometry and Physics considered in that way.

Einstein's reasoning about the indissociability between Geometry and Physics, which started from the difference between purely mathematical, axiomatic, Geometry, and practical Geometry applied to physical situations, and considered the first with respect to Poincaré's conception, was as a matter of fact shifted from axiomatic to physical Geometry. We actually get, in Einstein's description of the problem, $G_{pr} + P$, and not $G + P$, G_{pr} standing for practical Geometry, and being defined as Geometry G endowed with relations of coordination and congruence between its mathematical concepts and idealized physical objects, such as to define for the latter the notion of distance. Practical Geometry, G_{pr}, is, as we said earlier, nothing but a theory of the (idealized) physical space obtained through the idealization of physical bodies, i.e. the theory of distances for physical bodies, as thought independently from other physical properties. As such, G_{pr} is a part of Physics, which we could as well designate by P_d (i.e., Physics of distances), if we were to emphasize its relation with the rest of physics, P_p, (Physics of other, physical, properties), Physics as a whole being $P = P_d + P_p$, which is clearly indissociable in principle (and which could be better symbolized by $P_d \times P_p$).

Einstein's conception of physical Geometry, to which we shall come back later on, has borrowed from Poincaré something of his conventionalism but in a way which has modified Poincaré's conception *stricto sensu*. Einstein's purpose was indeed to oppose this part of conventionalism, which was in his view also a claim for rationalism, to empiricism, which considered that physical Geometry can be directly inferred from experiment.

In his translation of Poincaré's position, to which we shall turn now, Einstein was possibly influenced by considerations made by Schlick in his 1917 book on the theory of Relativity, concerning Poincaré's conception of the relativity of space[11]. Schlick recalled Poincarés consideration of a general modification of spatial dimensions of bodies occurring in a given universe, such that we would have no means of being aware of it, as everything in this universe would have its dimensions modified in the same way; and he added to it a corresponding physical transformation affecting the properties of really physical bodies, in such a way that the conclusion is the same, but based, this time, on a more plausible situation from the physical point of view. Schlick then spoke about the indissociability of Geometry and Physics, referring it to Poincaré's argumentation.

[11]Schlick 1917. See Paty (in press, chapter 6). On relativity of space, see Poincaré 1898 b, 1903 a, 1907.

3. Poincaré's True Conception of the Relation between Geometry and Physics

Poincaré's philosophy of Geometry[12] as he expounded it in particular in his famous 1891 and 1895 papers which constitute two chapters of *La Science et l'hypothèse*, is related to considerations on the physical world only in so far as it shows how the genesis of Geometry is obtained through man's experience of this physical world. It is not in these texts, but in later contributions, such as the chapter of the same book entitled "Experience and Geometry"[13], that he expressedly considers the physical character of the objects to which one relates Geometry when we wants to submit it to the judgement of experiment.

True, his previous considerations of the genesis of Geometry had led him to state that "the principles of Geometry are not experimental facts". But only in that last writing does he deal in detail with the relation of Geometry, as a branch of mathematics, considered as constituted, and not through its genesis, with concrete physical objects. He begins by making a radical distinction between Geometry considered as an axiomatic science and what is related with *practical experiment*: a distinction which Einstein, to some extent, will make too, but with a significant difference in vocabulary. Einstein will differentiate axiomatic Geometry, which is purely mathematical, and *practical or physical Geometry*, which is a physical science with a simplified object, i.e. a physical object with only geometrical properties[14]. This apparently slight difference is indeed an important one, having to do with the construction of physical theory, be it at the elementary level of a theory of the geometrical distances of standard objects (measuring rods) or at the more elaborated one of the theory of General Relativity.

In Poincaré's view, a fundamental aspect of axiomatic Geometry is that it can be integrally translated from a system of axioms and concepts to another one, and this property suffers no exception. On the contrary, practical experiment considers material objects for which we must always inquire about their relations with ideal notions, such as, for instance, the notion of distance, as they never coïncide exactly with them. Reasoning on straight line and distance, he infers from this that "it is impossible to imagine a practical experiment which could be interpreted in the euclidean system and could not be interpreted in the lobachevskian one...".

Furthermore, considering a physical system with respect to the question of Geometry entails considering the physical state of the bodies which constitute that system (i.e. temperature, electric quantities and so forth), the relative position of those bodies (they are defined from their mutual distances), as well as their absolute position and orientation in space. Poincaré's reasoning is actually directed toward the question of relativity of space, namely whether this one is maintained when we perform measurements of distances between bodies and express their results in terms of one Geometry or another. His conclusion is that we never can get outside of a given interpretative frame: "If the law is true in euclidean interpretation, it ought to be true also in the non-euclidean one". This conclusion meets with what he had inferred from his previous considerations about the genesis of Geometry: as a matter of fact, "experiments brings only knowledge of the

[12]Poincaré 1886, 1891, 1892, 1895, 1898 b, 1899, 1902 b, 1903 a, 1907, 1912 a and b.

[13]Poincaré 1899, included in chapter 5 of *La science et l'hypothèse.*

[14] Einstein 1921 b, 1949.

mutual relations of bodies; none of them is, or can be, about the relations of bodies with space, or about the mutual relations of the various parts of space"[15].

When we speak of the "geometrical properties of bodies", for Poincaré, we can never point at the metrical properties of space, and our experiments never deal with space, but on bodies. He gives for this statement the following illustration: consider solid bodies made of portions of straight lines assembled together, and sets of points taken on them and put in relation the ones with the others. It is possible to arrange the relative positions of points and rods in such a way that the motion of the body obeys euclidean group; or that, on the contrary, with a different arrangement, it obeys lobachevskian group. We obviously will not conclude from it that these experiments have shown that space is euclidean or lobachevskian. Experiment has been not on space, but on bodies considered as mechanical constructions. And Poincaré concludes that our constatations on the respective positions of material bodies are independent of the metrical properties of space, and actually can be performed without any notion of these properties: our experiments deal "not on space, but on bodies"[16]. The so-called geometrical properties of bodies are nothing else than our definitions.

As a result, Geometry, in Poincaré's conception, is completely disconnected from measurable properties of physical bodies. When he used to evoke, in his 1895 paper, the relations between rigid bodies and Geometry, it was only for the purpose of showing how rigid bodies are at the origin of the constitution of Geometry, and there would be no Geometry without their existence. But he never intended to consider a direct and quantitative relationship, for "the ideal, absolutely invariables solids" of Geometry are only a simplified image of natural solids, "very far from them"[17].

Geometrical space (as distinct from representative space which is the space of our sensations)[18] and bodies are not of the same nature (the first is an object of the understanding, the other ones are empirical objects), and no direct relation between them is possible: the concept of congruence applied to the correspondence between the figures of Geometry and those of solid bodies, which is one of the senses in which Helmholtz used it, would be, considering Poincaré's conception, devoid of meaning, and it is only definition which plays a role. Indeed, Poincaré never mentioned "congruence" in this sense. When he speaks of "congruence", it is always in the sense of the congruence of geometrical figures. For instance, in his 1902 article in which he analyzes Hilbert's work *On the foundations of Geometry*, he invokes congruence as characterizing "the displacement of an invariable figure"[19].

For Poincaré, the choice of a given congruence, which corresponds to the choice of a given metric, and defines a given geometry, is a matter of convention when one wants to apply it to physical space. This "conventionality of congruence" (to use Adolf Grünbaum's word[20]) entails the possibility to choose for physical space the metrics one wants, and to reformulate, according to an alternative metric, any physical theory. The choice of an alternative metric is, to him, of the same nature as the choice of an alternative system

[15]Poincaré 1902 a, chapter 5.

[16]*Ibid.* Actually this brings us to representative space, which is precisely the space of our sensorial experience.

[17]Poincaré 1895.

[18]Poincaré 1891, 1895.

[19]Poincaré 1902 b.

[20]Grünbaum 1963, ed. 1973, p. 119.

of physical units, and one has, for that reason, spoken of a "linguistic interdependance" to characterize the interdependence of Physics and Geometry according to Poincaré[21].

This expression however is misleading, as it seems to consider on an equal footing these two sciences, when we know that, in Poincaré's view, Geometry, once it has been constituted as a mathematical science, is totally independent from Physics (there is no"interdependence", but only a"one way dependence", i.e. a dependence of the physical formulation on the geometrical definitions). Indeed, Grünbaum, who uses the expression, subsequently endeavours to show that such an "extreme conventionalism" does not correspond to actual Poincaré's position, which he considers to be, on the contrary, that of a "*qualified* geometrical empiricist"[22]. His thesis is that Poincaré's strong statements in favour of conventionalism are context dependent, and that he exaggerated his own position, in order to refute both Russell's and Couturat's neo-kantism and Helmholtz empiricism. Re-reading Poincaré in this perspective, Grünbaum invokes the latter's use of the expression "by a series of *observations*, (...) *experience has proven* to me that [bodies'] movements form an euclidean group, (...) without having any preconceived idea concerning metric Geometry"[23]. He sees in it an empiricist uttering about the nature of the geometry of space, whereas Poincaré means exactly the contrary, as we have seen, precisely, with the example of systems of rods endowed with a mechanical agencement whose motion obeys an euclidean or lobachevskian group. Such a behaviour has to do, as Poincaré unambiguously describes it, not with the space in which these bodies are located, but with the mechanism that relates these bodies between them.

Grünbaum's idea, in fact, is that Poincaré's position, for which geometries are abstract and without relation with physical facts, are "uninterpreted", meets with that of logical empiricism in the claim that the question of the truth of Geometry is a matter of coordinative definitions. According to Grünbaum, Poincaré's polemics is against the attribution of a factual truth to congruence when it is in fact a matter of definition. But this is, actually, an interpretation and a reformulation of Poincaré's thought in the terms of a philosophy which would come after; indeed, this later philosophy founded itself partly on some of Poincaré's criticisms, in particular on those which asked for a precise definition of concepts. For Poincaré never spoke of "interpreted geometry", nor of congruence related to definitions of coordination: such conceptions were alien to it.

In his 1912 text "Space and time"[24], in which he indicates how one defines space from the consideration of bodies, these last ones being submitted to the "principle of physical relativity", Poincaré speaks indeed of the transport of a solid body on another one, of application of a solid body on a figure, by which one can define by convention the equality of figures, from which Geometry is born. But this"congruence" (a word he does not use here) is referred only to bodies between them, or to bodies with figures that are images of these bodies, and not between bodies on one side and geometrical quantities on the other. Poincaré recalls in this writing that Geometry, born from these considerations, is the study of the structure of the group formed by spatial transformations, i.e. the group of displacements of solid bodies: it still has not become a"science of space". He insists again on the difference between Physics and Geometry on evoking the "principle of physical

[21]Grünbaum 1963, *ibid.*, p. 119.

[22]*Ibid.*, p. 129. Emphasis is Grünbaum's.

[23]Poincaré 1902. Emphasis is Grünbaum's in quoting Poincaré.

[24]Poincaré 1912 b.

relativity", seing in the latter "an experimental fact" which entails the possibility of its revision, when, on the contrary, "Geometry must be immune from such a revision". (To preserve Geometry from revision, one must raise the principle of relativity to the rank of a reasonable convention). Let us observe *en passant* that if, relatively to Physics, Poincaré holds indeed an empiricism mixed with conventionalism (which, as an effect, bestows him on that point some kinship with logical empiricism), his position relatively to Geometry is quite different.

Geometry, according to Poincaré, does not for all that identify itself exactly with "axiomatical Geometry", to which Einstein will refer in his conference on "Geometry and experiment", considering besides it a "physical Geometry", as in Riemann and in Helmholtz. Poincaré does not consider a "physical Geometry", but only Geometry under its mathematical aspect. But, to him, even purely mathematical, Geometry maintains something which is related to its origin, to the operation of the understanding which generated it, and finally to these bodies which displacement gives rise to the study of their groups, this study being properly the object of Geometry. We can at least interprete in this way his dissatisfaction of the axiomatic definition of Geometry as proposed by Hilbert, when he points out – as we recalled it earlier – that this definition does not refer to the "natural concept" of congruence of figures in their displacement, which is, indeed, the intuitive image of the congruence of bodies. Axiomatic thus fails to get its postulates back "to their true psychological origin".[25]

Geometry, if we look carefully at Poincaré's argumentation, is used in our description of the physical properties of bodies, only as a definition (and this extends his conception of the axioms of Geometry, which "*are nothing else than disguised définitions*"[26]).

All what we have said about the difference of status between Geometry and Physics is well confirmed by the clear difference Poincaré establishes between a purely (mathematical) science as Geometry and a (physical) science which is as well theoretical as well as experimental as Mechanics[27]. At variance from Geometry, which can and must be thought independently from its origin and from experiment, the theoretical part of Mechanics, "conventional mechanics" (that of general principles), cannot be separated from "experimental mechanics" without mutilation: for "what will remain of conventional mechanics when it will be isolated will amount to very little, and will be in no way comparable to that splendid system of tenet [*corps de doctrine*] which we call Geometry"[28].

Poincaré was indeed far from expressing the idea that, with regard to experiment, we shall never consider Geometry (G) alone, but always Geometry combined with Physics of bodies (P), i.e. the indissociable pair ($G + P$): what can be empirically tested, for Poincaré, is simply physics (P)[29]. We must nevertheless observe that, if it has been

[25] Poincaré 1902 b.On the importance of the notion of the displacement of figures in Poincaré, see Vuillemin 1973.

[26] Poincaré 1891, in 1902 a, p. 76.

[27] Poincaré 1902 a, p. 152-153.

[28] *Ibid.*

[29] This difference of status between Geometry and Physics in Poincaré's conception forbids to reduce his conclusion on the impossibility to decide experimentally about Geometry, which is of a logical nature (because of the conventional character of axioms), to a mere factual question (that of the practical obstacles which preclude the elimination of the distorsions due to perturbations) as Grünbaum does (Grünbaum 1963, ed. 1973, p. 131). See above. Louis Rougier (whom indeed Grünbaum invokes in favour of his thesis) seems also to modify

possible to afford to Poincaré the idea of indissociability of Geometry and Physics in relation with decision from experiment, it is because of his conception, relative to bodies and to physical systems, of the association of the properties of spatial quantities (related with positions and directions, and which we express as geometrical properties) and physical properties properly speaking (those of states, internal to, and characterizing, physical systems, and they include dynamics). This conception lent itself to the above interpretation, but in the very peculiar way to which we are turning now.

If, for Poincaré, Physics and Geometry have no direct relations which would put them on the same level, because of their difference in nature, as we have seen, their connexion can only be of definition and of analogy. Analogies led to the genesis of Geometry from the experience of physical properties of bodies; definitions are the means of having Geometry entering in Physics.

Let us first emphasize again the role of definitions in the association of spatial and physical quantities. Poincaré observed, in his writing on "Experiment and Geometry"[30] that if, in Astronomy, we were to find that the parallaxes are larger than a certain limit, i.e. light is not propagating in a straight line, we would have the choice between "either to give up euclidean Geometry", either to modify the laws of Optics and admit that light does not propagate exactly in a straight line. The statement of such an alternative has been read as if Geometry and Physics were on the same level in relation with experiment, when, on the contrary, the choice which is given to us is nothing more than a choice in the definitions of our physical concepts. If "Geometry has nothing to fear from new experiments", it is not because of an"indissociability of Geometry and Optics", but because we are free to use the geometrical definition which we want for the path of a light ray when dealing with Optics.

Let us now come to an analogy, which gets at the same time into definition, and whose consideration by Poincaré might have been influencial to the interpretation of his conception in terms of "indissociability". It is the link he seems to establish between Geometry and dynamics, when he gives as an example of the sensorial genesis of the abstract representation of space, the case of a world consisting of a heated sphere with a given temperature distribution. (The law of temperature distribution through the sphere is $T = R^2 - r^2$, r being the distance from the center, R the radius of the sphere, T the absolute temperature; the dilatation coefficient is proportional to T, and the refraction index varies as $1/T$.) In such a world, Geometry will be defined as the study of the displacement of solid bodies that undergo distorsions according to the difference of temperature (differing from our own definition which, from our experience of our world, is the study of invariable solid bodies), and it will be, indeed, hyperbolic (lobachevskian) Geometry. The inhabitants of such a world would maintain, when brought to our world, their Geometry, and define accordingly in a different way their Physics, whereas if we

Poincaré's conception about Geometry, when he estimates that, for Poincaré, once conventions have been adopted, the facts expressed by science are necessarily true or false (Rougier 1920, p. 200). But, from what we have discussed, this is not the case, actually, for Poincaré, when this science is Geometry. This being said, conventions in the other sciences coexist, in Poincaré's thought, with the notion of empirical content and with the possibility of verification, as Rougier rightly emphasizes. In physics, Poincaré seems to conciliate conventionalism and empiricism, as we said above.

[30] Poincaré 1902 a, chapter 5.

were to come to their world, we would maintain euclidean Geometry and consider in Physics thermodynamical changes.

We see how Poincaré is concerned, in such an example, essentially by the properties of physical space that are indissociably related with the dynamics of bodies, Geometry as such being left untouched and kept within the definitions. To these definitions, it is the formation of our notions, through an elaboration which started from common experience, which will be determinant. The theory of physical space is Physics and in no way Geometry (which, as we remind, is the theory of the displacements of idealized bodies). But we can indeed consider that in Physics, for Poincaré, the theory of space (S) is indissociable from the theory of the physical properties of bodies, which in fact is Dynamics (D), so that the couple $S \times D$ is indissociable. This means that in front of experimental evidence for a given $S \times D$ couple, one would be free to choose to change either the (physical) theory of space, either the dynamics. But we see that, despite the analogy of the conclusions concerning the conventionality of our choices, one cannot identify the situation for the $S \times D$ couple with the $G \times P$ one, for S is not identified with G (Geometry is not the theory of space) and, indeed, $S \times D$ belongs entirely to Physics, P.

4. Poincaré's Geometrical Thought and Relativistic Mechanics

This last example can serve us as a transition to the question of Poincaré's approach to"relativistic Mechanics" through his study of "The dynamics of the electron"[31], which looks at first sight completely independent from his considerations about Geometry. The concepts of space (and time) and their relation to dynamics are at the core of what was to become Relativity, i.e. the reformulation of the Electrodynamics of moving bodies.

We shall not give here a detailed analysis of the respective paths of Poincaré and Einstein towards special Relativity, and in particular of their specific concerns with regard to the concepts we just mentioned, and we refer to another work[32]. Let us only recall the main features of their achievements and attitudes in this field.

Poincaré's and Einstein's respective conceptions about space (properties of distances), time (relativity of simultaneity), and on velocity (relativistic addition of velocities, the speed of light as a limiting velocity), concerning the mathematical formulation and the physical interpretation of these concepts as well as their relation to dynamics, were at the same time very close and very different.

Very close, because both of them came to exactly the same formulae of transformation, with an identical interpretation as to the truely ***physical*** character of the concepts considered in any (inertial) reference frame (i.e., in the usual case of two frames in relative motion, the one taken as at rest, and the one in motion), and also to the same relativistic form of Maxwell equations and of the Lorentz force[33]. We must add also that Poincaré

[31]Title of his 1905 paper simultaneous to Einstein's one (Poincaré 1905 b and c, Einstein 1905).

[32]Paty (in press), with references to the current litterature.

[33]Lorentz had not got fully covariant equations for electrodynamics, and had not the good velocity transformation. As for space and time, his view was different : according to him, the transformed quantities were mathematical fictions, for example his "local time" defined in the moving system. Furthermore there was a privileged inertial system, the system related with ether, at absolute rest. See Paty (1987, in press, and to be published).

had (even before Einstein) the idea of relativity of simultaneity which he analyzed in 1898 and then in 1904[34].

But their views were very different concerning the theoretical meaning of these results, and only Einstein can be credited of having developped a *theory* of Relativity, where the idea of *covariance* is basic and founding. Although the word was coined afterwards, it summarizes indeed the essential of Einstein's 1905 theory (and, so to speak, the "object" of this theory): covariance, as the condition put on physical quantities so that the principle of relativity is obeyed, entails the Lorentz formulae of transformation through a redefinition of space and time, and the covariant form of (electro-)dynamical laws. Poincaré also considered covariance, but not as the founding concept. It was entailed from Lorentz formulae of transformation, and these were a consequence of electrodynamical properties as evidenced experimentally (with a particular emphasis on Michelson-Morley experiment, at variance with Einstein)[35]. The concepts of space, and, separately, of time, were given their relativistic form through dynamics. There was no relativistic kinematics thought independently from dynamics in Poincaré's approach as well as in his later thoughts (and the writing of time as an imaginary fourth spatial component, first introduced by him in his 1905 paper, and which was to be taken from him by Minkowski, was to him a purely formal trick to get invariant quantities).

This "dynamical" thought of the concepts of space, time and velocity explains in a way the difference of Poincaré's approach from Einstein's reform of kinematics. One could however argue that, even with a dynamical origin and nature, time and space could be thought in a way similar to that of Relativity in Einstein's sense. Indeed, this is the case if we consider not Special, but General Relativity: Einstein's redefinition of time and space through metrics when he took into account the gravitation field is in continuity with the previous one, which appears as a special case with no field. We are thus led to look for another reason of the difference between Einstein's and Poincaré's results. Poincaré's geometrical thought, although it bears on a quite different object, will help us here to understand better his physical thought.

Although, when speaking of Poincaré, the $G + P$ and $E + D$ couples cannot be identified, as we have seen, it is possible to see in his conception an analogy between Geometry on one hand, and Physics on the other, when we have to consider different possible representations for them. In Physics, we are left with newtonian classical Mechanics on one side, and with relativistic (or, better, in Poincaré's terminology, "new" or electromagnetic) Mechanics with its Lorentz transformed space and time and its peculiar composition of velocities on the other. If we are to deal with problems of low velocity motions, or of celestial Mechanics, for which classical Mechanics suffices, we shall be content with this description, which is the simplest one and which we are used to in our daily world. On the other hand, if we deal with electromagnetic phenomena, or with high velocity motions, we shall adopt the "new Mechanics". But the modifications entailed by the latter do not oblige us in any way to modify our classical concepts for the description of our daily world. These representations are in no way absolute, they are relative. Each one is, as a whole, equivalent to the other, as we can put in correspondence every concept of the first to every concept of the second. So to speak, the classical and the "new" Mechanics are respectively, when we consider them according to Poincaré's way

[34]Poincaré 1898 a, 1904 a.

[35]Paty (in press).

of thinking, in a situation similar to euclidean Geometry with respect to non euclidean ones.

When, precisely, we look at these concepts themselves, we are led to a similar conclusion concerning the equivalent reference frames in relative motions. Let us make the analogy explicit. Poincaré conceived the relativity of motions in the following manner: inside each system of reference in relative motion, one is not conscious of the fact that times, lengths, forces or the various electromagnetic quantities are not the same as in the other frames. But this does not matter, as every system is coherent in itself: physical quantities which can be measured are those quantities defined in the system, and no one is truer in one system than in the other.

Such is in particular the concept of time, and Poincaré wrote, as soon as 1898, that "we have not a direct intuition of simultaneity, nor of the equality of two durations". He insisted, in this respect, on the psychological analysis of the idea of simultaneity, originated from the sensations we receive from events, and considered that simultaneity statements are reduced to rules that "make statements on natural laws the simpler possible"[36]. In his 1904 paper, he inquired about the physical meaning of Lorentz's local time, and considered the synchronisation of distant clocks in a way rather similar to Einstein's 1905 analysis[37]. He took first the clocks in relative rest, then in relative motion: in the last case, did he notice, the synchronisation condition is different from that at rest, for motion alters the interval of time needed to transmit optical signals; the new time, determined by taking this into account, equates to Lorentz's local time. And Poincaré concludes: "Clocks set up in this way will not show any more the true time, they will show what we may call local time, so that one of them will lag behind the other. *But it matters very little, because we shall have no means to be conscicnt of it.* All phenomena which will occur in A, for example, will lag behind, but all with the same rate, and the observer will not notice it, because his watch is lagging behind too; so that, as required by the principle of relativity, there will be no way to know whether he is at rest or in motion"[38].

From this we see clearly that, besides the dynamical origin of time considered in a given physical system, lies a specific conception of what "relative" (time, and system) is. It must be added that the principle of relativity for physical laws is taken, by Poincaré, in a similar meaning as for space (the relativity of space). There is no absolute motion (or time, or space) in the same way, and for the same reason, as there is no absolute position and direction[39]. Furthermore, the genesis of the concept of space (from compensations of motions by our body and muscles to external motions) implies relativity of motion[40]. In the relativity of systems and motions, Poincaré is content with the consideration of any of these systems taken in itself, as a whole, each one being not less and not more true than any other. He does not insist about the possibility of passing from one to the other, once the general transformation between them is set. He considers the relations of different physical quantities inside a system (they have the same structure in all systems), and not the relation of a given quantity as taken in different systems (except

[36]Poincaré 1898 a.

[37]Poincaré 1904 a.

[38]Poincaré 1904 a (my emphasis, M.P.).

[39]Poincaré 1902 a, chapter 5, p. 98-99, and chapter 7, p. 129.

[40]Poincaré 1895, in 1902 a, p. 83.

for the establishment of formulas of transformation)[41], notwithstanding his analysis of simultaneity. These relative systems constitute, so to speak, closed worlds endowed with adequate and self-sufficient representations.

This view is analogous to his conception of Geometries. The structure of physical quantities in a given system is thought in the same way as the structure of geometrical concepts in a given Geometry. And the same is true for the concepts of newtonian Mechanics, or for those of the "new Mechanics". Poincaré is aware that, if we wish to consider together Mechanics and Electromagnetism, one must perform corrections on the quantities of the first one. But this does not entail, in his view, a general modification of the laws and concepts, and those of Mechanics are still valid in its domain. If he thought so, it well probably is because he considered these concepts always through the mediation of dynamical laws: he did not admit a representation of physical concepts independent from dynamics, or transcendent to it.

This structural identity of his argumentation about the concepts in Mechanics (and in particular space-time concepts) and about world geometries looks obvious if we compare the formulations he gives independently for each case. But Poincaré himself gave an indication in favour of such a comparison, when he considered on an equal footing, in the chapter on "Classical Mechanics" in *La science et l'hypothèse*, relativity of space, of time, of simultaneity and the variety of possible Geometries. He said that notwithstanding relativity of time one continues by convention to speak of absolute time, and that Geometry being a convention of language, and letting us free to take it euclidean or non euclidean as well, we nevertheless consider it, provisionally, as euclidean[42].

In fact, the true key to Poincaré's thought of relativistic Mechanics can be found in the identity of the situations one observes, in his descriptions, between, on the one hand, in Physics, a system where velocity is given by its relativistic formula (and not any more by galilean addition) and, on the other hand, in Geometry, the dynamical world represented by a heated sphere. In both caes, the structure is given by a hyperbolic relation. Considered from the physical point of view, this structure is such that its fundamental quantity has a limiting value. In the case of relativistic dynamics, this limiting value is the constant velocity of light, c, related, precisely, to the relativistic law of composition of velocities (in Poincaré's 1905 work, it was a consequence of the formula, whereas in Einstein's one the constancy of c entails the formula). For the dynamical heated world, the law of transformation for the lengths (dilatation) is characterized by the limiting value of the fundamental quantity, i. e. the absolute zero of temperature.

The law of relativistic velocities on one side, the law of change in temperature (and of lengths) with distance on the other, are formally analogous. The inhabitants of the world with a hyperbolic Geometry are not conscious that the laws of their world differ from our world, because their knowledge of distances (considered geometrically) is depending on the dynamical law which governs these distances. This situation can be transposed without difficulty to the"space of relativistic velocities". Here also it is dynamics (through Lorentz's local time and transformation formulas which, to Poincaré, originate in dynamics) that dictates the law of transformation of velocities. But the laws as described by the relativistic composition of velocities are not fundamentally different from the laws

[41] If one dares say, this transformation is conceived as a translation, in the linguistic sense, more than describing a motion.

[42] Poincaré 1902 a, chapter 6, p. 112.

as expressed with galilean addition, in a newtonian world where the constant c is taken infinite. (And we nowadays know, indeed, that this is due to the fact that such a simple operation as a change of variable, velocity v into rapidity y, such that $y = \sinh \frac{v}{c}$, reestablishes for the new variable, rapidity, a law of addition: the composition law for velocities v and w, $\frac{v+w}{1+vw/c^2}$, reads, for the corresponding y and z rapidities, $y+z$. In both cases we are facing a world which exhibits unusual laws, but which is strictly equivalent to the world of our ordinary representation (euclidean for the case of Geometry, newtonian for the case of Relativity), and which is translatable into the latter.

Having done the comparison just sketched and drawn our conclusion, entailed by this comparison, of an identity of structure, in Poincaré's thought, between the problem of relativistic representations and that of Geometries, we find a justification of it written in Poincaré's own ink. He wrote, in 1904, after considering the peculiarities of electromagnetic dynamics: "From all these results, if they were to be confirmed, an entirely new Mechanics would emerge, which would be essentially characterized by the fact that no velocity could overpass the velocity of light, in the same way as no temperature could fall beneath the absolute zero of temperature. For an observer that would be drifted along *a translation he would not be aware of*, no velocity could as well overpass the velocity of light; and this would be a *contradiction*, if one did not remember that this observer would not use the same clocks as an observer at rest, but indeed clocks showing the "local time"[43].

Clearly, temperature law and velocity law are put in parallel, and the dynamical character of the first suggests a similar characterization for the second. The first underlined expression let us see, here again, that Poincaré is concerned by the independence of the descriptions of each reference frame with inertial motion. Every observer deals with his own space and time (and physical laws related to them), in his own system, and nothing more. According to the principle of relativity, there is, for sure, no physical means to decide whether this system is in motion. But this would not (and actually does not, even for Poincaré !) forbid him to communicate with another system having a different motion (for this is, indeed, the paper of transformation laws themselves). As for the word "contradiction", which I underlined too, its use by Poincaré to characterize a situation where both the motion of translation and the velocity of light are at sake, shows how close he was to the problem on which Einstein, for his own part, insisted in his 1905 Relativity paper, namely the difficulty to reconcile the principle of relativity and the principle of the constancy of the velocity of light[44].

When reading this excerpt, we have only to remember his 1895 article on "Space and geometry", to get into the analogy between the thermodynamical world and the world (or system) of relativistic velocities. Indeed, the heated world is a "relativistic world" in the following sense: if, in the description of this world, one replaces absolute temperature (which varies from 0 to infinity) by velocity (which varies from c to 0), one obtains a space (in fact a space-time, with a four-coordinate r) where lengths do contract (and time is determined as local time) as a function of the value of the velocity considered. We see how these two independent situations are similar. This sheds light on Poincaré's thought about Relativity: it is structured identically to his thought of Geometry.

[43]Poincaré 1904 a, in Poincaré 1905 a, p. 138-139. Emphasis is mine.

[44]See Paty (in press).

Let us summarize. On the one hand, for Poincaré, time and distances considered independently each on its side, are made physical trough their implication in dynamics; the link they are keeping between them is mediated through the link that each one is keeping with dynamics. This is the reason why Poincaré thought them separately, as in their classical acception, so that there does not exist such a thing as a"Poincaré's space-time", although it was Poincaré who introduced the symbolization of time as the fourth spatial dimension, which Minkowski was to systematize after him.[45] Each definition of space and time is relative to a self-consistent dynamical representation, like a world representation.

On the other hand, the modification undergone by Mechanics, and particularly by these spatial and temporal quantities, is analogous to the necessity of considering, beside euclidean Geometry, non euclidean ones. It corresponds to introducing a new representation which is effective for electromagnetic phenomena and for high velocity motions, but free from any claim to be universal. In Poincaré's view, various theoretical representations may coexist in Physics, and for each class of phenomena one is bound to choose the simplest and the most convenient one. Furthermore, the properties of space and time, in so far as they are physical quantities, are not objects of Geometry, but of Physics, and are especially related with dynamics.

Although Poincaré's work in Physics was the work of a theoretical physicist properly speaking (in contradistinction with physico-mathematician, whose interest is essentially formalization), and notwithstanding our last remark, on the dissociation of physical space and Geometry, we can tentatively conclude this analysis by saying, without exaggeration, that his interpretation of time and space of Relativity was governed by his thought of Geometry.

5. An Interpretation of Einstein's Evolution concerning the Relation between Geometry and Physics

We are thus left with the last problem we wanted to consider: the comparison of Einstein's thought about the relations between Geometry and Physics, and between space and dynamics, before and after the general Theory of Relativity.

The last step of Einstein's path towards the General Theory of Relativity has been when he realized that it would be impossible to generalize the principle of relativity from inertial to accelerated motions unless one drops euclidean Geometry, and, with it, a direct physical meaning attributed to coordinates in a reference frame. This idea came to him from a meditation of the problem of the behaviour of rigid bodies and rods under Lorentz contraction when circular motions are considered.

[45]He actually expressed it through the invariant $s^2 = x^2 + y^2 + z^2 - t^2$, where he choose the unities such as $c = 1$ (Poincaré 1905 b, p. 146). In the last part of his 1905 paper, devoted to gravitation, he wrote explicitly $x = t\sqrt{-1}$, and spoke of "four dimension space", stating that "Lorentz transformation is nothing else than a rotation of that space around the origin" (*ibid.*, p. 168). Let us recall that we also owe to Poincaré the first formulation of the method of the search for invariants of the Lorentz group, in order to obtain restrictions on the possible forms of an interaction (gravitation in this case, *ibid.*).

Ehrenfest's paradox[46] had shown that rigidity cannot be maintained in the usual sense, as a rotating body, which would be rigid in its own system of reference, as advocated by Max Born[47], ought to be deformed when seen from the system taken at rest, due to the fact that Lorentz contraction acts tangentially and does not act radially. Such a deformation ought to be the effect of physical forces, as shown by Max Planck and Max von Laüe[48]: accelerated motion would result in a deformation of the body, which would not be any more rigid, but at least elastically deformable, or eventually worn into pieces. At that stage, such a dynamical property of rotating bodies entailed, considered from Einstein's point of view[49], the impossibility of a generalisation of the relativity principle to all kinds of motions, including accelerations.

Einstein got his solution when he came to consider that relative accelerated motions require "another definition of the physical meaning of lengths and times". In his 1916 article on General Relativity[50], Einstein evokes the case of a rotating disk, and gives an interpretation of the difference of the tangential contraction and radial invariance in terms not of dynamical properties of the"rigid" body (which had been the way Born considered it), but of geometrical properties of the reference space (when seen from the system at rest, the ratio circumference/diameter for the rotating disk was less than π). These geometrical propertics of the reference frame were not for all that less physical. Such a shift in the description of the problem (which he repeated in all his further writings on the subject) is meaningful. For Einstein overcame the difficulty he met to extend the principle of relativity to accelerated motions by pointing out what he called a "limitation of the concept of rigid rods (and clocks)", through the use of gaussian coordinates for the description of the space-time continuum. In gravitation fields there do not exist such things as "rigid bodies having euclidean properties" and we are compelled to use non rigid bodies of reference.

If he expressed the problem in such terms, i. e. in terms of a critique of euclidean Geometry, and not in terms of a dynamical structure, it is because he had previously a strong and definite idea of the physical meaning of distances (and durations), which he in fact adquired already when he defined these concepts for Special Relativity[51]. This conception which he did put in practice in his considerations on the transformations of reference frames in motion, made itself explicit with reference to Geometry thereafter. In his further descriptions of Special Relativity he then would always give at the start definitions referred to Geometry, and use to retrospectively read Special Relativity in such a way. In his first book of popularization, published already in 1917, *The Special and General Theory of Relativity*[52], the first chapter is entitled, precisely, "The physical content of geometrical statements". In other important works he would further develop

[46]Ehrenfest 1908, 1909, following considerations about the truely physical character of Lorentz contraction, Ehrenfest 1907, Einstein 1907. See Pais 1981, Klein 1970, Stachel 1980, Dieks 1990.

[47]Born 1909, 1910, 1911.

[48]Planck 1910, von Laüe 1911; the problem is evoked also in Einstein 1911.

[49]As he explained it in his retrospective accounts.

[50]Einstein 1916. See also Einstein 1917, 1921 a.

[51]Einstein 1905. See our more detailed work on this and other Einstein's achievements: Paty (in press).

[52]Einstein 1917.

his analyses of what appears to be, precisely, the relationship of Geometry and Physics, which we have sumarized before[53].

Clearly, Einstein's notion of "physical Geometry" is inherited from Helmholtz who coined the term[54]. But, this concept, such as he made it explicit with General Relativity, is directly inspired also by Riemann whose conception of Geometry appealed directly to Physics: it is the last one that provides the metrics, i.e. the proper Geometry for physical space, as Geometry has become, with Riemann, the "science of space", and no more the "science of figures in space" it was before[55]. Interestingly enough, when Einstein explains the relation of Geometry with the spatial properties of bodies, he generally prefers to use the expression "practical Geometry" rather than "physical Geometry". His "practical Geometry" is in fact the same as Helmholtz's "physical Geometry", being defined from purely mathematical Geometry (axiomatic Geometry, for Einstein) by "adding to it", or "completing it with", relations of coordination that relate geometrical quantities (such as, for example, distance) to corresponding quantities considered for "practically rigid" bodies. "Practical Geometry" is applicable to the spatial considerations of both Relativities, the Special and the General ones. Perhaps in his view the use of the expression "Physical Geometry" would better describe the theory of physical space in the General Theory, as we are there in a situation closer to Riemann's conception, with a deeper connection between Geometry and Physics. Anyway, Einstein did not state it; and we shall content ourselves in observing that this "practical" concern results in defining a theory of an abstract and simplified object ("space of reference").

The concept of "body of reference" is pregnant in Einstein's thought through all his path since Special Relativity up to General Relativity. Even his emphasis of inertia, when he defined the relativity principle, expresses nothing but a property of the systems of reference that are admissible[56]. (On the contrary, Poincaré did not restrict so clearly his statements on the "principle of relativity" to inertial systems, the restriction being always considered a practical one, an effect of an approximation[57]). It thus appears that Geometry, as euclidean Geometry, was in fact implicated right at the beginning in Einstein's work, since the considerations about rods and clocks of Einstein's 1905 paper, in the terms of an embedding of Geometry (that of bodies of reference) and Physics.

When he justifies, for instance in his 1921 Princeton lectures, the abstract construction of four-dimensional space-time, Einstein argues that the three-dimensional euclidean space of pre-relativistic physics was the result of an abstract elaboration as well, referring to Poincaré's considerations in *La science et l'hypothèse* on the foundations of Geometry from the study of displacements of bodies based on our own body's experience. At variance with Poincaré, Einstein expresses this construction of our geometrical notions in terms of a "space of reference", which is an abstraction constructed from bodies. Such a concept, absent in Poincaré, who used to speak only of the motions of bodies (be them idealized), is a kind of an intermediate between the purely abstract space of mathematical Geometry and material bodies which are the objects of our experience. It is this "space

[53]Einstein 1921 a, b, 1926, 1949 c, 1954, etc.

[54]Helmholtz 1868, 1870.

[55]This suggestive opposition of the two objects of Geometry has been proposed by Houzel (Houzel 1989). See also Boi (1989).

[56]Einstein 1917, chapter 4.

[57]See for instance Poincaré 1912 a.

of reference" that determines practical or physical Geometry as the *theory* of *that space*. Let us note that, from this point of view, it corresponds to a riemannian conception of Geometry as theory of space, which Helmholtz as well as Poincaré did not share, each one for different reasons.

We are able to see that Einstein's own elaboration of these concepts (*space of reference*, *practical* or *physical Geometry*, is a genuine one which borrows elements from Riemann, Helmholtz and Poincaré, and is not an alinement on Poincaré's philosophy of Geometry. Furthermore, this elaboration shows its ability to integrate the conceptions on space, and implicitly on Geometry, that Einstein did put in practice before his General Relativity, and which indeed conditioned already the Special Theory and his particular approach to the problem of the physical meaning of space (and time) coordinates. Thus, his separation of kinematics from dynamics in the Special Relativity appears as a simplification which was legitimated by the purpose he had in mind at that time, and which can be described as an approach to the theory conceived as determined strictly by the consideration of its object. (This type of approach being characteristic of Einstein's "scientific style".) And we recall that this object was, to summarize, *covariance* in the sense of inertial transformations, and in no way *dynamical properties* of physical systems (what it was for Poincaré)[58]. When the object aimed at will be changed, i.e. when covariance will be taken in the general sense, and the problem will show itself to be of a dynamical nature, this simplification will not be held anymore. We cannot help to make the diagnosis of a deep continuity in Einstein's path and thought in the field of Relativity and concerning Geometry, despite the strong differences which we recalled at the beginning.

It thus appears from the comparative examination of Poincaré's and Einstein's contri butions to the Special Theory of Relativity, that, for both of them, their respective ways towards Relativity was strongly influenced by their conceptions concerning the relation between Mathematics, and in particular Geometry, and Physics. But although Poincaré thought Physics with his geometer's mind, Einstein thought Geometry trough its use in buiding Physics, a view that he would maintain and refine afterwards.

References

Boi, Luciano 1989. Conception 'dynamique' en géométrie, idéalisation et rôle de l'intuition, Communication au Colloque *Rationalisme scientifique et philosophique*, UNESCO, Paris, 1989 (prétirage).

Born, Max 1909. Die Theorie des starren Elektrons in der Kinematik des Relativitätsprinzips, *Annalen der Physik* (4) 30, 1909, 1-56; "Berichtigung", *Ibid.*, 840. Also *in* Born 1963, vol. 1, p. 105-160.

Born, Max 1910 a. Die träge Masse und das relativitätsprinzip, *Annalen der Physik* 28, 1911, 571-584; also in Born 1963, vol. 1.

Born, Max 1910 b. Ueber die Definition des starren Körpers in der Kinematik des Relativitätsprinzip, *Physikalische Zeitschrift* 11, 1910, 233-234.

Born, Max 1911. Elektricitätstheorie und Relativitätsprinzips, *Physikalische Zeitschrift* 12, 1911, 569-575; also *in* Born 1963, vol. 1.

Born, Max 1963. *Ausgewählte Abhandlungen*, Vandenhoeck & Ruprecht, Göttingen, 1963, 2 vols.

[58] Paty (in press).

Carnap, Rudolf 1921. *Der Raum. Ein Beitrag zur Wissenschaftlehre*, Iena, 1921; Berlin, 1922.

Carnap, Rudolf 1924. Dreidimensionalität des Raumes und Kausalität, *Annalen der Philosophie* ,1924, 105-130.

Carnap, Rudolf 1925. Ueber die Abhängigkeit der Eigenschaften des raumes von denen der Zeit, *Kanstudien* 30, 1925, 331-345.

Cassirer, Ernst 1921. *Zur Einstein'schen Relativitätstheory*, Bruno Cassirer, Berlin, 1921. Engl. transl. by William Curtis Swabey and Mary Collins Swabey: *Einstein's theory of relativity considered from the epistemological standpoint, in* Cassirer E., *Substance and function and Einstein's theory of relativity*, Open Court, Chicago, 1923; re-ed., Dover, New York, 1953, p. 347-460.

Dieks, Dennis 1990. Physics and geometry: the beginnings of relativity theory, *Proceedings of the 8 th General Conference of the European Physical Society (History of Physics Section)*,1990 (to be published).

Ehrenfest, Paul 1907. Die Translation deformierbarer Elektronen und der Flächensatz, *Annalen der Physik* 23, 1907, 204-205. Also *in* Ehrenfest 1959.

Ehrenfest, Paul 1908. Gleichförmiger Rotation starrer Körper un Relativitätstheorie, *Annalen der Physik* 26, 1908, 24-25. Also *in* Ehrenfest 1959.

Ehrenfest, Paul 1909. Gleichförmige Rotation starrer Körper und Relativitätstheorie, *Physikalische Zeitschrift* 10, 1909, 918. Also *in* Ehrenfest 1959, p. 154.

Ehrenfest, Paul 1959. *Collected scientific papers*, edited by M. Klein, North Holland, Amsterdam, 1959.

Einstein, Albert 1905. Elektrodynamik bewegter Körper, *Annalen der Physik*, ser. 4, XVII, 1905, 891-921. Engl. transl., On the electrodynamics of moving bodies, *in* Einstein et al. 1923 (ed. 1952, p. 34-65).

Einstein, Albert 1907. Bemerkung zur Notiz des Herrn P. Ehrenfest: *Translation deformierbarer Elektronen und der Flächensatz, Annalen der Physik*, ser. 4, XXIII, 1907, 206-208.

Einstein, Albert 1911. Zum Ehrenfesten Paradoxon, *Physikalische Zeitschrift* XII, 1911, 509-510.

Einstein, Albert 1916. Die Grundlage der allgemeinen Relativitätstheorie, *Annalen der Physik*, 3 ser. 4, XLIX, 1916, 769-822. Engl. transl., The foundation of the generalized theory of relativity, *in* Einstein et al. 1923 [ed. 1952, p. 109-164].

Einstein, Albert 1921 a. *Vier Vorlessungen über Relativitätstheorie*, Vieweg, Braunschveig, 1922. Engl. transl. by Edwin Plimpton Adams, *The Meaning of relativity: four lectures delivered at Princeton University* (Stafford lectures, given at the University of Princeton, may 1921), Princeton University Press, Princeton (N.J.), 1922. Re-ed. augm. with Appendices, Chapman, London, 1973.

Einstein, Albert 1921 b. Geometrie und Erfahrung (Conference given at the Priussian Acad. of Sciences, Berlin, jan. 1921, the 27 th), *Preussische Akademie der Wissenschaften, Sitzungsberichte*, 1921, part 1, 123-130. Engl. transl., Geometry and experience, in Einstein 1954 b (ed. 1981), p. 227-240.

Einstein, Albert; Lorentz, Hendryk A.; Minkowski, Hermann; Weyl, Hermann 1923. *The principle of relativity*, with notes by Arnold Sommerfeld, transl. by W. Perrett and G.B. Jeffery, Methuen, London, 1923; re-ed., Dover, New York, 1952.

Einstein, Albert 1926 e. Geometria no euclidea y fisica, *Revista mathematica hispano-america* (Buenos-Aires), ser. 2, 1926, 72-76. (Comm. to the Acad. of sciences of Buenos Aires, april 1925, the 16 th).

Einstein, Albert 1954 a. Relativität und Raumproblem, engl. transl., Relativity and the problem of space, in Einstein 1954 b, p. 350-356.

Einstein, Albert, 1954 b. *Ideas and Opinions*, Engl. transl. by Sonja Bargmann, Laurel, New-York, 1981.

Grünbaum, Adolf 1963. *Philosophical problems of space and time*, Knopf, New York, 1963. Second, enlarged, ed., Reidel, Dordrecht, 1973.

Helmholtz, Hermann L.F. von 1868. Ueber die Tatsachen, die der Geometrie zugrunde liegen, *Nachrichten von der königlichen Gesellschaft der Wissenschaften zu Göttingen* , n° 9, 1868. Engl. transl., On the facts underlying geometry, *in* Helmholtz 1978, p. 39-71.

Helmholtz, Hermann L.F. von 1870 . Ueber der Urspring und die Bedeutung der Geometrische Axiome (Göttingen, 1870). Ré-éd. *in* Helmholtz 1882-1895. Trad. angl., On the origin and significance of the axioms of geometry, *in* Helmholtz 1978, p. 1-38.

Helmholtz, Hermann L.F. von 1978. *Epistemological writings.* The Paul Hertz/Moritz Schlick centenary edition of 1921; newly transl. by Malcolm F. Lowe. Ed by Robert S. Cohen and Yehuda Elkana, Reidel, Dordrecht and Boston, 1978.

Houzel, Christian 1991a. Histoire de la théorie des parallèlles, *in* R. Rashed (ed.) *Mathématique et philosophie de l'Antiquité à l'âge classique*, Ed. du CNRS, Paris, 1991, p. 163-179.

Houzel, Christian 1991b. Interprétations diverses des géométries non-euclidiennes de Gauss à Poincaré, *in* Boi L. ; Flament D. et Salanskis J.M. (eds), this volume.

Klein, Martin J. 1970 b. *Paul Ehrenfest*, vol. 1: *The making of a physicist*, North-Holland, Amsterdam, 1970.

Laüe, Max von 1909. Die Wellestrahlung einer bewegten Punktladung nach dem Relativitätsprinzip, *Annalen der Physik* 28, 1909, 436-442.

Laüe, Max von 1911. Zur Diskussion über den starren Körper in der Relativitätstheorie, *Physikalische Zeitschrift* 12, 1911, 85-87.

Pais, Abraham 1982. *Subtle is the Lord. The science and life of Albert Einstein*, Oxford University Press, Oxford, 1982.

Paty, Michel 1987. The scientific reception of relativity in France, *in* Glick, Thomas (ed.), *The comparative reception of relativity*, Reidel, Boston, 1987, p. 113-167.

Paty, Michel 1989. Interprétation et construction dans le rapport des mathématiques à la physique, *Fundamenta scientiae* 10, 1989 , 35-55.

Paty, Michel (in press). *Einstein philosophe (La physique comme pratique philosophique: relativité, quanta, épistémologie)*, Presses Universitaires de France.

Paty, Michel (to be published). Trois conceptions des rapports entre les mathématiques et la physique: Lorentz, Einstein, Poincaré, in *L'application des mathématiques. Travaux de l'équipe REHSEIS*, Paris.

Planck, Max 1910. Gleichförmige Rotation unfd Lorentz-Kontraktion, *Physikalische Zeitschrift* 11, 1910, 294.

Poincaré, Henri 1886. Sur les hypothèses fondamentales de la géométrie, *Bulletin de la Société mathématique de France* 15, 1886-1887, 203-216. Also in Poincaré 1950-1965, vol. 11, pp. 79-91.

Poincaré, Henri 1891. Les géométries non-euclidiennes, *Revue générale des sciences pures et appliquées* 2, 1891, 769-774. Also in Poincaré 1902 a (chap. 3).

Poincaré, Henri 1892. Lettre à M. Mouret sur les géométries non euclidiennes, *Revue générale des sciences pures et appliquées* 3, 1892, 74-75. Also in Poincaré 1902 a (parts of chap. 2 and 4).

Poincaré, Henri 1895. L'espace et la géométrie, *Revue de métaphysique et de morale* 3, 1895, 631-646. Also in Poincaré 1902 a (chap. 4).

Poincaré, Henri 1898 a. La mesure du temps, *Revue de métaphysique et de morale* 6, 1898, 1-13. Also in Poincaré 1905 a, chapter 5.

Poincaré, Henri 1898 b. On the foundations of geometry, *The Monist* 9, 1898-1899, 1-43.

Poincaré, Henri 1899. Des fondements de la géométrie: à propos d'un livre de M. Russell, *Revue de métaphysique et de morale* 7, 1899, 251-279. Also in Poincaré 1902 a (in chap. 5: Expérience et géométrie).

Poincaré, Henri 1900. Relations entre la physique expérimentale et la physique mathématique, *in Rapports présentés au Congrès international de physique de 1900*, Paris, 1900. Also in

Poincaré 1902 a (chap. 9: Les hypothèses en physique and 10: Les théories de la physique moderne).

Poincaré, Henri 1902 a. *La science et l'hypothèse* (1902), Flammarion, Paris, 1902; 1968.

Poincaré, Henri 1902 b. Compte-rendu de Hilbert: *Les fondements de la géométrie*, *Bulletin des sciences mathématiques* 26, 1902, 249-272; Rectification, *ibid.* , 27, 1903, 115. Republished as Les fondements de la géométrie, *in* Poincaré 1950-1965, vol. 11, pp. 92-113.

Poincaré, Henri 1903. L'espace et ses trois dimensions, *Revue de métaphysique et de morale* 11, 1903, 281-301; 407-429. Also *in* Poincaré 1905 a (chap. 3: La notion d'espace, and 4: L'espace et ses trois dimensions), ed. 1970, p. 55-76, 77-100.

Poincaré, Henri 1904. L'état actuel et l'avenir de la physique mathématique, *Bulletin des sciences mathématiques* 28, 1904 (déc.), 302-324 (Lecture at the International Congress of Arts and Sciences, Saint-Louis, Missouri, sept. 1904, 24 th). Also *in* Poincaré 1905 a (chap. 7: L'histoire de la physique mathématique, 8: La crise actuelle de la physique mathématique, and 9: L'avenir de la physique mathématique), 1970 ed., pp. 123-128, 129-140, 141-147].

Poincaré, Henri 1905 a. *La valeur de la science*, Flammarion, Paris, 1905; 1970.

Poincaré, Henri 1905 b. Sur la dynamique de l'électron, *Compte-rendus des séances de l'Académie des sciences* 140, 1905 (session of june the 5 th.), pp. 1504-1505. Also *in* Poincaré 1950-1965, vol. 9, **3** pp. 489-493.

Poincaré, Henri 1905 c. Sur la dynamique de l'électron (received july 1905, 23 th), *Rendiconti del Circolo matematico di Palermo* XXI, 1906, p. 129-176. Also *in* Poincaré 1950-1965, vol. 9, pp. 494-550.

Poincaré, Henri 1907. La relativité de l'espace, *L'année psychologique* 13, 1907, 1-17. Also *in* Poincaré 1908, Livre 2, chap. 1.

Poincaré, Henri 1908. *Science et méthode*, Flammarion, Paris, 1908.

Poincaré, Henri 1910. La mécanique nouvelle, *Revue d'électricité* 13, 1910, 23-28.

Poincaré, Henri 1912 a. L'espace et le temps (Lecture at London University, may 1912), *Scientia* 12ème année, vol. XXV, 1912, 159-170. Also *in* Poincaré 1913, éd. 1963, pp. 97-109.

Poincaré, Henri 1912 b. Pourquoi l'espace a trois dimensions, *Revue de métaphysique et de morale* 20, 1912, 483-504. Also *in* Poincaré 1913, p. 133-157.

Poincaré, Henri 1912 c. Les rapports de la matière et de l'éther, *Journal de physique théorique et appliquée*, 8ème série, 2, 1912, 347-360. Also in Poincaré 1913, éd. 1963, p. 68-83 and *in* Poincaré 1950-1965, vol. 10, p. 669-682.

Poincaré, Henri 1913. *Dernières pensées*, Flammarion, Paris, 1913; réed. 1963.

Poincaré, Henri 1924. *La mécanique nouvelle. Conférence, mémoire et note sur la théorie de la relativité*, Introduction de M. Edouard Guillaume, Gauthier-Villars, Paris, 1924.

Poincaré, Henri 1950-1965. *Oeuvres*, Gauthier-Villars, Paris, 11 vols., 1950-1965.

Reichenbach, Hans, 1920. *Relativitätstheorie und Erkenntnis a-priori*, Springer, Berlin, 1920. Engl. transl. by Maria Reichenbach, *The theory of relativity and a-priori knowledge*, University of California Press, Berkeley, 1965.

Reichenbach, Hans 1922. Der gegenwärtige Stand der Relativitätsdiskussion, *Logos* 10, n° 3, 316-378. Engl. transl., The present state of the discussions on Relativity. A critical investigation, *in* Reichenbach 1978, p. 3-47.

Reichenbach, Hans, 1924. *Axiomatik der relativischen Raum-Zeit-Lehre*, Braunschveig, Vieweg, 1924. Trad. angl. par Maria Reichenbach, *Axiomatisation of the theory of relativity*, University of California Press, Berkeley, 1969.

Reichenbach, Hans, 1928. *Philosophie der Raum Zeit Lehre*, de Gruyter, Berlin, 1928. English Translation. Maria Reichenbach et John Freund, *The philosophy of space and time*, Dover, New-York, 1957.

Reichenbach, Hans 1949. The philosophical significance of the theory of Relativity, in Schilpp 1949, p. 289-311.

Reichenbach, Hans 1978. *Selected writings*, edited by Robert S. Cohen and Maria Reichenbach, 2 vols, Reidel, Dordrecht, 1978.

Riemann, Bernhard 1854. Ueber die Hypothesen, welche der Geometrie zugrunde liegen [A memoir presented on june 1854, the 10 th, at the Philosophical Faculty of Göttingen], *Abhandlungender königlischen Gessellschaft der Wissenschaften zu Göttingen*, vol. 13, 1867; also in Riemann 1902, p. 272-287.

Riemann, Georg Fiedrich Bernhard 1876. *Gesammelte mathematische Werke und wissenschaftlicher Nachlass*, ed. by par Richard Dedekind and Heinrich Weber, Leipzig, 1876; 2 nd ed., 1892.

Riemann, Georg Fiedrich Bernhard 1902. *Gesammelte mathematische Werke. Nachträge*, ed. by M. Noether et W. Wirtinger, Leipzig, 1902 [a supplement to Riemann 1896.].

Riemann, Georg Fiedrich Bernhard 1953. *Collected works*, ed. par H. Weber et R. Dedekind, with supplement by M. Noether and W. Wirtinger, engl. transl., New York, 1953.

Rougier, Louis 1920. *La philosophie géométrique de Henri Poincaré*, Alcan, Paris, 1920.

Schilpp, Paul-Arthur 1949. *Albert Einstein: philosopher-scientist*, The library of living philosophers, Open Court, Lassalle (Ill.), 1949. Ré-ed. 1970.

Schlick, Moritz 1917. *Raum und Zeit in der Gegenwärtigen Physik. Zu Einfuhrung in das Verständnis der Allgemeinen Relativitätstheorie*, Berlin, 1917. Engl. transl. (on the 3rd. germ. ed., modif. for 4 th. ed.) by Henry L. Brose and Peter Heath, Space and time in contemporary physics. An introduction to the theory of relativity and gravitation, *in* Schlick 1978, vol. 1, p. 207-269.

Schlick, Moritz 1921. Kritizistische oder empiritische Deutung der neuen Physik, 26, 1921, 96-111. Engl. transl. by Peter Heath, Critical or empiricist interpretation of modern physics, *in* Schlick 1979, vol. 1, p. 322-334.

Schlick, Moritz 1922. Die relativitätstheorie in der Philosophie, *Verhandlungen der Gessellschaft Deutscher Naturforscher und Aertze. 87 Versammlung, Hundertjahrfeier*, Leipzig, 1922, p. 58-69. Engl. transl., *in* Schlick 1979, vol. 1.

Schlick, Moritz 1978. *Philosophical papers*, edited by Henk L. Mulder and Barbara F. B. van de Velde-Schlick, translated by Peter Heath, Wilfrid Sellars, Herbert Feigl and May Brodbeck, 2 vols., Reidel, Dordrecht, 1979.

Stachel, John 1980. Einstein and the rigidly rotating disk, *in* Held, A. et al. (ed.), *General relativity and gravitation, one hundred years after Albert Einstein*, vol. 1, Plenum, New York, 1980.

Vuillemin, Jules 1972. Poincaré's philosophy of space, *Synthese* 24, 1972, 161-179.

Transport parallèle et connexions en Géométrie et en Physique

Jean-Pierre Bourguignon

Centre de Mathématiques U.R.A. 169 du C.N.R.S. Ecole Polytechnique F-91128 PALAISEAU Cedex (France)

Le thème de cet exposé[1], *"Transport parallèle et connexions en Géométrie et en Physique"*, est vraiment au cœur de la problématique de ce colloque consacré à "Un siècle de Géométrie" car c'est entre 1830 et 1930 que s'est élaborée la notion de connexion, cette Longue Marche alliant étroitement Physique et Mathématiques. Presque tous les grands géomètres de cette période ont contribué à cette transformation. Je n'ai donc aucun effort à faire pour qu'apparaissent dans mon exposé des noms de grands mathématiciens comme Carl Friedrich Gauss, Bernhard Riemann, et Elie Cartan, qui complètent le titre de ce colloque. Pour faire cette présentation, deux options s'ouvraient à moi :

- m'en tenir à une démarche historique,
  prendre le point de vue (qui était presque obligatoire il y a quelques années seulement) de présenter les notions de la façon la plus récente et moderne, en ne faisant que quelques allusions à l'Histoire.

J'ai choisi la première solution. Peut-être est-ce pour ne pas faire comme on m'a appris ? Une raison moins passionnelle tient à ma volonté de mettre en évidence les diverses formes sous lesquelles est apparue la notion de connexion avant de s'imposer comme centrale[2]. Il y a, je crois, peu d'exemples de concepts mathématiques qui ont, comme celui de connexion, mis un siècle et demi à s'imposer. Cet exposé va en effet montrer comment cette notion disparaît dès qu'elle apparaît, comment, peu après son introduction, elle se transforme en un puissant outil de calcul, puis se subordonne à un autre concept, pour enfin triompher en existant vraiment pour elle-même. Pour terminer ma démonstration, j'ai d'ailleurs demandé aux organisateurs de me permettre de présenter des résultats postérieurs à 1930 car il a fallu attendre les années 1970 pour que les connexions soient vraiment étudiées pour elles-mêmes, et la reconnaissance par les physiciens de leur pertinence dans la modélisation des interactions fondamentales est pour beaucoup dans ce phénomène. Cette épopée permet de corriger une idée fausse selon laquelle l'apparition des concepts devrait suivre une marche inexorable. Cette quête a aussi conforté ma conviction qu'il est nécessaire d'apporter une attention plus grande aux conditions d'élaboration des concepts pour en faciliter l'accès, y compris dans le cadre

[1] Recherche soutenue par le programme européen C.E.E. G.A.D.G.E.T. SC1-0105-C.

[2] Je tiens à appeler l'attention des lecteurs sur l'article [10] qui contient une excellente discussion de beaucoup des points abordés dans cet article.

de l'enseignement. Avant d'entrer dans le vif du sujet, je tiens à exprimer ma gratitude aux organisateurs du colloque pour la confiance qu'ils m'ont témoignée en me proposant de faire ce retour aux sources. Je dois à l'hospitalité d'Ohio State University d'avoir pu trouver le temps nécessaire à l'étude de l'ensemble des articles historiques que je présente ; que tous les collègues qui ont rendu ce séjour possible et fructueux y trouvent une expression de ma gratitude. La préparation matérielle de ce texte a été facilitée par l'aide technique que j'ai reçue d'Eugène Durenard et de Madame Dezetter ; qu'ils en soient remerciés.

I. Avant la Naissance des Connexions

Pour commencer, remontons un peu dans l'Histoire, ... jusqu'à Isaac Newton.

Il est bien connu que Newton n'a pas seulement eu des contributions fondamentales à la Mécanique et à la Physique, mais aussi aux Mathématiques. Pour formuler les équations fondamentales de la Mécanique sous leur forme la plus standard, à savoir exprimer que la résultante des forces appliquées à un système mécanique, à une particule par exemple, est proportionnelle à son accélération, Newton a dû développer un calcul d'un nouveau type : le Calcul Différentiel dont le concept fondamental est celui de *dérivée d'une fonction*.

La notion d'*accélération* n'est pas des plus concrètes. Celle de *vitesse instantanée* suppose déjà une démarche assez abstraite, faisant intervenir l'ensemble des positions du mobile au cours d'un intervalle de temps entourant un instant donné pour n'en retenir qu'une information *infinitésimale*, la dérivée de la fonction position à cet instant. Pour définir l'accélération, il est nécessaire d'aller un cran plus loin encore et de considérer la dérivée *seconde* : étant données les vitesses instantanées d'une particule le long d'une trajectoire (qu'il est naturel d'attacher à la position d'un mobile x de masse d'inertie m à l'instant t où elle est calculée), il faut transporter parallèlement les vitesses instantanées prises en des points voisins et calculer le vecteur vitesse instantanée de cette nouvelle courbe *abstraite*. Ceci conduit à la forme classique des équations de la Mécanique

$$\mathrm{F} = m \frac{d^2\mathrm{x}}{dt^2} .$$

Ces développements remontent à trois siècles, et sont le sous-bassement des *Philosophiae naturalis Principia Mathematica* de Newton parus en 1687.

Le pas suivant, qui est loin d'être négligeable, a été franchi par Joseph Louis de Lagrange dans [16]. Ce dernier a généralisé le Calcul des Variations d'Euler au cas de systèmes de coordonnées arbitraires. Lagrange a parfaitement compris une notion assez subtile[3], celle de *quantité d'accélération*, et c'est là une de ses contributions importantes à la Mécanique. Ainsi il a pu énoncer ce que l'on appelle aujourd'hui *les équations d'Euler-Lagrange* qui étendent les équations de Newton aux systèmes conservatifs repérés dans des systèmes de coordonnées quelconques : pour un système régi par un lagrangien L dépendant des positions par les coordonnées q^i et des composantes $\dot{q}^i$ du vecteur vitesse instantanée, la quantité d'accélération $m\,d^2\mathrm{x}/dt^2$ est remplacée par l'expression de Lagrange ayant pour composantes

[3] ... même aujourd'hui, elle nécessite qu'on y réfléchisse un peu.

$$\frac{d}{dt}\left(\frac{\partial L}{\partial \dot{q}^i}\right) - \frac{\partial L}{\partial q^i} .$$

Cette expression doit être égalée aux forces appliquées au système. Dans le cadre de notre discussion, le point important est que, pour définir cette notion, on a recours à une réflexion géométrique qui est rendue nécessaire par la nature subtile de la dérivation seconde lorsqu'on travaille dans un système de coordonnées quelconques. En effet, en prenant un point de vue moderne, i.e. en se mettant sur une variété différentielle (ce qui est le cadre naturel pour modéliser des systèmes mécaniques ou physiques un peu généraux), les vecteurs tangents en deux points distincts de l'espace de configuration (que sont les vecteurs vitesses instantanées en deux instants distincts) sont des quantités a priori incommensurables. On rencontre donc une difficulté pour définir l'accélération qui suppose qu'on compare les vecteurs-vitesses en des points distincts. Lagrange parvient à la surmonter en utilisant une donnée supplémentaire, le *lagrangien* L qui est une fonction dépendant des positions et des vitesses : dans le cas d'une particule dans un potentiel V, on a $L = E_c - V$ où E_c désigne l'*énergie cinétique* de la particule qui est une fonction quadratique de la vitesse. Ceci est fait de façon un peu implicite dans [16], mais Lagrange ne semble pas intéressé[4] à développer les concepts nécessaires pour analyser complètement le processus par lequel cette équation se forme, ce qui aurait pu le mettre sur la voie de la notion de connexion. Continuons à explorer la période qui précède immédiatement celle retenue pour la conférence. L'étape suivante a un rapport explicite avec la Géométrie, ce qui n'était pas le cas pour les travaux de Lagrange. Je voudrais brièvement présenter les *"Disquisitiones Generales circa superficies curvas"* (cf. [9]) qui furent publiées par Gauss en 1827, mais dont les résultats essentiels étaient connus de lui bien avant. Dans cet article de Géométrie, Gauss utilise les courbes géodésiques qui sont des plus courts chemins pour établir des propriétés géométriques alors que chez Lagrange les géodésiques interviennent essentiellement pour leur intérêt mécanique. Pour développer la théorie des surfaces, Gauss utilise la propriété fondamentale suivante : *il existe des paramétrages de toute surface par des systèmes de coordonnées qui se déduisent des vecteurs-vitesses dans le plan tangent en un point* (ces coordonnées sont depuis appelées *coordonnées gaussiennes*). Gauss met notamment en évidence l'influence de la courbure sur la géométrie d'une surface ; il montre en particulier comment l'expression de la métrique dans ces coordonnées géodésiques fait seulement intervenir la racine carrée de la courbure et comment la courbure est déterminée complètement par la métrique et n'est pas simplement une donnée *extrinsèque*, i.e. dépendant de la forme de la surface dans l'espace ambiant. C'est son fameux *Theorema Egregium*. Grâce à cette étude, il comprend notamment l'influence de la courbure sur les propriétés angulaires des triangles (cette motivation était directement liée à son travail en Géodésie). C'est la première utilisation un peu systématique des géodésiques pour faire une Géométrie générale, mais Gauss, lui non plus, n'éprouve pas le besoin d'introduire un procédé de comparaison entre espaces tangents.

---

[4] Ces termes exacts sont les suivants : *"Les méthodes que j'expose ne demandent ni constructions, ni raisonnements géométriques ou méchaniques, mais seulement des opérations algébriques, assujetties à une marche régulière et uniforme"*.

II. Une Première Occasion Manquée : La Naissance de la Géométrie de Riemann

La généralisation en dimension supérieure de ce que Gauss a fait pour les surfaces est due à Riemann. C'est le contenu de sa dissertation inaugurale (cf [25]) qui était un exercice imposé. Ce document fondamental, à la présentation duquel Gauss a assisté en 1854, n'a été publié que beaucoup plus tard en 1868, après la mort de Riemann.

Si, dans ce texte (qui n'est pas un article scientifique au sens ordinaire du terme), Riemann développe un grand programme et donne des formules fondamentales pour le développement de la Géométrie appelée aujourd'hui *riemannienne*, il ne fait qu'y esquisser les démonstrations sans énoncer de résultats de façon systématique et linéaire. C'est dans ce document que Riemann introduit les ***étendues à n dimensions*** (qu'on appelle aujourd'hui ***variétés différentielles de dimension n***) et considère des expressions différentielles quadratiques, les ***métriques riemanniennes***, qui permettent d'y mesurer la longueur des courbes[5]. Il se sert alors, comme le faisait Gauss, des plus courts chemins pour construire un système de coordonnées géodésiques : dans un voisinage d'un point donné a, un point q est repéré par le vecteur-vitesse de la géodésique issue de a et atteignant le point q au bout du temps 1. Mais ces courbes n'apparaissent pas comme définies par des équations (dans le texte de Riemann, il n'y a en fait aucune équation). Il se préoccupe par contre de mesurer en quoi les espaces généralisés qu'il a introduits dévient de l'espace ordinaire (qu'il propose d'ailleurs d'appeler *plat*). Dans les coordonnées géodésiques l'expression de la métrique n'est plus donnée par des constantes, mais par un développement commençant par l'élément de longueur au point considéré. Le terme suivant contient une quantité plus complexe dont Riemann identifie la dépendance par rapport aux directions et qui se réduit pour chaque sous-surface à la courbure introduite par Gauss : c'est la ***courbure de Riemann***. Des développements plus explicites se trouvent dans [24][6]. En notations modernes, la formule de Riemann est donnée par

$$g(q) = \sum_{i,j=1}^{n} g_{ij}(a)\, dx^i\, dx^j - \frac{1}{6} \sum_{i,j,k,l=1}^{n} R_{ikjl}\,(q^k\, dx^i - q^i\, dx^k)(q^l\, dx^j - q^j\, dx^l) + \dots$$

où g désigne le produit scalaire généralisé et où les q^i sont les coordonnées du point q dans le système de coordonnées géodésiques centré en a. Dans cette formule apparaît clairement que, ***dans les coordonnées géodésiques, la métrique riemannienne est osculatrice à la métrique euclidienne.*** Au premier ordre, l'écart ne dépend que des éléments de surface engendrés par le vecteur directeur de la géodésique le long de laquelle on se déplace et par le vecteur à mesurer. (Rappelons qu'à cette époque-là le calcul extérieur n'existait pas.) A ce point, nous n'avons toujours ni les connexions, ni le transport parallèle !

---

[5] Noter qu'il ne propose cette étude que comme un début, et envisage explicitement de considérer ensuite le cas où la longueur serait donnée par la racine quatrième d'une expression de degré 4 des éléments différentiels.

[6] A cause de son état de santé, il n'a pu développer complètement son argumentation ce qui semble lui avoir coûté le prix qu'il briguait.

III. Les $\left\{ {}^{j\,k}_{\;i} \right\}$ d'E.B. Christoffel

Le mathématicien qui a vraiment introduit la notion de connexion est Edwin Bruno Christoffel : dans l'article [7] paru une année après la publication du texte de Riemann[7], il s'est intéressé aux *conditions que doivent vérifier des expressions différentielles de degré 2 pour pouvoir se transformer l'une dans l'autre par un changement de variables.*

Le point de vue de Christoffel est strictement celui d'un mathématicien intéressé par la *théorie des invariants*. Il résoud un problème de *classification*. Il part donc de deux expressions $\sum_{i,j=1}^{n} g_{ij}(x^k)\,dx^i\,dx^j$ et $\sum_{\alpha,\beta=1}^{n} g_{\alpha\beta}(y^\gamma)\,dy^\alpha\,dy^\beta$, dans lesquelles les quantités (g_{ij}) et $(g_{\alpha\beta})$ sont des matrices de fonctions inversibles sur leur domaine de définition, et il cherche une transformation f qui échange ses deux expressions (ce sont donc les expressions de la même métrique riemannienne dans deux cartes reliées par le difféomorphisme local inconnu f). Il est à noter que dans [6] il considère le problème plus général de l'équivalence d'expressions homogènes de degré n ce qui souligne que la Géométrie riemannienne (i.e. définie par un produit scalaire à coefficients variables) n'avait pas encore établi la suprématie que nous lui connaissons maintenant.

Dans [7], Christoffel introduit les symboles $\left\{ {}^{j\,k}_{\;i} \right\}$ de la façon suivante. Si les variables y et x vérifient la relation $y = f(x)$, alors $dy^\alpha = \sum_{i=1}^{n}(\partial f^\alpha/\partial x^i)\,dx^i$ où $(\partial f^\alpha/\partial x^i)$ désigne la *matrice jacobienne* de f. L'idée fondamentale est de partir de la relation différentielle

$$\sum_{i,j=1}^{n} g_{ij}(x^k)\,dx^i\,dx^j = \sum_{\alpha,\beta=1}^{n} g_{\alpha\beta}(y^\gamma)\,dy^\alpha\,dy^\beta$$

et de la dériver une fois en supposant les coordonnées (x^k) et (y^γ), et leurs différentielles, reliées par la transformation f. Ce faisant, on fait donc apparaître des dérivées partielles secondes de la transformation f que l'on cherche, mais aussi des dérivées partielles premières des coefficients de la métrique.

Christoffel remarque alors que, pour isoler une dérivée partielle seconde de la transformation f, il est commode d'introduire les coefficients[8] notés aujourd'hui

$$\left\{ \begin{matrix} i \\ j\;k \end{matrix} \right\} = \frac{1}{2}\sum_{l=1}^{n} g^{il}\left(\frac{\partial g_{jl}}{\partial x^k} + \frac{\partial g_{kl}}{\partial x^j} - \frac{\partial g_{jk}}{\partial x^l}\right)$$

où (g^{ij}) désigne la matrice inverse de la matrice (g_{ij}). Un fait technique tout à fait important pour plus tard est la non-dégénérescence de la métrique, car dans l'écriture des symboles de Christoffel apparaît l'inverse de la matrice de la métrique. Cela lui donne l'équation fondamentale

$$\frac{\partial^2 f^\alpha}{\partial x^i \partial x^j} = \sum_{l=1}^{n}\left\{ \begin{matrix} l \\ i\;j \end{matrix} \right\}\frac{\partial f^\alpha}{\partial x^l} - \sum_{\beta,\gamma=1}^{n}\left\{ \begin{matrix} \alpha \\ \beta\;\gamma \end{matrix} \right\}\frac{\partial f^\beta}{\partial x^i}\frac{\partial f^\gamma}{\partial x^j}\ .$$

---

[7] Christoffel fait explicitement référence dans son texte aux travaux de Riemann (en mentionnant d'ailleurs Dedekind). Je ne sais par contre pas si des notes de Riemann avaient circulé parmi les mathématiciens de l'époque.

[8] L'automatisme maintenant habituel de calcul sur la position des indices est perturbé dans l'article de Christoffel. En effet il ne met pas les indices en positions opposées dans l'expression de la métrique, ni ne se préoccupe de la variance de la matrice inverse des coefficients de la métrique.

A aucun moment, Christoffel ne s'intéresse à la relation qui pourrait exister entre ces quantités et des courbes particulières dans l'espace. En particulier, aucune remarque n'est faite pour dire que ces coefficients sont naturellement présents dans l'équation des géodésiques $t \mapsto (\gamma^i(t))$ que l'on obtient en appliquant les équations de Lagrange à l'énergie E_c que définit la métrique riemannienne g, équation qui s'écrit

$$\frac{d^2\gamma^i}{dt^2} + \sum_{j,k=1}^{n} \left\{ {i \atop j\ k} \right\} \frac{d\gamma^j}{dt}\frac{d\gamma^k}{dt} = 0 \,.$$

Ces symboles jouent un rôle important dans la résolution du problème étudié par Christoffel et dont il donne une solution presque complète. Comme condition nécessaire à l'équivalence des deux métriques, il obtient notamment que l'évaluation dans les deux cartes des composantes de la courbure de Riemann, qui se calculent à partir des symboles de Christoffel par la formule

$$R_{ijk}{}^{l} = \frac{\partial \left\{ {l \atop j\ k} \right\}}{\partial x^i} - \frac{\partial \left\{ {l \atop i\ k} \right\}}{\partial x^j} + \sum_{m=1}^{n} \left(\left\{ {m \atop j\ k} \right\}\left\{ {l \atop i\ m} \right\} - \left\{ {m \atop i\ k} \right\}\left\{ {l \atop j\ m} \right\} \right) ,$$

doivent se correspondre par la matrice jacobienne qui est déterminée par les deux expressions métriques, i.e. qu'elles satisfont l'égalité

$$\sum_{l=1}^{n} R_{ijk}{}^{l} \frac{\partial f^\delta}{\partial x^l} = \sum_{\alpha,\beta,\gamma=1}^{n} \frac{\partial f^\alpha}{\partial x^i}\frac{\partial f^\beta}{\partial x^j}\frac{\partial f^\gamma}{\partial x^k} R_{\alpha\beta\gamma}{}^{\delta} \,.$$

Christoffel établit un certain nombre d'identités pour ces quantités à quatre indices (antisymétries, et ce qu'on appelle aujourd'hui la première identité de Bianchi). Mais il va aussi plus loin, et montre que le même type d'identité doit être satisfait pour certaines expressions qui sont des *dérivées modifiées de la courbure*. Ces expressions qui seront identifiées ultérieurement comme les dérivées covariantes de la courbure sont un peu compliquées. Christoffel voit cependant qu'il y a un procédé général de formation d'objets nouveaux puisqu'il applique ce procédé de dérivation aux composantes de la métrique et mentionne que les expressions obtenues sont identiquement nulles. Christoffel trouve ainsi une famille infinie de conditions pour que les deux expressions différentielles soient équivalentes, à savoir que les composantes de la courbure et de ses dérivées covariantes soient mises en correspondance par la matrice jacobienne. Sa démonstration du caractère suffisant (cf. [8]) qui repose sur l'étude des conditions de compatibilité du système différentiel n'est pas tout à fait correcte. Il remarque qu'un nombre fini de ces conditions suffit pour assurer l'équivalence dans le cas générique. En présence d'un groupe d'automorphismes non trivial (i.e. lorsque la métrique a des isométries), il a conscience que la situation est plus complexe. En fait la solution finale de ce problème sera donnée dans le cas analytique par Elie Cartan (cf. [3]) et, dans le cas général, par Shlomo Sternberg dans les années 50. C'est donc l'aspect *"outil de calcul"* qui a poussé Christoffel à introduire les symboles qui portent son nom. Bien que mis en œuvre, le concept proprement dit de connexion (dans sa version *dérivation covariante*) n'est donc pas étudié par Christoffel.

IV. G. Ricci et le Calcul Différentiel Absolu

C'est Gregorio Ricci-Curbastro qui a vraiment reconnu que les formules introduites par Christoffel permettaient de créer un outil d'une grande importance théorique. Il a présenté cela dans l'article [22] publié en 1888 (voir aussi [21]), qui est resté relativement inconnu. Grâce à ce nouvel outil, il est en effet possible de sortir des difficultés créées par le caractère non intrinsèque des opérations de dérivation partielle effectuées dans des systèmes de coordonnées locales.

L'article [23], qui est généralement considéré comme la référence sur ce point, s'intitule *"Méthodes de calcul différentiel absolu et leurs applications"* et est paru aux Mathematische Annalen en 1900. Ricci l'a écrit en collaboration avec un des ses élèves, Tullio Levi-Civita. Il est important de souligner la modernité de ce texte qui est *l'acte de naissance et le manuel fondamental du Calcul Tensoriel.* Une de leurs motivations en écrivant cet article était de donner une démonstration de l'invariance par changement de coordonnées de l'opérateur de Laplace-Beltrami (qui généralise le laplacien ordinaire aux variétés riemanniennes).

Le point de vue qui y est adopté est tout à fait intrinsèque, et les notions de *covariance* et de *contravariance* clairement définies. Les auteurs s'y préoccupent abondamment de ce qu'une équation établie dans un système de coordonnées soit encore valable dans un autre système de coordonnées. C'est d'ailleurs de cette analyse que les auteurs tirent le nom d'*absolu* qu'ils donnent à cette extension du Calcul Différentiel. La dérivée covariante d'un champ de vecteurs X est un champ d'endomorphismes dont les composantes $(X^i{}_{;j})$ se déduisent des composantes (X^i) de X dans un système de coordonnées x^i par la formule

$$X^i{}_{;j} = \frac{\partial X^i}{\partial x^j} + \sum_{k=1}^{n} \left\{ {i \atop j\ k} \right\} X^k .$$

Le lien avec les géodésiques associées à cette métrique est naturellement fait dans [23], et il est montré que leur quantité d'accélération (i.e. la dérivation covariante du vecteur-vitesse de la géodésique le long de lui-même) est nulle.

L'article est bâti comme un cours dans lequel les préoccupations physiques sont pratiquement à l'égal des préoccupations mathématiques. Passons en revue les titres des différents chapitres : I. Algorithme du Calcul différentiel ; II. La Géométrie Intrinsèque comme instrument de calcul ; III. Applications analytiques ; IV. Applications géométriques ; V. Applications mécaniques ; VI. Applications physiques.

C'est donc sous la forme de dérivation covariante que le concept de connexion s'est formalisé pour la première fois. Cette nouvelle dérivation se déduit de la métrique riemannienne (qu'il était alors de coutume d'appeler la *forme fondamentale* par extension du cas des surfaces). Ainsi développé, ce nouveau Calcul allait pouvoir devenir un outil de base pour la Physique Mathématique.

V. L'Aventure Mathématique de la Relativité Générale

La naissance de la théorie de la Relativité Générale a constitué une occasion extraordinaire pour développer le Calcul Différentiel Absolu, et pousser à l'approfondissement de la notion de connexion. La création de cette théorie, qui a indiscutablement été l'œuvre d'Albert Einstein, s'est échelonnée sur plusieurs années. Le point de départ est bien entendu la Théorie de la Relativité Restreinte. Dans ce cadre, l'espace-temps est un espace vectoriel lorentzien dans lequel les rayons lumineux sont des génératrices des cônes de lumière et où les particules libres se meuvent suivant des lignes droites du type temps. Je donne deux références pour la Relativité Générale. La première, [12], est, à tort à mon avis, considérée comme une tentative avortée. Il est intéressant de noter qu'elle a deux parties : l'une, écrite par Einstein, porte sur les aspects physiques, et l'autre traitant des aspects mathématiques a, elle, été écrite par son ami Marcel Grossmann. Einstein y justifie le saut épistémologique fondamental de la Relativité Générale consistant à *concevoir les effets de la Gravitation comme des modifications de la Géométrie de l'espace-temps.* Il y propose donc que le tenseur métrique (lorentzien puisque nous sommes dans l'espace-temps) serve de potentiel de gravitation. Ce sont les inhomogénéités du champ de gravitation qui accélèrent les particules. C'est seulement dans le deuxième référence, [13], qu'il donne les équations de champ correctes (après une course effrénée avec David Hilbert !). Le coût mathématique de cette démarche est considérable puisqu'il lui faut utiliser les théories mathématiques les plus avancées de son temps. Une des sources de la Relativité Générale est la réflexion menée par Einstein sur l'*identification qu'il convient de faire sur la base des expériences physiques entre masse d'inertie et masse gravitationnelle.* Quand on adopte le point de vue de l'inertie, on est naturellement amené à étudier les trajectoires des particules tombant librement, et à utiliser ces courbes pour définir une Géométrie, par exemple en les considérant comme des géodésiques pour une dérivation covariante. A ce point, nous sommes donc confrontés à deux familles d'objets mathématiques : *une connexion* définie par l'intermédiaire de ses géodésiques, d'autre part *une métrique lorentzienne* correspondant à la distribution de masse gravitationnelle. Dans le prolongement de la Relativité Restreinte, Einstein les relie alors comme suit : *les vecteurs-vitesses des photons, particules de masse nulle tombant librement, doivent être partout tangents au cône de lumière.* Dès lors, la connexion définie par ces géodésiques est nécessairement celle que Christoffel et Ricci et Levi Civita avaient déterminée à partir des coefficients de la métrique[9] Ainsi, *alors qu'apparaissait l'occasion d'introduire de façon autonome une dérivation covariante, celle-ci est immédiatement subordonnée à la métrique lorentzienne.* Noter que le caractère non tensoriel des symboles de Christoffel prend ici une signification physique : le fait que, le long d'une géodésique, il existe un système de coordonnées où ces symboles s'annulent correspond au fait que le passage à des systèmes de référence accélérés crée des effets inertiels et gravitationnels. Du point de vue des connexions, l'aventure de la Théorie de la Relativité Générale a été une nouvelle occasion perdue : *la notion de connexion n'a pas encore acquis un statut pour elle-même.*

---

[9] Noter que, dans la définition des symboles de Christoffel, seule la non-dégénérescence de la métrique intervient. Il est donc possible d'étendre aux métriques lorentziennes les considérations introduites jusque là seulement pour les métriques riemanniennes.

VI. La Naissance du Transport Parallèle

C'est dans [18] écrit par Levi Civita en 1917 qu'apparaît la notion de *transport parallèle* ce qui constitue une étape très importante dans l'autonomisation de la notion de connexion.

C'est là que le processus géométrique par lequel on peut définir l'accélération dans un espace général est vraiment analysé de façon propre. Levi-Civita envisage la résolvante de l'équation différentielle du premier ordre définissant les champs de vecteurs, dits *parallèles*, dont la dérivée covariante le long d'une courbe est nulle. Grâce à cette résolvante, on peut en effet comparer les vecteurs tangents en deux points distincts et par là-même définir le vecteur accélération d'une courbe comme dans un espace vectoriel : en transportant parallèlement les vecteurs-vitesses le long de la courbe jusqu'à un point de référence, on définit une courbe dans l'espace vectoriel tangent en ce point qui est un espace fixe. Les *géodésiques* sont donc *les courbes dont le vecteur-vitesse est transporté parallèlement sur lui-même*.

Levi Civita fait une analyse assez poussée de ce que peut être une notion de parallélisme. Cela le conduit à travailler dans un espace vectoriel ambiant pour pouvoir le comparer au parallélisme ordinaire. Il interprète notamment *le transport parallèle le long d'une courbe tracée sur une sous-variété d'un espace euclidien munie de la métrique induite comme le roulement sans glissement le long de cette courbe*. C'est Hermann Weyl qui va vraiment se libérer de l'idée de transport parallèle conçu en termes géométriques traditionnels. Dans [28], [29] et dans les versions successives de [30] (livre fondamental pour les relations entre Mathématiques et Physique), Weyl définit une notion générale de *connexion affine* et développe une tentative d'unification de la Gravitation et de l'Electromagnétisme à laquelle il donne le nom de *Weltgeometrie*. Il interprétait le champ scalaire nécessaire pour représenter le potentiel électromagnétique comme le facteur conforme d'une métrique (il s'agit donc d'un potentiel réel). Cette tentative était suggérée par les propriétés d'invariance conforme de l'Electromagnétisme. Mais cette théorie n'a pas pu être validée physiquement.

Pourtant, sur le plan mathématique, ces recherches ont permis à Weyl de bien dégager la notion de transport parallèle associée à une dérivation covariante générale. Elle a aussi été chez Weyl une puissante motivation pour *développer la Géométrie Conforme*. Mais, peut-être encore plus important pour l'avenir, il pressent à cette occasion l'existence d'équations qui admettent une invariance plus générale que l'invariance par un groupe de transformations de dimension finie. C'est la première apparition d'une *Théorie de Jauge* ("Eichinvarianz" dans le texte). Dans [31], après l'avènement de la Mécanique Quantique, Weyl reprend ces idées d'un point de vue physique en donnant une nouvelle version de sa Théorie de Jauge. Il prend cette fois le facteur de phase imaginaire pur et il décrit très précisément comment ce changement est lié à la considération du groupe multiplicatif des nombres complexes de module 1 (dont l'axe imaginaire est l'algèbre de Lie), alors que la construction précédente était reliée au groupe multiplicatif des nombres réels. Cette approche a été par la suite considérée comme une alternative à la présentation traditionnelle des équations de Maxwell (voir aussi [19]). Dans une des sections suivantes, nous verrons combien, convenablement généralisé au cas de groupes de Lie compacts non nécessairement abéliens, ce point de vue s'est révélé fécond.

VII. L'Apport d'Elie Cartan

Elie Cartan a évidemment joué un très grand rôle dans l'émergence des concepts géométriques modernes, et les connexions sont au cœur de son œuvre. On sait maintenant que l'aboutissement sera *la Théorie des Espaces Fibrés*, et la notion de connexion dans ces espaces. Il ne fait cependant pas de doute qu'on trouve tout cela en germe dans les travaux d'Elie Cartan. A propos des connexions, on pourrait citer beaucoup d'articles d'Elie Cartan ; je n'en cite que deux, [1] et [2]. Dans le cadre de la Théorie de la Relavité Générale, Elie Cartan a fait beaucoup de tentatives pour libérer la métrique et la connexion du lien fort qui existait entre elles. Ses efforts n'ont pas été couronnés de succès comme le montre par exemple l'échec de la Théorie du Parallélisme Absolu. On peut mentionner l'étude fine de la notion de *torsion* d'une connexion qu'il a menée à ce propos.

On doit à Elie Cartan une autre notion importante attachée aux connexions : celle de *groupe d'holonomie*. Il s'agit du sous-groupe du groupe linéaire de l'espace tangent en un point qui est engendré par les transformations résultant du transport parallèle d'un vecteur le long de tout lacet issu de ce point. Cette notion introduit en fait une extension tout à fait intéressante de la notion de groupe d'invariance, puisqu'au lieu de considérer des propriétés conservées par des transformations, on s'intéresse ici à des notions infinitésimales qui restent bien définies même lorsqu'elles sont transportées parallèlement le long de n'importe quel chemin. Cette approche est l'acte de naissance de certaines Géométries appelées génériquement du nom de *Géométries de Chemins*. On peut noter que certaines Géométries spéciales (comme les Géométries Kählériennes ou Quaternionniennes) peuvent être détectées par le biais de leur groupe d'holonomie. Les articles d'Elie Cartan sur les *espaces (localement) symétriques* ont joué un rôle considérable dans le développement de la Géométrie moderne car ces espaces se sont révélés être des modèles extrêmement précieux à cause à la fois de leur grande diversité et de la possibilité de les décrire complètement de façon algébrique. Cette classe d'espaces peut être distinguée de beaucoup de façons équivalentes, à première vue difficiles à rapprocher : l'une d'entre elles consiste à demander que *le transport parallèle laisse invariant le tenseur de courbure de Riemann* comme l'indique le titre de [2].

VIII. Les Connexions dans la Théorie Moderne des Espaces Fibrés

Le pas suivant nous fait dépasser 1930. Il a été franchi avec la théorie d'Ehresmann (cf. [11]) qui a donné finalement un contenu réellement géométrique à la notion de connexion. Pour simplifier, nous allons nous restreindre à la famille des espaces fibrés *principaux*. On se place sur un espace P sur lequel un groupe de Lie G agit sans point fixe de telle sorte que toutes les orbites sont difféomorphes. Il est dès lors naturel d'introduire $M = P/G$, l'espace des orbites, sur lequel P se projette naturellement ce qui donne naissance à la fibration $P \to M$ (i.e. la contre-image d'un ouvert suffisamment petit de M est le produit de cet ouvert par la *fibre-type*, ici un espace difféomorphe à G). Le groupe G s'appelle dans ce contexte le *groupe structural*. Dans ce cadre *une connexion* n'est alors rien d'autre que *la donnée en tout point p de P d'un sous-espace de l'espace tangent* *en p à P transverse à l'espace tangent à la fibre* (appelé *espace horizontal* puisqu'on représente le plus souvent l'espace tangent à la fibre comme vertical). Comme le groupe G agit sur P, il est naturel de demander en plus que cette donnée soit équivariante sous l'action du groupe, i.e. qu'au point $p.g$ (pour $g \in G$) le sous-espace horizontal soit l'image du sous-espace horizontal au point p par l'action de g. Avec cette définition, une connexion apparaît comme un objet purement géométrique, et l'existence d'une telle représentation est importante : ainsi nous ne sommes pas contraints de travailler avec des fibrés de repères ou des fibrés qui leur sont associés comme le fibré des vecteurs tangents. Toutes les notions attachées à celle de connexion peuvent s'interpréter dans ce cadre (cf. [19]). Ainsi le transport parallèle est le résultat du relèvement horizontal d'une courbe (ce qui se traduit, pour une géodésique, par le fait que le relèvement de son vecteur-vitesse est à tout instant horizontal). Le groupe d'holonomie prend aussi une signification géométrique intéressante comme sous-groupe opérant dans la fibre qui permet une *réduction* du groupe structural du fibré. De même la courbure apparaît dans ce contexte comme l'*obstruction à l'intégrabilité de la distribution horizontale définissant la connexion*, d'où une interprétation purement géométrique de la platitude.

Il est à noter que les préoccupations physiques étaient totalement absentes chez Ehresmann. Ce point vaut la peine d'être souligné car il montre que *des cheminements différents par des physiciens et des mathématiciens peuvent aboutir à des objets très similaires.* Certains physiciens mathématiciens avaient proposé très tôt des constructions de cette nature. Ainsi une théorie, développée en 1919 par Theodor Kaluza dans [14], puis reprise en 1926 par Oscar Klein dans [15], visait à unifier l'Electromagnétisme et la Gravitation en considérant des métriques spéciales dans un espace à cinq dimensions. La construction de ces métriques faisait déjà appel à un supplémentaire de l'espace tangent à l'espace élargi. Les dérivations covariantes peuvent être aussi vues comme des *opérateurs différentiels du premier ordre à symbole injectif.* Ce point de vue, développé notamment par Jean-Louis Koszul, s'est révélé important dans l'étude des structures complexes sur les espaces totaux d'espaces fibrés et dans l'introduction de fibrés de coefficients qui sont indispensables dans les diverses formes du Théorème de l'Indice. Cela a d'ailleurs amené Daniel Quillen à donner une généralisation algébrique de la notion de connexion sous le nom de *superconnexion.* Ces travaux ont été amplifiés par Jean-Michel Bismut. Dans la section suivante, nous développons encore une autre facette des connexions.

IX. Le Triomphe des Théories de Jauge

Il y a aujourd'hui plusieurs raisons indépendantes pour lesquelles les connexions sont reconnues comme des objets fondamentaux à la fois par les mathématiciens et par les physiciens. La publication en 1954 de [33] écrit par Chen Ning Yang et Robert Mills a représenté un saut qualitatif dans la géométrisation de la Physique. La Théorie de Yang-Mills n'a pas eu un succès immédiat, et surtout ce n'est que plus tard qu'elle a été identifiée comme une théorie s'appuyant fondamentalement sur des connexions sur un espace fibré[10]. L'objectif initial de Yang et Mills était de développer un modèle classique des interactions fortes (celles qui sont responsables de la cohésion des noyaux) bien que

[10] Il semble que plusieurs physiciens mathématiciens comme Andrzej Trautmann aient compris cela bien plus tôt, mais ils sont restés isolés.

ces interactions soient fondamentalement de nature quantique. Ils proposèrent donc à cet effet un lagrangien en espérant que les extrêmales de l'action qu'il définit permettraient de donner les premiers termes d'un développement perturbatif dont on pourrait tirer des informations quantiques. *Le lagrangien proposé par Yang et Mills n'était rien d'autre que la norme L^2 de la courbure d'une connexion* sur des fibrés ayant pour base l'espace-temps et pour groupe structural le groupe non abélien SU_2. A cette époque, ce groupe était considéré comme le groupe de symétrie associé au nombre quantique de spin isotopique ; depuis lors, d'autres nombres quantiques ont été mis en évidence pour les interactions fortes ce qui a conduit à prendre SU_3 comme groupe de symétrie des interactions fortes. Cet article proposant de prendre l'espace des connexions comme espace des paramètres pour une Théorie des Champs est remarquable de clairvoyance. Ce modèle a eu finalement plus de succès quand il a été repris par Steven Weinberg et Abdus Salam (cf. [26] et [27]) pour expliquer les *interactions faibles* (responsables de la radioactivité β) au point de permettre l'unification de ces interactions avec l'Electromagnétisme sous le nom d'interaction *électrofaible*. Une étape importante pour le succès de ces théories a été, en 1971, la preuve par Gerhart 't Hooft du fait que *ces théories pouvaient être renormalisées*, i.e. qu'on peut définir un procédé cohérent pour éliminer les termes infinis apparaissant dans certaines intégrales qui permettent d'estimer les perturbations quantiques. Aujourd'hui, le succès de ces approches, dites *de jauge*, est presque total. Elles servent de cadre de pensée à plusieurs branches de la Physique Théorique. Pour la Théorie des Particules Elémentaires, la conception suivante est maintenant bien installée. On distingue

- *les champs représentant des particules de matière*, qui, au niveau classique, ont des fonctions d'ondes identifiées à des *sections de certains fibrés* dont le groupe structural reflète les interactions dans lesquelles cette particule est engagée ;
- des *champs d'interaction* correspondant à des *particules virtuelles d'échange* qui sont toujours des bosons (comme les photons, ou les vecteurs de l'interaction faible que sont les bosons Z^0, W^+ et W^- dits *intermédiaires*), dont la version classique correspond à des *connexions* sur les fibrés mentionnés précédemment. On voit combien il était fondamental de disposer de la notion de connexion sur des fibrés de groupe structural un groupe de Lie n'opérant pas sur l'espace-temps. *C'est donc dans le cadre de théories physiques que, pour la première fois, l'espace de toutes les connexions est apparu comme un objet digne d'intérêt.* Sur une variété de dimension 4 orientée, l'espace des minima de la fonctionnelle de Yang-Mills (convenablement normalisé en le quotientant par le groupe de jauge) est un objet que les physiciens souhaitaient vivement comprendre vers la fin des années 70 en liaison avec le calcul de perturbations quantiques. De nombreux mathématiciens (parmi lesquels Michael F. Atiyah, Vladimir Drinfeld, Nigel Hitchin, Yuri Manin, Isadore M. Singer, Clifford Taubes, Karen Uhlenbeck) ont contribué à mieux comprendre cet espace de modules. En 1982, Simon Donaldson a complètement renversé la problématique en montrant comment *cet espace de connexions minimisantes contient des informations subtiles sur la structure différentielle de la variété de base.* Ce fut le point de départ d'une révolution de la topologie différentielle des variétés de dimension 4,... et d'un intérêt accru d'autres mathématiciens pour les connexions. Simon Donaldson (et d'autres) ont étendu la portée de cette approche à l'étude de divers *espaces de modules*, comme ceux des *fibrés stables* qui intéressent beaucoup les géomètres

algébristes, ou les ***espaces de représentations*** de certains groupes en les identifiant à des espaces de connexions holomorphes ou plates.

X. L'Apport des Connexions dans l'Etude de Propriétés Globales

Dans les dernières décennies, d'autres démarches totalement indépendantes ont aussi amené les mathématiciens à étudier les connexions pour elles-mêmes. Faisons un petit saut en arrière, et regardons comment les connexions ont participé à une des évolutions fondamentales de la Géométrie dans ce siècle, à savoir *le passage du local au global.* Une des étapes décisives de cette évolution a été la description de certaines classes de cohomologie dites *caractéristiques* en terme de courbure grâce à la fameuse *théorie de Chern-Weil.* Le point de départ en fut la découverte en 1944 par Shiing-Shen Chern[11] (cf. [4]) d'une nouvelle preuve du théorème de Gauss-Bonnet exprimant la caractéristique d'Euler-Poincaré par une intégrale de courbure. La construction de base en était une *transgression* de la forme de courbure d'une surface, i.e. l'opération consistant à la remonter en la différentielle extérieure de la forme de connexion sur le fibré en cercles tangents. Une nouvelle fois, les connexions sont introduites pour disparaître dans le résultat final! En 1974, S.S. Chern et James Simons (cf. [5]) ont étendu la théorie de Chern-Weil aux classes caractéristiques secondaires avec, cette fois, dépendance en la connexion. La quantité introduite par Chern et Simons en dimension 3 a d'ailleurs trouvé sa place dans certaines Théories des Champs contemporaiens sous le nom de *lagrangien de Wess-Zumino.* L'opération de transgression est aussi fondamentale dans la cohomologie B.R.S. qui permet la description de certaines *anomalies* rencontrées en quantifiant certaines théories. L'étude de ces objets a encore été stimulé récemment par le point de vue développé en 1989 par Edward Witten dans [32] sur le nouvel invariant des nœuds que constitue le polynôme de Jones. Son détour utilise une Théorie des Champs pour le lagrangien de Wess-Zumino et pour chaque connexion la trace du transport parallèle le long du nœud. Ces deux exemples montrent l'actualité des recherches menées autour des connexions, et la place prise par l'espace de toutes les connexions dans la Géométrie Moderne. Bien que le cheminement historique ait été tortueux, nous pouvons dire aujourd'hui que *la notion de connexion occupe une place centrale en Géométrie et en Physique.*

[11] A ce propos, notons que la première personne qui a exprimé un nombre caractéristique d'une variété de dimension 4 par une intégrale de courbure est en fait un physicien, Cornelius Lanczos, qui, étudiant dans [17] les équations de champ engendrées par les lagrangiens quadratiques en la courbure, a constaté qu'un lagrangien particulier n'engendrait aucune équation de champ : c'était celui qui exprimait la caractéristique d'Euler-Poincaré en dimension 4.

Bibliographie

[1] E. Cartan, *Sur les variétés à connexion affine et la théorie de la relativité générale*, Ann. Sci. Ec. Norm. Paris **40** (1923), 325-412.

[2] E. Cartan, *Sur les espaces de Riemann dans lequel le transport par parallélisme conserve la courbure*, Rend. Acc. Lincei **3** (1926), 544-547.

[3] E. Cartan, *Leçons sur la Géométrie des espaces de Riemann*, Deuxième édition, Gauthier-Villars, Paris (1946).

[4] S.S. Chern, *A simple intrinsic proof of the Gauss-Bonnet formula for closed Riemannian manifolds*, Ann. Math. **45** (1944), 747-752.

[5] S.S. Chern, J. Simons, *Characteristic forms and geometrical invariants*, Ann. Math. **99** (1974), 48-69.

[6] E.B. Christoffel, *Über die Transformation ganzer homogener Differentialausdrücke*, Monatsber. Kgl. Preuss. Akad. Wiss. Berlin (1969), 1-6.

[7] E.B. Christoffel, *Über die Transformation der homogenen Differentialausdrücke zweiten Grades*, J. reine angew. Math. **70** (1869), 46-70.

[8] E.B. Christoffel, *Über ein die Transformation der homogenen Differentialausdrücke zweiten Grades betreffendes Theorem*, J. reine angew. Math. **70** (1869), 241-245.

[9] P. Dombrowski, *150 years after Gauss' Disquisitiones Generales circa Superficies Curvas*, Astérisque **62** (1979).

[10] J. Ehlers, *Christoffel's work on the equivalence problem for Riemannian spaces and its importance for modern field theories of Physics*, in *E.B. Christoffel's Festschrift*, Aachen (1980), 526-542.

[11] C. Ehresmann , *Les connexions infinitésimales dans un espace fibré différentiable*, in *Colloque de Topologie*, Bruxelles (1950), 29-55.

[12] A. Einstein, *Die Grundlagen der allgemeinen Relativitätstheorie*, Ann. Phys. **49** (1916), 769-822.

[13] A. Einstein, M. Grossmann, *Entwurf einer allgemeinerten Relativitätstheorie und einer Theorie der Gravitation, I. Physikalischer Teil*, Z. Mat. Phys. **62** (1913), 225-244 ; *idem, II. Mathematischer Teil*, ibidem, 245-261.

[14] T. Kaluza, *Zum Unitätsproblem der Physik*, Sitzungber. Preuss. Akad. Wiss. Phys., Math. Klasse (1921), 966-972.

[15] O. Klein, *Quantentheorie und fünf-dimensionale Relativitätstheorie*, Z. Phys. **37** (1926), 895-906.

[16] J.L. Lagrange, *Mechanique Analitique*, Paris (1788).

[17] C. Lanczos, *A remarkable property of the Riemann-Christoffel tensor in four dimensions*, Ann. Math. **39** (1938), 842-850.

[18] T. Levi-Civita, *Nozione di parallelismo in una varietà qualunque et conseguente specificazione geometrica della curvatura riemanniana*, Rend. Circ. Mat. Palermo **42** (1917), 173-204.

[19] A. Lichnerowicz, *Théorie globale des connexions et des groupes d'holonomie*, Cremonese, Roma (1955).

[20] F. London, *Quantenmechanische Deutung der Theorie von Weyl*, Z. Phys. **42** (1927), 375-389.

[21] G. Ricci,*Sui parametri e gli invarianti delle forme quadratiche differenziali*, Ann. Mat. Pura Appl. **XIV** (1886).

[22] G. Ricci, *Delle derivazioni covarianti e controvarianti*, Stud. ed. Univ. Padova, Padova (1888).

[23] G. Ricci, T. Levi-Civita, *Méthodes de calcul différentiel absolu et leurs applications*, Math. Ann. **54** (1900), 125-201.

[24] B. Riemann, *Commentatio mathematica, qua respondere tentatur quaestioni ab Illma Academia Parisiensi propositae* (1861), in *Riemann's Gesamm. Math. Werke* XXII, 2. Aufl. (1892), 391-423.
[25] B. Riemann, *Über die Hypothesen, welche der Geometrie zu Grunde liegen*, in *Riemann's Gesamm. Math. Werke* XIII, Abhandl. Kgl. Gesell. Wiss. Göttingen XIII (1868), 272-287.
[26] A. Salam, *Gauge unification of fundamental forces*, Rev. Modern Phys. **92** (1980), 525-536.
[27] S. Weinberg, *Conceptual foundations of the unified theory of weak and electromagnetic interactions*, Rev. Modern Phys. **92** (1980), 515-524.
[28] H. Weyl, *Gravitation und Elektrizität*, Sitz. Kgl. Preuss. Akad. Wiss. (1918), 465.
[29] H. Weyl, *Reine Infinitesimalgeometrie*, Math. Z. **2** (1918), 384-411.
[30] H. Weyl, *Raum, Materie, Zeit*, Springer, Berlin (1918).
[31] H. Weyl, *Elektron und Gravitation*, Z. Phys. **56** (1929), 330-352.
[32] E. Witten, *Quantum Field Theory and the Jones polynomial*, Commun. Math. Phys. **121** (1989), 351-399.
[33] C.N. Yang, R.L. Mills, *Conservation of isotopic spin and isotopic gauge invariance*, Phys. Rev. **96** (1954), 191-196.

3

Formalism and Intuition

De la Géométrie Formelle à l'Algèbre Abstraite

Hourya Sinaceur

Institut d'Histoire et Philosophie des Sciences et des Techniques, URA 1079 du CNRS, 13, rue du Four 75006 Paris

Chacun sait l'importance des *Grundlagen der Geometrie* de Hilbert, non seulement dans l'histoire de la géométrie proprement dite, mais, plus généralement, dans l'histoire des méthodes mathématiques et dans l'histoire de leur philosophie, c'est-à-dire des idées relatives à leurs concepts fondamentaux comme ceux de nombre et d'espace. C'est d'ailleurs le fait de ce triple impact qui constitue à la fois le caractère décisif et l'ampleur de cette importance. On a beau retrouver, après coup, et mettre en valeur les acquis techniques, méthodiques ou philosophiques qui ont, durant le XIXe siècle, anticipé ou préparé tel aspect ou telle conséquence du point de vue représenté par les *Grundlagen*, ce point de vue n'en doit pas moins sa radicale nouveauté, sa "modernité", à sa capacité d'intégrer le général au technique et la philosophie mathématique à la pratique du mathématicien. Hilbert a réussi cette intégration par la méthode axiomatique conçue comme instrument d'analyse, d'organisation et de fondement des disciplines mathématiques. Les *Grundlagen* sont un monument de la pensée dont on sait bien, même sans y être jamais entré, qu'il consacre l'avènement de cette méthode.

I. La Géométrie comme Langage Formel

Ce monument couronne tout un train de résultats partiels ou locaux différents par leur nature et de portée inégale. On insiste généralement sur le rôle significatif de la géométrie projective ou sur le caractère précurseur des *Vorlesungen über neuere Geometrie* de M. Pasch [1882]. On souligne aussi[1] l'influence des discussions sur les "points imaginaires", sur le principe de dualité ou sur les géométries non euclidiennes; on mentionne les travaux quasi contemporains de F. Klein, F. Enriques, G. Veronese, G. Peano, M. Pieri, A. Padoa, etc. Mais toutes les parentés ponctuelles que l'on découvre *a posteriori* à l'œuvre de Hilbert, et qui confirment d'ailleurs sa force à déterminer notre compréhension de ce qui l'a précédée aussi bien que de ce qui l'a suivie, ne peuvent estomper une différence essentielle, de nature plus philosophique que technique comme l'a souligné Freudenthal [1957]. Hilbert *est le premier à avoir pleinement considéré l'espace comme un concept mathématique*[2] et non comme le lieu de notre expérience,

[1] Nagel 1939.

[2] Bien entendu, une telle conception a été largement préparée par la géométrie de Riemann avec ses multiplicités à n dimensions et la classification des géométries par Klein d'après leur groupe de transformations.

et la géométrie comme une science formelle et non comme un ensemble de propositions *relatives à* "la réalité", soit qu'elles l'expriment de façon idéalisée comme l'ont pensé Aristote ou Helmholtz, soit qu'elles en organisent *a priori* notre intuition comme l'a affirmé Kant, soit qu'elles la structurent selon des relations qui constituent les hypothèses préalables de notre savoir comme l'a soutenu Riemann. Pour la première fois, il n'y a plus lieu de chercher la nature du rapport de la géométrie à "la réalité", en fonction d'ailleurs de l'idée qu'on se fait de cette réalité, parce que ce rapport est brisé.

Les premières lignes, si justement célèbres, des *Grundlagen* nous apprennent que les "éléments" (*στοιχεῖα*) de géométrie *ne représentent pas* des entités réelles, composants simples de corps complexes. Ce sont, dit Hilbert qui n'évite cependant pas totalement les métaphores réalistes, des "choses" quelconques[3], objets dépourvus d'individualité d'une collection, éléments certes, mais en un sens "déréifié". Nous leur donnons les noms de "points", "droites" ou "plans", mais cette désignation ne se justifie pas par une référence à des objets réels ou réalisables. Elle s'autolégitime sans sortir de la sphère du langage, en instituant un nouveau type d'"objets", lesquels sont précisément dépourvus des caractéristiques propres au concept d' objet : stabilité et détermination relatives, puisqu'il s'agit, en fait, de ce qu'en mathématiques ou en logique on appelle des "variables". Ainsi l'axiomaticien n'ancre son travail ni dans des objets réels ni dans des objets idéels. C'est que les "objets" géométriques ne sont pas donnés au départ, comme il était supposé pour les "éléments", mais obtenus au terme de constructions plus ou moins complexes. Ce sont, non pas les points ou les lignes, éléments premiers, mais les structures invariantes à certains égards par rapport auxquelles points et lignes jouent le rôle de variables, le terme structure renvoyant aussi bien à un ensemble d'énoncés compatibles qu'à l'ensemble des modèles qui le réalisent. Bref, déréifiées toutes deux, les notions d'élément et d'objet sont, en plus, disssociées l'une de l'autre, en sorte que les éléments de la géométrie ne peuvent plus être pensés comme des objets, que ceux-ci soient pris pour des êtres simples, lieu d'une intuition réaliste, ou pour des constructions individualisées, résultat d'une activité mathématique. Les éléments de la géométrie ce sont, dans les *Grundlagen*, les variables d'un langage formel, plus exactement les symboles de variables de ce langage, qui comprend, par ailleurs, des symboles de constantes, nécessaires pour marquer un point origine sur une droite orientée ou dans un plan, par exemple, et des symboles de relations ou de fonctions : ceux de l'égalité, de l'ordre, de l'addition, de la multiplication, etc. Puisque les noms des éléments ne sont pas des noms d'objets, il est clair que nous n'avons besoin de *définir* ces éléments ni de façon "réelle" ni de façon "nominale". Quand nous parlons de "point", nous n'associons pas de signification particulière à ce mot, nous indiquons simplement son statut logique de contenu d'un symbole *indéfinissable*. En toute rigueur, point peut être remplacé par chaise ou cuillère comme on l'a souligné plaisamment.

Il est pourtant remarquable que Hilbert et les mathématiciens, en général, conservent l'usage des noms traditionnels même quand ils leur font subir, comme c'est le cas ici, une véritable mutation sémantique, les vidant de toute signification immédiate et convenue pour en faire des symboles de signification variable et déterminée seulement par

---

[3] En 1904, dans sa conférence au 3e congrès international des mathématiciens (publiée en 1905), Hilbert précisera qu'il entend par "chose" (Ding) un "objet de notre pensée " (ein Gegenstand unseres Denkens), corrélat du signe ou du symbole que nous écrivons pour en tenir lieu (p. 176).

les règles d'utilisation explicitement énoncées. Le goût ou la tendance pour cet usage complètement métaphorique, qui s'affirme aussi bien dans le choix de termes nouveaux : anneau, filtre, gerbe, etc., s'explique sans doute par la nécessité de greffer une intuition formelle sur l'intuition matérielle. Il traduit, en tout cas, un fait épistémologique structurel. Puisqu'ils ne sont pas des noms d'objets, des termes comme point, nombre, espace, etc. ne sont-ils pas, comme le propose Granger [1988], des "formes a priori d'expression", c'est-à-dire des termes dont la permanence s'accomode de significations différenciées en fonction de l'interprétation considérée et du degré de sophistication de la connaissance mathématique? Granger nous paraît tirer là, pour une perspective kantienne de la connaissance, une *conséquence* remarquable de ce fait que la géométrie n'est plus considérée comme une épure de la réalité, ni comme une forme de notre intuition, mais comme la *constitution d'un langage*. Il y aurait dans notre échaffaudage de concepts perpétuellement remaniés par le travail du mathématicien des invariants verbaux prêts à recevoir les mille et une inflexions sémantiques accumulées par ce travail.

De même que la morphologie du langage géométrique comporte des indéfinissables, sa syntaxe comporte des indémontrables, traditionnellement appelés "axiomes". Ceux-ci étaient censés, dans la géométrie conçue comme "science de l'espace", énoncer des vérités évidentes par elles-mêmes. Hilbert les traite comme des hypothèses suffisantes pour la déduction des théorèmes. Ce qui l'intéresse c'est non pas de savoir *si* les axiomes sont *vrais* – la question ne se pose pas – mais *quels* axiomes sont *indispensables* à la démonstration de tel théorème. Hilbert a classé les axiomes de la géométrie en cinq groupes : appartenance, ordre, congruence, parallèlles, continuité. Il montre, par exemple, que le théorème de Pappus exige l'usage des axiomes de congruence si l'on veut se passer des axiomes de continuité, et réciproquement. A la question de la vérité des axiomes, il a substitué celles de leur compatibilité et celle de leur indépendance. Ces deux questions, de nature logique, sont résolues par construction de modèles algébriques aussi bien que géométriques. On observe ainsi dans les *Grundlagen* une solidarité profonde entre les exigences logiques, les problèmes géométriques et les méthodes algébriques. Et sous l'aspect de *modèle*, une construction géométrique n'a ainsi *pas plus* de "réalité" qu'une construction algébrique.

Si la géometrie a préparé longtemps à l'avance (comme l'a montré E. Nagel) la naissance de la logique mathématique moderne, on voit bien dans les *Grundlagen* que c'est l'alliance de celle-ci avec l'algèbre qui a donc fait définitivement basculer la géométrie dans le formel. Lorsqu'il faut vaincre les résistances d'un Frege, qui refuse de voir dans la géométrie un système formel, alors qu'il n'avait lui-même pas hésité à formaliser l'arithmétique (pour la réduire à la logique), Korselt, un élève de Hilbert, invoque précisément l'exemple de l'algèbre dont les lettres désignent des nombres "de manière ambiguë" et dont les lois "ne fournissent que des règles auxquelles se conforment les interprétations qu'on a en vue et doivent se conformer celles qu'on n'a pas en vue"[4]. C'est dire que, du point de vue de la déduction logique, il n'y a pas de différence entre la plus ancienne des sciences symboliques, l'algèbre, et la nouvelle géométrie. Une idée qu'avait eue Leibniz dans ses essais de "caractéristique géométrique", mais qui n'avait pas résisté à la séparation catégoriale dominante entre l'intuitif et le formel et à la division des mathématiques en sciences pures et sciences appliquées. Pour la majorité des mathématiciens du XIXe siècle, même pour des esprits formalistes comme Grassmann[5],

[4] Korselt 1903, p. 407.
[5] Grassmann 1844, 2e édition, 1878, p. XXIII.

même pour quelqu'un comme Klein qui avait promu, à la suite de Riemann, le concept d'espace abstrait et mis en évidence le caractère structural des différentes géométries et la possibilité d'en donner une expression algébrique, *la* géométrie fait partie des sciences mathématiques appliquées. C'est donc seulement avec les *Grundlagen* qu'elle apparaît comme une science pure, "ein reiner Lehrbegriff" comme dit Korselt[6]. C'est alors qu'elle devient la pierre de touche pour un nouveau type de rapport entre l'intuitif, ou ce que l'on prend pour tel, et le formel. Intuitif et formel apparaissent relatifs l'un à l'autre et étroitement liés dans un mouvement d'engendrement réciproque, dans ce que Gonseth appelait, dans son ample commentaire de l'axiomatique hilbertienne, une "dialectique de l'espace"[7]. Ainsi, faire ressortir l'architecture formelle d'une science déductive ne vise pas à comprimer ou refouler l'intuition, mais à l'assouplir, l'amplifier, lui ouvrir un champ de possibles, un champ d'expérimentation dans le formel. La géométrie est à la fois intuitive et formelle, saisie concrète des objets et pensée abstraite de leurs relations logiques, contenu (Inhalt) et système[8]. Comme y insistait Hilbert, la méthode axiomatique est non seulement "logiquement inattaquable", mais aussi mathématiquement "féconde"[9]. Et comme le notera Hermann Weyl, pourtant gagné par les arguments de l'intuitionnisme hollandais, elle est devenue, dans les mains des élèves de Hilbert, "un instrument de la recherche mathématique concrète"[10].

II. L'Algèbre Réelle

Illustrons ce témoignage direct de Hermann Weyl en relevant la part des *Grundlagen* dans la constitution de ce que l'on appelait, dans les années 1930, l'algèbre abstraite. Hans Freudenthal (1957) a observé deux choses. 1°) Le modèle de géométrie non archimédienne construit au §4 du chapitre II conduit au concept abstrait de *corps* dans la mesure où les éléments considérés ne sont pas des nombres mais des fonctions obtenues à partir d'une indéterminée x par combinaison finie de cinq opérations : les quatre opérations rationnelles et l'opération $x \mapsto \sqrt{1+x^2}$. 2°) Hilbert a envisagé au chapitre III le concept abstrait de *corps ordonné* sous l'expression "système complexe de nombres". On peut, en fait, établir une connexion plus précise avec l'algèbre abstraite en montrant le rôle des *Grundlagen* dans l'élaboration par Artin et Schreier de l'axiomatique des corps ordonnables et des corps réels clos[11]. Ce rôle s'est exercé par le biais d'au moins deux aspects.

1°) D'abord, le problème posé dans la première édition des *Grundlagen* (chapitre VII) et reposé au congrès international des mathématiciens de 1900 : une fonction rationnelle f à coefficients rationnels, *définie*, c'est-à-dire qui ne prend de valeur négative pour aucun système de valeurs réelles de ses variables, peut-elle être décomposée en carrés de fonctions rationnelles à coefficients rationnels des mêmes variables? C'est le fameux dix-septième problème de Hilbert qui se présente dans les *Grundlagen* à l'occasion de la question

[6] Korselt 1903, p. 405.

[7] Gonseth 1945-55. Voir mes observations dans Sinaceur 1991 b.

[8] Hilbert/Cohn-Vossen, Vorwort, p.V.

[9] Hilbert 1922, p. 161.

[10] Weyl 1932, p. 349.

[11] Artin-Schreier 1926. La deuxième partie de Sinaceur 1991a est tout entière consacrée à ce mémoire.

suivante : quels points peut-on construire en utilisant, avec la règle, un transporteur de distances appelé "empan" à la place du compas? Ou, en termes algébriques : quels points obtient-on par combinaison finie des cinq opérations de tout à l'heure, les opérations rationnelles correspondant au tracé de droites, l'opération $x \mapsto \sqrt{1+x^2}$ correspondant au report de segments et d'angles?

La solution était largement tributaire des travaux de Hilbert sur les formes algébriques et la théorie des nombres algébriques[12], ainsi que de résultats arithmétiques plus anciens sur les sommes de carrés. Hilbert insiste sur le fait que la théorie des nombres algébriques montre "la raison intime"[13] des rapports analogiques entre arithmétique et algèbre. Aussi est-il naturel de songer à généraliser à un corps quelconque k de nombres algébriques le théorème de Lagrange selon lequel tout nombre entier positif se décompose en somme de quatre carrés d'entiers relatifs[14]. Cette généralisation exige d'envisager des nombres "totalement positifs" : ceux dont les correspondants dans tous les corps conjugués réels[15] de k sont positifs – si aucun des corps conjugués de k n'est réel, tous les nombres de k sont totalement positifs. Mais comme la théorie des nombres algébriques induit une théorie analogue des corps de *fonctions* algébriques, il est encore naturel de chercher à généraliser le résultat aux fonctions polynomiales ou aux fractions rationnelles à coefficients rationnels. Si l'on s'intéresse aux fonctions des cinq opérations sus-indiquées, on retombe sur les constructions à la règle et à l'empan, dont Hilbert a montré qu'elles sont permises par l'ensemble des axiomes d'incidence, d'ordre, de congruence et des parallèles et indépendantes des axiomes de continuité[16].

C'est pour résoudre le dix-septième problème de Hilbert que fut imaginée la théorie des corps réels. Ceux-ci sont définis comme les corps où une somme $\sum_1^n x_i^2$ de carrés d'éléments est nulle seulement si chacun des éléments x_i est nul. Artin et Schreier montrent que ces corps sont exactement les corps ordonnables, c'est-à-dire ceux qui peuvent être ordonnés par une relation d'ordre total compatible avec les opérations d'addition et de multiplication. On dit plus volontiers aujourdhui que ce sont les corps qui possèdent un "cône positif", c'est-à-dire une partie P stable pour l'addition et la multiplication et telle que, pour tout élément x non nul du corps, $x \in P$ ou $-x \in P$, un ordre étant univoquement déterminé par un cône positif. Le lien établi entre l'existence d'un ordre sur un corps K et les sommes de carrés d'éléments de ce corps permet d'identifier celles-ci avec les éléments totalement positifs de K [17], ces derniers étant définis comme éléments positifs pour toute relation d'ordre compatible avec la structure de corps, c'est-à-dire comme éléments appartenant à tous les cônes positifs de K. L'analogie est patente entre corps ordonnables (ou réels) et corps de nombres, et le dix-septième problème a une solution positive pour les fonctions définies à coefficients dans un corps réel ordonnable d'une manière unique comme c'est le cas des corps de nombres rationnels, réels algébriques

[12] Hilbert 1888, 1893, 1897.

[13] Hilbert 1897, préface.

[14] Hilbert 1899, chapitre VII, §2, dans Hilbert 1971, p. 167-168.

[15] Rappelons la définition de corps conjugué d'un corps de nombres algébriques k. Soit t un élément primitif de k. L'équation algébrique de plus bas degré m, à coefficients rationnels, satisfaite par t possède $m-1$ autres racines. k' est conjugué de k s'il est engendré par l'une de ces m-1 racines. Un corps conjugué est réel s'il ne contient que des nombres réels.

[16] Hilbert 1971, théorème 63, p. 160.

[17] Artin 1927, proposition 1.

et réels[18]. A vrai dire, dans la démonstration des deux lemmes de spécialisation dont Artin a besoin préalablement, on remarque qu'il prend soin de se placer dans un corps contenant celui des fractions rationnelles $\mathbb{R}(X_1, X_2, \cdots, X_n)$ et conservant les propriétés algébriques de $\mathbb{R}$. Il s'agit d'un corps *réel clos*, c'est-à-dire d'un corps réel admettant un ordre unique et tel que tout élément positif a une racine carrée et que tout polynôme de degré impair a une racine[19]. On s'apercevra ensuite que le dix-septième problème a une solution affirmative pour les fonctions définies à coefficients dans un corps réel clos quelconque et que la démonstration peut être considérablement simplifiée en utilisant des résultats de théorie des modèles, plus précisément la propriété de modèle-complétude de la théorie élémentaire des corps réels clos[20].

2°) A joué également un rôle efficient dans l'élaboration de la théorie d'Artin et Schreier la conjonction des deux circonstances suivantes : la construction d'un modèle de géométrie indépendante des deux axiomes de continuité et le choix pour les nombres réels d'une axiomatique *parallèle* à celle de la géométrie linéaire, avec les *mêmes* axiomes d'ordre et de continuité [21]. Comme pour la géométrie linéaire ou plane, Hilbert sépare nettement, pour l'arithmétique des nombres réels, les axiomes de continuité de tous les autres axiomes qu'il subdivise en deux groupes, ceux qui énoncent les propriétés de l'addition et de la multiplication et ceux qui établissent la compatibilité de la relation d'ordre avec ces deux opérations. En construisant un modèle où certaines constructions géométriques – celles que l'on peut faire grâce à la règle et à l'empan ou grâce aux cinq opérations ci-dessus mentionnées – ne nécessitent pas l'usage des axiomes de continuité, Hilbert induit une question analogue pour le corps ordonné des nombres réels : quels problèmes posés en termes de nombres réels peuvent être résolus indépendamment de la continuité, c'est-à-dire indépendamment de la topologie de $\mathbb{R}$? Quels problèmes peuvent être résolus par des méthodes purement algébriques? Plus précisement : l'ordre du corps des nombres réels ou d'un corps formellement analogue, c'est-à-dire d'un corps réel clos, est-il séparable d'une structure topologique qui fonde, elle, l'idée de continuité?

Dedekind [1872] avait montré, il est vrai, que la structure d'*ensemble* ordonné de $\mathbb{R}$ peut être définie indépendamment du concept de limite et ouvert ainsi la voie à l'idée que la topologie de $\mathbb{R}$ n'intervient pas forcément dans tous les problèmes impliquant les nombres réels. Mais la théorie algébrique de $\mathbb{R}$ comme *corps* ordonné est l'œuvre d'Artin et Schreier. Ceux-ci montrent qu'un corps réel clos ne peut être ordonné que d'une unique façon et que cette unique relation d'ordre est définissable à partir des opérations d'addition et de multiplication et de la relation d'égalité, c'est-à-dire définissable à partir de la structure algébrique de corps. En conséquence, tous les théorèmes de l'analyse

[18] Artin 1927, proposition 4.

[19] Il est équivalent de dire qu'un corps réel clos est un corps réel dont aucune extension algébrique propre n'est réelle. Le corps des nombres réels et celui des réels algébriques sont réels clos, mais, bien entendu, le corps des nombres rationnels ne l'est pas.

[20] Robinson 1955. On dit qu'une théorie est modèle-complète si toute inclusion de modèles de cette théorie est une inclusion élémentaire, c'est-à-dire étant donnés deux modèles quelconques M et M' de ladite thorie tels que $M \subset M'$, M et M' vérifient les mêmes énoncés élémentaires (formulables dans un langage logique du premier ordre).

[21] Hilbert pose pour axiomes de continuité l'axiome d'Archimède et un axiome dit d'inextensibilité imposant aux modèles qui vérifient l'ensemble des autres axiomes du continu arithmétique ou géométrique d'être maximaux ; il montre que ces deux axiomes sont indépendants l'un de l'autre.

réelle, dont l'énoncé ou la démonstration impliquent la relation d'ordre $\geq$ et s'appuient sur l'hypothèse de continuité des fonctions réelles de variables réelles considérées, peuvent être formulés et démontrés de manière purement algébrique. C'est le cas du théorème de Bolzano que toute fonction continue sur un intervalle réel et qui prend, pour deux valeurs a et b de l'intervalle, des valeurs de signe opposé, s'annule forcément entre a et b, du théorème de Rolle qu'entre deux zéros d'une fonction continue sur un intervalle réel il existe au moins un zéro de la fonction dérivée, du théorème de Sturm qui permet de déterminer exactement le nombre de racines réelles d'une fonction polynomiale (donc continue), etc. Tous ces théorèmes peuvent être formulés pour des polynômes et des fractions rationnelles à coefficients réels ou dans un corps réel clos quelconque et démontrés sans utiliser l'hypothèse que polynômes ou fractions sont des fonctions continues. Tous ces théorèmes, d'abord connus dans le cas de $\mathbb{R}$, sont en réalité valables pour un corps réel clos quelconque. C'est pourquoi Artin et Schreier les désignent comme "théorèmes de l'algèbre réelle", délimitant du même coup les frontières d'une nouvelle discipline, l'algèbre réelle, dénomination certes plus imagée que celle, équivalente, de théorie des corps réels clos. Pour le dire en un mot, l' algèbre réelle est cette partie de l'analyse réelle qui peut être traitée sans faire intervenir la topologie ni la métrique de $\mathbb{R}$.

Cette nouvelle discipline, qui n'eût sans doute jamais vu le jour sans l'immense rayonnement de l'enseignement de Hilbert et la fortune des *Grundlagen der Geometrie*, apparaît aujourd'hui comme l'ancêtre fondateur d'au moins quatre branches mathématiques. Elle constitue, en effet, le premier exemple d'étude de structures algébriques ordonnées, celles-ci tendant à affirmer de plus en plus l'autonomie de leur domaine propre; elle est à l'origine des récents développements (depuis la fin des années 1960) de la théorie algébrique abstraite des formes quadratiques à coefficients sur un corps quelconque[22]; elle a fortement influencé l'épanouissement, dans les années 1950, de la théorie des modèles[23]; elle a contribué enfin à la naissance de la toute nouvelle géométrie algébrique réelle ou géométrie algébrique développée sur un corps de base supposé réel clos.

Bibliographie

Artin E., Schreier O. 1926 : Algebraische Konstruktion reeller Körper, *Abh. math. Sem. Hamb.* 5, 85-99. Dans Artin 1965, 258-272.

Artin E. 1927 : Über die Zerlegung definiter Funktionen in Quadrate, *Abh. math. Sem. Hamb.* 5, 100-115. Dans Artin 1965, 273-288.

Artin E. 1965 : *The Collected Papers* (éd. S. Lang et J. Tate), Addison, Wesley Pub. C°. Rééd. 1982, Springer-Verlag.

Dedekind R. : Stetigkeit und irrationale Zahlen, *Gesam. math. Werke*, III, 315-334. Trad. frç. J. Milner et H. Sinaceur, Paris, 31 rue de Navarin, 1978.

Freudenthal H. 1957 : Zur Geschischte der Grundlagen der Geometrie, *Nieuw Archief voor Wiskunde* (4), V, 105-142, P. Noordhoff N.V., Groningen.

Gonseth F. 1945-55 : *La géométrie et le problème de l'espace*, 6 volumes, Neuchâtel, Editions du Griffon.

[22]On trouvera dans Sinaceur 1991a, troisième partie, une présentation synthétique de certains de ces développements.

[23]Un aperçu en est donné dans Sinaceur 1991a, quatrième partie.

Granger G.-G. 1988 : Sur l'idée de concept mathématique naturel, *Revue internationale de philosophie*, vol. 42, n° 167, 474-499.

Grassmann H. 1844 : *Ausdehnungslehre*, Leipzig, dans *Gesam. Werke*, I, Leipzig 1894.

Hilbert D. 1888 : Über die Darstellung definierter Formen als Summe von Formenquadraten, *Math. Ann.* 32, 342-350. Dans Hilbert 1932-1935, II, 154-161.

Hilbert D. 1893 : Über ternäre definite Formen, *Acta Math. 17*, 169- 197. Dans Hilbert 1932-1935, II, 345-366.

Hilbert D. 1897 : Theorie der algebraischen Zahlkörper, *Jahresbericht D. M. V. 4*, 175-546. Dans Hilbert 1932-1935, I, 63-361. Trad. frç., Paris, Hermann, 1913.

Hilbert D. 1899 : *Grundlagen der Geometrie*, Teubner. 7e édition, 1930; 10e édition avec des suppléments de P. Bernays, 1968.

Hilbert D. 1905 : Über die Grundlagen der Logik und der Arithmetik, *Verhandlungen des dritten internationalen Mathematiker-Kongresses*, Heidelberg 1904, Leipzig, Teubner, 174-185.

Hilbert D. 1922 : Neubegründung der Mathematik, *Abh. math. Sem. Hamb. 1*, 157-177. Dans Hilbert 1932-1935, III, 157-177.

Hilbert D. 1932-35 : *Gesammelte Abhandlungen I, II, III.* Berlin. Reprint, New York, Chelsea, 1965.

Hilbert D. 1971 : *Les fondements de la géométrie*, édition critique de P. Rossier, Paris, Dunod.

Hilbert D. et Cohn-Vossen S. 1932 : *Anschauliche Geometrie*, Berlin, Springer.

Korselt A. 1903 : Über die Grundlagen der Geometrie, *Jahresbericht D. M. V. 12*, 402-407.

Nagel E. 1939 : The formation of modern conceptions of formal logic in the development of geometry, *Osiris*, 7, 142-222.

Pasch M. 1882 : *Vorlesungen über neuere Geometrie*, Berlin, Springer. 2. Auflage, mit einem Anhang : Die Grundelegung der Geometrie in historischer Entwicklung von Max Dehn, 1926.

Robinson A. 1955 : On ordered fields and definite functions, *Math. Ann. 130*, 257-271. Further remarks on ordered fields and definite functions, *Math. Ann.* 130, 405-409. Dans *Selected Papers* (éd. H.J. Keisler, S. Körner,W.A.J. Luxemburg, A.D. Young), North-Holland, I, 355-374.

Sinaceur H. 1984 : De D. Hilbert à E. Artin: les différents aspects du dix septième problème et les filiations conceptuelles de la théorie des corps réels clos, *Arch. hist. ex. sc. 29*, n° 3, 267-287.

Sinaceur H. 1991a : *Corps et modèles, Essai sur l'histoire de l'algèbre réelle*, Paris, Vrin, collection Mathesis.

Sinaceur H. 1991b : La dialectique de l'espace, à paraître aux Editions Masson dans les *Actes* du colloque *La philosophie ouverte de F. Gonseth l'épreuve de la géométrie*, Genève, 14-15 Décembre 1990.

Weyl H. 1932 : Topologie und abstrakte Algebra als zwei Wege des mathematischen Verständnisses, *Unterrichtsblätter für Math. und Naturwiss. 38*, 117-188. Dans *Gesam. Abh.* (Hrsg. von K. Chandrasekharan), Springer-Verlag, III, 348-358.

Le Principe de Dualité : sa Signification Historique et Epistémologique

Ludovico Geymonat

Dipartimento di Filosofia, Università degli Studi di Milano, Milan, Italie ; et via Visconti d'Aragonna 25, 20133 Milan, Italie

Tous les mathématiciens savent que l'expression "principe ou lois de dualité" a été introduite en géométrie par le mathématicien-philosophe Gergonne dans trois articles publiés en 1824-25, 25-26, 26-27 dans la revue *Annales de Mathématiques pures et appliquées* que Gergonne lui-même avait fondée et qu'il dirigea jusqu'en 1834 (en réalité, il la dirigea d'une façon très autoritaire, jusqu'à modifier certains articles qu'on lui avait envoyés, sans l'autorisation des auteurs, ou, pis encore, jusqu'à ne pas publier certains articles et, profitant de cette non-publication, utiliser personnellement les résultats qu'y étaient démontrés.

Le principe de Gergonne (dualité entre point et droite sur le plan, ou entre point et plan dans l'espace) suscita tout suite un grand nombre de discussions. Avant-tout, une polémique avec Poncelet, à propos de la priorité de la découverte de ce principe car Poncelet avait parlé de principe de réciprocité dans le même sens que celui de principe de dualité. Une autre question qui a été débattue concerne la justification de ce principe: doit-il être justifié par des considérations analytiques ou par des considérations proprement géométriques? En fait, on avait déjà noté ce principe à propos de la polarité par rapport à une courbe de second degré, ou d'une surface de ce même degré, et Gergonne lui-même avait fait certaines applications particulières du principe en question. Mais, à l'époque dont nous avons parlé, il y avait une remarquable différence entre la façon de présenter le principe de dualité de Gergonne et de Poncelet. Ce dernier le justifie sur la base de considérations analytiques de portée limitée ; au contraire, Gergonne a cru pouvoir le justifier sur la base de réflexions philosophiques qui prétendaient en démontrer la validité naturelle, générale et absolue. Ces réflexions étaient profondes, sans doute, mais confuses et par conséquent pas utilisables en géométrie.

Après quelques années, le principe fut repris et reélaboré par deux mathématiciens allemands, Möbius et Plücker. Möbius essaya de le démontrer par des considérations analytiques qui utilisent pour la première fois un système de coordonnées homogènes. Par ce moyen, il arrive d'une part, à un résultat qui a un caractére de portée bien plus générale que les résultats obtenus par Poncelet, d'autre part, à un raisonnement bien plus rigoureux que celui de Gergonne. Plücker est le mathématicien auquel on attribue, d'habitude, la première démonstration vraiment rigoureuse du principe de dualité. Dans son mémoire ayant pour titre *Über ein neues Coordinatensystem* (1829) il reconnait que le mérite d'avoir inventé le principe de dualité va à Gergonne et à Poncelet, et pourtant, dans d'autres articles de peu postérieurs, il ouvre une polémique très explicite contre ces deux auteurs, en affirmant que c'est lui, Plücker, qui est arrivé à ce principe par une voie totalement autonome, indépendante de leurs travaux et pour tout dire avant leurs publications.

En rappelant toutes ces discussions, nous pouvons observer que la découverte du principe de dualité marque un point très important dans l'histoire de la géométrie, parceque c'est à partir de là que date le rapide développement de la géométrie projective au XIX siècle.

En effet cette géométrie a traversé plusieurs phases et nous voulons ici en rappeler deux. La première phase est répresentée par les mathématiciens qui oscillaient entre deux thèses : la thèse selon laquelle la géométrie projective doit se fonder sur la géométrie analatyque (pas seulement sur celle de Descartes, mais aussi sur d'autres, par exemple sur celle de Möbius ou celle de Plücker) et la thèse selon laquelle la géométrie projective doit se fonder sur des méthodes purement géométriques. La seconde phase est répresentée par les mathématiciens qui, au contraire, exigent de fonder la géométrie projective sur des bases purement géométriques, sans l'aide de considérations analytiques.

Cette seconde phase, que l'on appelle la "phase puriste" est liée surtout au nom du mathématicien allemand Cristian von Staudt, qui publia en 1847 une oeuvre fameuse, la *Geometrie der Lage*, à laquelle il a fait suivre, une dizaine d'années plus tard, des ajouts eux aussi très importants. Pour caractériser l'oeuvre de Staudt, je veux reporter ici l'introduction de la *Geometrie der Lage* :

> "Ces derniers temps on a distingué, bien à raison, la géométrie de position de la géométrie métrique, et toutefois, l'on démontre ordinairement par des considérations métriques aussi des propositions dans lesquelles il n'est question d'aucune grandeur. Dans l'oeuvre présente j'ai cherché à faire de la géométrie de position une science indépendante pour laquelle le concept de mesure ne soit pas nécessaire."

Sur quels fondements sera donc baseé la géométrie de Staudt? Avant tout sur certains concepts qui étaient déjà présents dans la géométrie d'Euclide, c'est-à-dire sur le concept de point, de droite, de plan, etc ; deuxièment sur certaines relations entre les entités indiquées par ces concepts. Ces relations peuvent être divisées en deux catégories: relations d'appartenence (par exemple entre points et droites, entre droites et plans, etc.) et relations de correspondance (par exemple entre les demi-droites qui partent d'un point et sont sur un plan, et les points dans lequels ces demi-droites rencontrent une autre droite qui appartient au même plan et qui ne passe pas par l'origine des demi-droites).

En ce qui concerne le principe de dualité, von Staudt comprend avec clarté que ce principe dérive de la symétrie formelle avec laquelle se présentent les concepts de point et de droite, ou de point et de plan dans les relations dont nous avons parlé.

Certains spécialistes de la pensée de von Staudt affirment qu'il n'a pas fait une rigoureuse axiomatisation de la géométrie projective ; d'autres, au contraire, pensent qu'il a fait cette axiomatisation, et je suis de cette opinion. Les uns et les autres sont quand même d'accord que la *Geometrie der Lage* de Staudt a donné une réelle contribution pour faire passer la géométrie projective d'une phase aproximative à une phase rigoureusement axiomatisée. C'est ainsi qu'après Staudt presque tous les géomètres ont compris que leur science doit être exposée en forme axiomatique et que par conséquent les questions qui se réferent à la dualité doivent être exposées déjà dans les premiers passages de la géométrie.

Si l'on compare la géométrie des points et des droites sur le plan avec la géométrie duale des droites et des points (ou par analogie celle des points et des plans dans l'espace avec celle duale des plans et des points) on peut les considérer comme deux interprétations différentes du même modèle et, par conséquent, les théorèmes qui en dérivent et les théorèmes duaux comme deux interprétations des mêmes théorèmes démontrés dans le modèle. Tout ceci prouve que le principe de dualitè conduit à la théorie des modèles, qui est un des arguments fondamentaux non seulement de la science mathématique mais de toute la science contemporaine.

Finalement, en 1896-97 sortent les *Leçons de géométrie projective* du mathématicien-philosophe Federigo Enriques ; il s'agit des *Leçons* qu'Enriques avait tenues, cette année-là, à l'Université de Bologne. Tout le monde considére ces *Leçons* comme le premier ouvrage de géométrie projective vraiment moderne.

Selon Enriques la géométrie est la science du concept d'espace, concept qui naît de l'ordre des choses du monde extérieur. Cette science se base sur un système d'axiomes, ou propositions fondamentales, que l'on admet de façon arbitraire, à la seule condition qu'ils ne soient pas contradictoires l'un avec l'autre. Toujours selon notre auteur, le principe de dualité dérive de la symétrie avec laquelle figurent les mots point et droite dans le plan ; et par conséquent, si nous changeons arbitrairement les anciens axiomes, mais que nous conservons la même symétrie entre les nouveaux, le principe de dualité continuera à être valide dans la science qui se base sur ces axiomes modifiés. Après Enriques, Gino Fano, qui a été mon professeur de géométrie supérieure à l'Université de Turin en 1930-31, s'est occupé lui-mème du principe de dualité, en particulier dans l'espace .

Dignes d'étre remarqués sont aussi les travaux des mathématiciens américains Veblen et Young (1910) qui, les premiers, considérent ouvertement les axiomes comme définitions implicites des concepts géométriques de point, droite, etc. Ils généralisent le principe de dualité en l'appliquant à l'espace à n dimensions. Aujourd'hui nous savons qu'un principe analogue vaut aussi pour l'algèbre de Boole et pour d'autres systèmes algébriques.

En conclusion, le développement de la géométrie projective a démontré que le principe de dualité n'a rien de mystérieux, rien de philosophique comme pensait Gergonne, mais se base sur une propriété formelle des axiomes. Il s'agit d'un résultat très important, obtenu par l'axiomatique et la logique.

The Formal and the Transcendental in Mathematics

Gilles Gaston Granger

Collège de France, 11, place Marcelin Berthelot, 75231 Paris, cedex 05

The classical philosophical problem we propose addressing here remains one of the most general and one of the most significant ones that the mathematician impressed by the extraordinary perenniality of his science and the prodigious success of its applications to the world of phenomena can still raise today. It is a question that clearly concerns geometries especially, with their apparently priviledged relations to the sensible world, but also concerns mathematics in general as well. Just what is the nature and scope of the *objectivity* of mathematics? Objectivity is naturally to be taken here in an altogether radical sense of validity as knowledge - or, perhaps, in the sense of the creation of objects. It is from this point of view that we use the originally kantian dichotomy here between the two terms "logical" and "transcendental". We'll begin then by justifying the sense of the word transcendental as used here. After that we'll advance a certain idea of the status of logic and of its role in mathematics and finally, beginning with points of view already sketched, we shall end with the examination of a conception of mathematical objects and of their relation to other types of objecthood.

What does "Transcendental" mean here?

In his *Critique of judgment* Kant writes : "A transcendental principle is a principle by which is represented the *a priori* universal condition under which alone things (Dinge) can become in a general manner objects of our knowledge." (K. U. Einleitung, V, xxix).

We begin with the kantian notion, though without the pretension of faithfulness to it that would be demanded of an historian. While preserving what we believe is essential to it, we shall nevertheless redefine it in the light of later scientific developments.

The transcendental function of a concept will thus come to be defined as establishing *the conditions of the possibility of considering as objects* the entities to which it refers. Two remarks should be made in this connection.

1) Kant provided a safeguard against the thought of illusory objects : it lies in the necessity of connecting concepts to the sensible, without which he recognized no possibility of *experience*. Mathematical objects themselves are thus constructed as a development of the spatial and temporal *a priori* forms of *sensible intuition*.

2) The kantian transcendental conditions are clearly distinguished from conditions that would depend on the empirical, psychophysiological nature of our minds. They

are, nevertheless, ascribed to a *subjectivity*, in some sense anonymous, and rightly called transcendental.

We shall not dispute the essential character of these two aspects within kantian philosophy ; but we think that the nature of scientific thought as we have it today fails to confirm them. To guarantee against the formation of pseudo concepts and the consideration of pseudo objects, we therefore replace the restriction to the sensible and the idea of subjectivity by the thesis of a *fundamental and inalienable correlation between the "objectal" and the operational.* To the kantian formula :

> "thoughts without content are empty ; intuition without concepts are blind" (K. R. V. 75),

we shall substitute : "the thought of operations without objects is sterile ; the thought of an object without a system of operations is opaque." The position of an entity as object at once requires the functioning of a *system of operations*, whether its rules be explicit or only implicit. We should like to show then that the transcendental function of mathematical entities depends on the essential character and particular clarity of the object/operation correlation they present. One last remark, however, as regards Kant. In the *Critique of Pure Reason*, transcendental is opposed to "metaphysical" when the principle

> "represents the *a priori* conditions on which only objects whose concept must be empirically given can be further determined *a priori*" (*Ibid.*)

and is opposed to "formal" when, in speaking of logic, he says that "general and pure" logic is

> "a canon of understanding and reason, but only with respect to what is formal in their use, whether the content be empirical or transcendental" (K. R. V. A. 53).

It is this second dichotomy that interests us here. Though Kant still call "logical" a knowledge of the rules of the pure thought of an object not excluding all, but only empirical, content, we retain rather of the logical *stricto sensu* its radically formal character that distinguishes it from a transcendental discipline.

The Formal and the Logical.

With the notion of logic delimited in this way, we can bring out by contrast with the purely formal aspect just what the nature of a content proper to mathematics is. But first of all we should underscore the relativity of the form/content dichotomy. What is form at one level may become content at a higher level of organization so that characterization as formal makes sense only if a content is contraposed and, in some sense, there are degrees of the formal. It would be appropriate here to evoke the notion of "theme" introduced by J. Cavaillès and which has been commented on in a preceeding lecture. Doesn't the "transversal" - or "vertical" passage effected in "thematization" consist precisely in rendering explicite as a new type of object - and as new content - that which at first appeared to be purely operational? The "paradigm", in emptying the statements of their content, now mobilized as "variables" (what Cavaillès calls "the moment of the variable"), brought to light a form operating on that content ; the "theme" then takes that form as point of application of a new operational system : the additions and multiplications of arithmetic become more general entities subject to the laws of higher degree of "universal" algebra ; the operation of integration becomes the new entity : "linear functional", the object of a new analysis. The real dichotomy is that of the pair operation/object. A higher operational level determines as new object that which was operational in act.

The logical aspect *stricto sensu* of a thought will be defined as the extreme, absolute term of that dichotomy : when the content characterizing the object-moment is reduced to the operational system itself, zero degree content. The logical "object" is nothing then but the shadow of an object, a blank, a fulcrum of operation. This is what happens with the "propositional calculus". Wittgenstein's assertion in the *Tractatus* that there are no logical constants may be understood from this angle ; constants - that is to say, logical operators - and propositional objects are but two faces of a same thought. Likewise, in the formulation of this calculus as a "Lindenbaum algebra", the object is explicitly reduced to two states (presence, absence ; true, false ; supposed, not-supposed) whose reality is reduced to the translation of an operational efficacy.

As a matter of fact, two equivalent interpretations of the calculus already glimpsed by Leibniz are also legitimate and correspond to highlighting either the objectal or the operational aspect : the calculus of *classes* (though without the class-element,and finite-infinite distinctions) and the calculus of propositions. On this last interpretation the calculus is seen as regulating in an elementary way the *acts* of all thought of objects, independently of any specification. This logic *stricto sensu* is at once theory of the form of the object *überhaupt*, and theory of entailment of elementary statements functioning now as the fundamental *metatheory* for every other formal level of the manipulation of objects of thought.

Its fundamental character can be seen, moreover, in the particular formal properties resulting from the perfect adequation of the operational system and the object to which it applies. We can speak in this case of a strict *duality*, that is to say, of a reciprocity of determination corresponding, in a way, to a simple reversal of points of view. What the metatheorical privilege of the calculus translates (non-contradiction, completeness, decidability) loses its universal guarantee as the object of the calculus becomes richer and takes on some consistency, no matter how little. With the object "first order predicate" decidability is already lost...

The Mathematical Object.

Just what is the status of mathematics relatively to this pure logical element? It is true that the concepts employed in mathematics are such as to allow of a complete account of its deductive steps regulated by logic ; but mathematics has *specified objects*. The problem we are dealing with is not primarily that of the demarcation of logic and mathematics ; but it does seen both meaningful and important to recognize the way in which mathematics constitues this objecthood that logic is manifestely incapable of. To this end, let us try to specify some of the traits of the mathematical object.

1. In the first place, the mathematical object is characterized by the appearance of "formal content", not found in logic. This appearance of "content" whose origine is not to be found among the *givens* of the sensible has as consequence the loss of the perfect object/operation adequation : loss of universal decidability by a finite algorithm as soon as the object takes on a specification at the level of the predicate calculus ; the more decisive loss of completeness and of the possibility of establishing non-contradiction, once number makes its appearance in elementary arithmetic. "Content" here means properties of the object somehow escaping the confines of the demonstrative operational system, even though the object in question was introduced as a correlate of that system. Semantics, one could say, comes into its own here, breaking away from syntax.

An especially clear example of this trait is to be seen in the dialectic of the *extension of operational fields* which is one of the modalities of mathematical progress. It turns out not to be possible to fully exploit the potentialities of an operational system within the field of objects constituted by the system ; operational integrity is then restored through an extension of the system of objects ; generally speaking, a proper part of the new system will then be an isomorphic image of the former one. Examples of this are numerous, from the invention of complex numbers down to that of distributions.

2. From the point of view just mentioned, mathematics would begin with the first order predicate calculus and would take on its full sense with elementary arithmetic subject to Gödel's theorems. We must, however, underscore another aspect of the properly mathematical object, doubtlessly closely associated with the first though distinct from it. This is the consideration of the *infinite as object* as opposed to the simply *virtual* infinite which is only the possibility of *indefinite* operational iteration. This latter is already present in the purely logical calculus of propositions as a freedom to operate, in no way implying the introduction of an object or a procedure explicitly posited as *infinite*. That mathematics "begins with the infinite" was the definition of Cavaillès. It will be remarked that it is precisely this introduction of the infinite that gives rise to the formulation of the intuitionist requirements and gives significance to the difficulties raised by "non-predicative" definitions. (When an exhaustive enumeration of the elements of a set is possible there is no real problem, and for one thing, the intuitionist logic will here be no different from the classical one). Let us add in passing, without developing the point, that one of the central expressions of the introduction of the infinite is the axiom of choice

If it is thematization of the infinite that must be recognized as marking the properly mathematical moment of objective thought, it is doubtlessly because it is with this acquisition that we definitely accept the impossibility of having complete control over the objects, and is the proliferation of formal contents. From this point of view, intuitionist mathematics does not consist in a total rejection of that acquisition, but expresses a particularly acute awareness of its conditions and reinforces the requirements, of kantian inspiration, concerning the validation of the thought of objects. In the same way, the metamathematical debates about the degrees of constructivity of concepts correspond to the concern more or less vividly felt by mathematicians (and especially by logicians) for limiting the liberty one takes in "ideally" manipulating objects that lie beyond the purview of effective operations.

3. The mathematical imagination thus creates systems of objects submitting only, on the one hand, to logic *stricto sensu* which regulates its tactical steps, and, on the other, to the necessity of a possible explicitation of the operational system correlative to said objects. We have used the word "creation" ; and while it is admittedly a conditioned creation, its amazing fruitfulness cannot fail to suggest that it is at least in some manner a result of art. But aren't there, at the same time, some mathematical objects that might be said to be "natural"?

We don't mean here notions that are first as to their psychogenesis and cultural history, although this empirical naturalness is certainly that most often associated with an intrinsically natural character. We shall call "natural" those mathematical objects that by themselves have a privileged pregnancy we think related to two traits which at first blush could seem incompatible.

a) - an essential "incompleteness" that leaves their exploration indefinitely open and makes it so that the knowledge of all their aspects is never exhausted by the capture of some of them in the axiomatics (fixing the operational system).

b) - a sort of "accomplishment", connected to the complexity of the abstract structures that can be dissociated from them, and that gives them, in the domain of mathematical beings, the character of relatively concrete entities. They are themselves their own models.

If our science were more advanced and certain in these areas, this naturalness could undoubtedly be tied to the biological conditions of the relation of an organism to its environment. We shall not deal with this question that belongs rather to biopsychology and psychogenetics. But from another point of view, the "natural" concepts are not the simple reflection of an empirical reality from which they would have been abstracted, nor are they the expression of a *a priori* conditions of the thought of objects of *sensible* experience. The notion of spatiality, for instance, is an example of a "natural" mathematical concept ; but in this sense a Lobachevskian space is just as natural as Euclidean one, for it constitutes just as rich and complex a variant of a same constitutive notion of any representation of a world of objects. The idea of real number, seen as idea of the "numerical line", is as much a natural concept as that of whole number is. In our sense, the naturalness of mathematical objects is not assimilable to their immediacy or convenience nor to our spontaneity in applying them to experience. What contrasts with natural objects are not deviant or overdetermined ones but those that are constructed essentially by *reduction*, *abstraction*, and *generalization* : the notion of group, the notion of vectorial "space", the notion of Category, for instance...

To interpret this distinction, it must be recognized that the transcendental function of mathematics has two orientations : one toward the constitution, in rigorous symbolism, of many-facetted and in some sense self-sufficient objects which are its "natural" objects ; the other, towards bringing out, through surgical dissociation, the "elementary" structures susceptible of appearing to be abstract aspects of the former. We have here two modalities of mathematical "existence", or, more precisely, of showing, or even of discovering, mathematical existence. Of course, once explicitly brought to light, both natural and reduced concepts are on an exactly equal footing, and the reduced concepts, even though appearing more fundamental or general in the architectonic of structures, have no genuine priority : "natural" notions such as that of "space" or whole number necessarily reappear as "naive", meta-operational instruments in the construction of the more abstract objects.

4. If our conception of the mathematical is admitted, there still remains the problem of understanding how the autonomous transcendental function of mathematical thought is linked with the possibility of its *application* to the empirical world. There was talk of the "miracle" of adequation of mathematical objects and natural phenomena, and it was found to be, *a priori*, "unreasonable" ... There were two classical solutions advanced that can be schematized as follows. A "Platonic" solution (the adjective certainly does not do justice to Plato here) : the *mathemata* are essences of all things and it is through them that we can attain to the reality of things. An empiricist solution : the *mathemata* are abstract forms derived from experience. But the development of mathematics leads to the rejection of both these theses.

We shall say rather that mathematics aims at constructing the totality of forms of possible objects, and not only of objects constructed in sensible intuition (as is witnessed by the importance of "teratological" beings). We should perhaps state more precisely : the totality of forms of objects *constructible in symbolic intuition.*

All in all, this would be the Leibnizian thesis on condition that the possibles not be restricted to logic *stricto sensu* (which for Leibniz was subject to the sole principle of identity) and that the set of principles corresponding to the Leibnizian "architectonic" principles be incorporated into mathematics itself : properly mathematical production creates objects that are admittedly imaginary, but that have *formal content* not derivable through pure logic. "Free" mathematics, to use Cantor's expression, gets its fecundity from the recognition and exploitation for the constraints engendered by the emergence of such content. It is true that some from among the multitude of forms of objects mathematics engenders have been sought explicitly with a view to modelling phenomena, and it is in this sense that the empirical sciences pose problems and suggest solutions for the mathematician. And some, though the fruit of an abstract exploration undertaken by the purely creative imagination, turn out unexpectedly to be useful for the physicist long after their elaboration. But the most by far come to be and remain without external application, simply as the *a priori* forms of possible objects, rich in content that owes nothing to the experience of our senses. That is why the transcendental function of mathematics is truly creative, indefinitely creative ; and that is also why it incessantly confronts the philosopher, in the starkest way, with the enigma of the relation of representation to reality.

Un Panorama des Mathématiques

René Thom

Institut des Hautes Études Scientifiques, Le Bois-Marie, 35, route de Chartres
91440 Bures-sur-Yvette

A l'origine de toute grande discipline scientifique, on peut déceler ce que j'ai proposé d'appeler une *aporie fondatrice.* On entend par là une opposition fondamentale, une contradiction de base, qui, au cours des temps, reçoit des solutions "'fantasmatiques" ; toute solution de ce type permet un certain développement de la discipline ; ces solutions sont donc, lors de la phase d'extension associée, considérées comme ayant une validité définitive ; mais, au bout d'une certaine durée, le développement même de la discipline, l'ampleur croissant des résultats aboutissent à remettre en question la solution initiale, dont on reconnait alors le caractère arbitraire et illusoire. Il n'est pas difficile d'isoler ce qui, en Mathématique, joue ce rôle : c'est l'opposition Continu-Discret.

La solution fantasmatique actuellement en vigueur est celle qui consiste à engendrer le continu à partir du discret. Nous faisons ici allusion, bien entendu, à la définition des nombres réels à partir des rationnels par les deux procédés concurrents des coupures de Dedekind et de la complétion des suites de Cauchy (procédés qui sont ici équivalents, bien qu'ils cessent de l'être dans un cadre "catégoriel" plus général). Si on traite ces solutions de "fantasmatiques", c'est, évidemment en raison de leur caractère hautement non-constructif, qui les a fait rejeter des intuitionnistes. Mais il faut reconnaître que pour les besoins de la mathématique – besoins à la fois de la pure mathématique et des sciences appliquées, ces solutions n'ont pas, jusqu'à présent, donné naissance à des difficultés majeures ; dans l'optique actuelle, on peut donc considérer le continu géométrique de dimension un, la droite réelle, comme engendrée par l'un ou l'autre de ces procédés. Les continus pathologiques (droite transfinie) élaborés comme contre-exemples au problème de Souslin n'ont jusqu'ici trouvé aucun répondant "naturel". Jusqu'à plus ample informé, on peut donc considérer la construction classique des réels comme pertinente. Mais il ne faut pas se dissimuler que si l'on abandonne l'aspect proprement technique de la mathématique pour une vue plus philosophique, on ne peut que donner au continu, par ses capacités de réalisation intuitive immédiate, une assise ontologique infiniment plus solide que les constructions itératives infinies de la théorie des Ensembles et de l'Algèbre

La seule difficulté du continu, c'est que, à lui seul, on n'en peut rien dire ; c'est une entité proprement ineffable ; il faut y faire des marques, y distinguer des points pour qu'un discours puisse s'instaurer à son sujet. donc il faut meubler le continu de discontinuités pour que la science commence. Mais dès qu'on a des discontinuités, on peut les soumettre à des opérations : opérations ensemblistes (union, intersection, complémentaire) ou même à des opérations de caractère géométrique comme la translation dans $\mathbb{R}$. L'aspect constructif, opératoire des mathématiques est nécessairement lié à son aspect discret, discontinu. Car les opérations ne peuvent être effectués simultanément, elles ont lieu chacune dans un segment bien défini du temps. L'itération indéfinie d'opérations crée des objets pour lesquels il est souvent difficile, voire impossible, d'obtenir des représentations intuitives; A partir d'objets relativement raisonnables et représentables, l'itération indéfinie d'opérations décolle de l'univers de la représentation (décollage sémantique) pour créer des êtres difficilement maîtrisables. Tel est bien le drame du constructivisme en Mathématique. On ne peut se contenter d'une attitude finitiste, et il faut considérer des suites infinies d'opérations, lesquelles vont engendrer des objets à caractère fantomatique ou délirant. Tel a été le cas des transfinis cantoriens, théorie promise à ses débuts à un avenir paradisiaque – selon la célèbre formule de David Hilbert –, mais qui finalement apparaît maintenant comme une sorte de gigantesque impasse que le courant central des mathématiques a soigneusement évitée Il peut cependant arriver qu'une itération infinie d'opérations conduise conduise à un objet acceptable intuitivement ; l'exemple le plus spectaculaire du phénomène est donné par le paradoxe éléate d'Achille et de la Tortue. L'expression infinie

$$S = 1/2 + 1/4 + 1/8 + \ldots + 1/2^n + \ldots$$

admet une interprétation intuitive globale (Achille dépasse la tortue), à savoir le segment [01]. Pour les modernes, le contenu du paradoxe pourrait ainsi se formuler – Alexandre Koyré avait d'ailleurs suggéré une telle interprétation – : l'infini dénombrable, du seul fait qu'il peut se plonger dans le continu selon la formule (S), en retire une légitimité ontologique. C'est là un cas particulier de l'assertion générale suivante : un objet mathématique défini par une infinité d'opérations ne peut être considéré comme réellement existant que si cet objet peut être "naturellement" plongé dans le continu, l'itération ayant elle-même une représentation continue.

Si donc l'on considère le problème classique du statut ontologique des entités mathématiques, il est difficile d'échapper à l'impression que toutes ces entités n'ont pas, en quelque sorte, même droit à l'existence. Peu de mathématiciens se rallient actuellement à la vision formaliste qui fut celle de Russell-Hilbert. On admettra qu'il existe une sorte de "tronc commun" à la Mathématique, constitué essentiellement des constructions qui, partant des entiers naturels, via les entiers relatifs et les rationnels, aboutissent aux nombres réels, et de là au Calcul Différentiel classique. Sur ce tronc vont se greffer diverses extensions liées, soit à la générativité algébrique, soit à la générativité dimensionnelle (augmentation du nombre de dimension des espaces considérés) ; les êtres ainsi construits seront plus ou moins naturels, moins ou plus artificiels. Il est commode – sinon très rigoureux – de se représenter l'ontologie des être mathématiques selon une métaphore géométrique: le "tronc commun" serait une sorte de variété W plongée dans un espace euclidien ambiant. L'itération indéfinie d'opérations, agissant sur les objets de W, peut être assimilée à la propagation dans un "plan tangent" en un point w de W ; ce faisant, le plan tangent décolle de la variété W : c'est le décollage sémantique décrit plus haut. Il peut arriver que certaines de ces constructions, par une harmonie mystérieuse, retombent dans la variété W. Elles s'en trouvent de fait légitimées. Par contre, le continu lui-même échappe à une définition constructive... Sans doute, on peut identifier le continu à l'ensemble des suites dénombrables binaires de deux symboles 0, 1 (). Mais les difficultés liées à la définition intrinsèque du cardinal d'un ensemble, manifestées dans le théorème de Löwenheim- Skolem, ainsi que dans la démonstration de l'indécidabilité de l'hypothèse du continu (P. Cohen) montrent à quel point ces identifications ensemblistes du continu sont fantomatiques. Un espace continu donné dans l'intuition a toujours quelque structure de plus, un supplément de propriétés : une topologie, une métrique, une dimension ..., et ces donnée supplémentaires permettent souvent d'y définir opérations ou constructions.

L'ambition initiale de Bourbaki, classifier tous les objets mathématiques grâce à la notion d'échelle de structures n'est plus guère de mise aujourd'hui. Il y a peu en commun entre les trois grands types de structures : structures d'ordre, structures topologiques, structures algébriques. Les premières sont essentielles dans la définition des nombres réels ; les secondes dans la distinction : Ouvert-Fermé. Quant aux troisièmes, elles sont loin de répondre à l'idée formaliste d'un système défini par un choix arbitraire d'axiomes. En fait, le nombre des systèmes d'axiomes conduisant à des structures algébrique dûment répertoriées comme significatives (groupes, anneaux, corps ... etc) est sévèrement limité, et c'est, je crois, une tâche essentielle de toute saine vision des mathématiques que de comprendre, d'expliquer pourquoi tel ou tel système d'axiomes est infiniment plus fécond que tel autre. Selon moi, il importe de revenir à une vision plus spatiale que logiciste des mathématiques. Ce qui importe ce n'est pas d'éprouver si oui ou non une proposition est vraie, mais de savoir se retrouver, se localiser dans un espace...

A cet égard, les idées que j'ai récemment proposées relatives aux notions de saillance et de prégnance, notions constitutives, selon moi, de toute ontologie intelligible me paraissent pouvoir également jouer un rôle pour élucider la structure "génétique" interne des mathématiques. En mathématique, il y a un "substrat" universel, c'est le continu, et ce substrat s'identifie avec une prégnance diffuse qui le remplit et le constitue, et qui va assurer l'intelligibilité des objets (légitimer leur statut ontologique) ; mais cette prégnance ne se propage qu'en s'investissant sur des formes saillantes (par exemple les caractères d'imprimerie symboles de l'Algèbre). La distinction Prégnance-Saillance recoupe l'opposition Ouvert-fermé de la Topologie (toute forme saillante est un fermé ; la propagation de la prégnance fondamentale (le Continu) se fait par adhérence à des symboles saillants, et la forme des axiomes doit être telle qu'elle se prolonge en un plongement continu du mot pris comme axiome. Par exemple, l'axiome $xyx^{-1}y^{-1} = e$ des groupes commutatifs s'écrirait comme le rectangle

$$\begin{array}{ccc} & \underline{y^{-1}} & \\ x \downarrow & \mathrm{e} & \uparrow x^{-1} \\ & \overrightarrow{y} & \end{array} \quad ,$$

où l'identité (e) est symbolisée par la prégnance du continu ouvert qu'est l'intérieur du rectangle. La combinatoire engendre par concaténation spatiale des lettres d'un alphabet fini un gigantesque objet, le monoïde libre $M(a_i)$ qui est en quelque sorte l'univers de toutes les structures algébriques. Toute structure algébrique de présentation finie est un quotient du monoïde libre par un sous-ensemble Q d'éléments "nuls", qu'on envoie sur l'identité (e). Il faut considérer que (Q) est le domaine des éléments investis par une prégnance spatiale (c) engendrée par ces formes-sources que sont les axiomes. (Ce sont les éléments qu'il faut "tuer", proies d'un chasseur omniscient!). On pourrait presque, à cet égard, emprunter une métaphore à la Mécanique des Fluides : la générativité libre du monoïde $M(a_i)$ s'apparente à la diffusion illimitée d'un fluide initialement concentré en un point : phénomène typiquement irréversible. Les structures algébriques quotient du monoïde $M(a_i)$ apparaissent alors comme des sous-ensembles cycliques où le flux de diffusion revient sur lui-même par identification "translocale" de points distants dans $M(a_i)$. Cette métaphore explique donc le caractère "translocal" des axiomes, qui décrit le caractère plus ou moins lié de la propagation de la prégnance annihilante (q). Un peu comme un attracteur qui, en Dynamique Qualitative, est un régime asymptotique stationnaire, présente nécessairement une dynamique réversible (l'effet du temps y est annulé), les structures algébriques sont les "compactifications" autour desquelles peut s'"enrouler" la générativité libre du monoïde. (De fait, l'attracteur le plus simple, en Dynamique, après le point, est le cercle (orbite périodique) un groupe topologique). La déduction logique peut elle-même être considérée comme la propagation, dans l'univers formel des propositions, de la prégnance Vérité ; on voit à cet égard que la prégnance Vérité a une propagation plus contrainte que son opposé, la Fausseté (du Faux, on peut déduire le Vrai). Dans la présentation synthétique des mathématiques que nous donnerons en annexe, on s'efforcera d'ordonner les grandes théories génératives selon le caractère plus ou moins contraint de leur générativité (la générativité la moins contrainte étant la générativité libre, sans relations) : les groupes à croissance exponentielle sont beaucoup moins liés que les groupes à croissance polynomiale.

On voit l'importance – centrale à nos yeux – de la notion de *propagation* en Mathématique, puisqu'elle couvre un immense champ allant de la Logique formelle, à la Physique. En fait, théoriser un champ expérimental, c'est "réduire l'arbitraire de sa description", et ceci exige de définir des opérateurs d'extension, de prolongement qui permettent d'étendre la donnée expérimentale définie sur un ouvert D à un ouvert D_1 plus grand du substrat. Toute la prédiction, en particulier, exige un algorithme permettant de déduire d'une connaissance du passé une connaissance de l'avenir. Mais la propagation n'est qu'un aspect d'une problématique d'une grande généralité, à savoir celle liant *Local et Global en Mathématique.*

On peut dire sans hésitation que les principaux théorèmes de l'Analyse – en assez petit nombre finalement, cinq à six peut-être – ont à faire avec la problématique Local-Global. Cette problématique peut être envisagée dans deux sens :

A) Le sens Global $\Longrightarrow$ Local. Étant donné un être mathématique défini globalement par une procédure constructive (par exemple défini par une équation $F(x_i) = 0$), en déduire les propriétés locales de cet espace au voisinage d'un de ses points. On citera ici deux théorèmes extrêmement importants : le théorème des fonctions implicites, qui permet parfois de reconnaître si l'ensemble défini par $F = 0$ est lisse (sans singularités) ; et la formule de Taylor, qui donne une forme approchée de l'équation locale à partir de l'équation globale. Ces procédures de localisation, aussi connues en Géométrie Algébrique (et Analytique moderno sensu) permettent de décrire la notion de "singularité" d'un espace en un de ses points.

On pourrait penser que ces procédés de localisation, qui restreignent à un petit voisinage l'objet étudié, sont de peu d'intérêt (à cause de la "perte d'information" qu'ils entraînent). En fait, pour comprendre comment est fait un espace, il faut d'abord se faire une idée aussi exacte que possible de ses singularités. D'ailleurs, dans le cas d'un espace analytique, la connaissance des singularités suffit souvent à déterminer l'espace entier. Comme l'a écrit Gaston Julia : "Regardez les singularités il n'y a que cela qui compte ..."

B) Le sens Local $\Longrightarrow$ Global. Il s'agit ici de construire un être global à partir de conditions locales imposées en certains (ou tous) ses points. Le procédé typique est ici le prolongement analytique, qui permet, étant donné un germe de fonction analytique en un point 0 d'une variété analytique, de prolonger ce germe à tout le "domaine d'holomorphie" de la fonction ; on a donc là un mode d'extension qui est strictement déterministe, et qui reste de nature locale. C'est le procédé d'extrapolation universel de la Physique, qui permet la prédiction quantitative.

Le prolongement analytique d'un germe d'ensemble analytique (ensemble défini localement par des équations analytiques) exige que ce germe soit donné plongé dans un espace analytique. Si le germe est donné intrinsèquement (par son algèbre locale), le prolongement analytique n'est plus uniquement déterminé (exemple de $\mathbb{C}$ dans $\mathbb{C}^2$). Néanmoins il existe alors pour le germe une "déformation plate universelle" qui permet parfois une extension locale de l'espace associé au germe. C'est là le principe de l'algorithme en Théorie des Catastrophes Élémentaires pour une extension locale qualitative de certains germes de fonctions.

Enfin on peut s'efforcer de construire un être global assujetti à des contraintes locales de manière non-déterministe ; on peut alors se heurter à des impossibilités de nature globale, usuellement désignées sous le vocable d'obstructions. Par exemple, si l'on a une 1-forme ω dans un ouvert U de l'espace euclidien qui y est fermée ($d\omega = 0$), un lemme de Poincaré nous dit que ω est localement la différentielle d'une fonction réelle $dV = \omega$; peut-on synthétiser toutes ces présentations locales $\omega_i = dV_i$ en une différentielle unique $\omega = dV$ où V est définie sur tout U ? La réponse est positive si U est simplement connexe ; sinon, le problème peut n'admettre aucune solution du fait d'une obstruction située dans un groupe de cohomologie $H^1(U, \pi_1(U))$. La topologie algébrique et la théorie de l'homotopie ont été créées essentiellement pour résoudre ce type de questions. On notera aussi dans ce genre de problèmes le rôle joué par le théorème de Stokes, qui est un puissant moyen d'exprimer globalement l'effet global de conditions locales partout valides.

Pour définir les procédés de propagation en Mathématique, la générativité discrète des opérations ne suffit pas. Dès l'invention du Calcul, on a pu constater que la structure additive des réels servait à représenter l'addition des longueurs, l'addition des temps, et aussi celle des vitesses ; c'est-à-dire que la droite réelle est un espace qui agit dans lui-même par translation, nous donnant ainsi le prototype d'un groupe topologique abélien : premier cas particulier d'un groupe continu ou *groupe de Lie* ; ces espaces, en quelque sorte, se propagent dans eux-mêmes de manière canonique. De plus l'engendrement d'un groupe de Lie à partir de son Algèbre de Lie donne l'image d'une propagation globale instantanée ne nécessitant pas – comme le prolongement analytique classique –, l'existence d'un espace ambiant. Outre leur intérêt mathématique propre (qui est considérable), les groupes de Lie sont indispensables en Physique pour définir les symétries fondamentales (internes et externes) des particules élémentaires. Ainsi de la générativité continue du temps, quand elle s'applique dans un espace (variété) différentiable : si on suppose la diffusion du temps contrôlée localement par la position, on obtient l'importante notion de système *dynamique autonome* – paradigme le plus parfait du déterminisme scientifique : en effet, c'est l'exemple type d'une évolution déterministe "locale", i.e. excluant l'action à distance. La Science Moderne s'est toujours efforcée d'exclure les actions à distance – non locales, mais elle n'y est jamais totalement parvenue. En Mécanique Quantique notamment, le caractère non local des évolutions temporelles décrites par la théorie est patent. Cet ultime élément de "pensée magique", réside donc dans des éléments translocaux, mais mathématiquement contrôlés sous forme des grandes symétries de la Physique.

On a représenté en annexe un tableau (T) donnant une vue globale des mathématiques.

Description du tableau (T).

Le courant central des Mathématiques prend naissance avec l'arithmétique (théorie des entiers naturels $\mathbb{N}$) ; de $\mathbb{N}$ le besoin de symétriser l'addition conduit aux entiers relatifs $\mathbb{Z}$, puis aux rationnels $\mathbb{Q}$. On peut sur cette base développer la théorie des nombres, et une bonne part de l'algèbre classique. La théorie des nombres reste à beaucoup d'égards une sorte de noyau dur des mathématiques. Les problèmes qui y sont posés sont souvent extrêmement difficiles ; certains – comme la conjecture de Fermat – ont résisté aux efforts des mathématiciens depuis plusieurs siècles. Du fait de la nature discrète de $\mathbb{N}$ et de $\mathbb{Z}$, ces questions ne se laissent pas facilement spatialiser. C'est seulement tout récemment qu'on a pu appliquer à certaines questions d'analyse diophantienne les techniques en topologie m-adique de l'Algèbre homologique (théorie des schémas d'A. Grothendieck ...). Mais même là, beaucoup de spécialistes estiment que certains progrès récents (telle la démonstration de l'irrationalité de $\zeta(3)$ par Roger Apery) sont plus dus à l'usage de méthodes élémentaires astucieusement conduites qu'à des techniques de haute sophistication ...

Le courant central des mathématiques, dès la constitution des rationnels, se hâte vers la définition des réels. On sait comment la Géométrie, avec la diagonale du carré, a imposé l'introduction des irrationnels : la conquête du continu, de la mesure de l'espace, exigeait cette extension. Le courant essentiel passe ensuite par l'apparition du Calcul Différentiel avec la théorie des fonctions d'une variable réelle. La pluri-dimensionalité des espaces oblige alors à introduire les fonctions de plusieurs variables, les dérivées partielles, la notion de différentielle, la formule de Taylor ... Avant de poursuivre l'évolution de ce courant central, évoquons les principales branches des Mathématiques Discrètes, toutes situées dans le tiers gauche du tableau.

Il y a tout d'abord le cas de la théorie des Ensembles et de la Logique mathématique. Il s'agit là de théorie à générativité trop faiblement liée pour présenter un intérêt mathématique intrinsèque. Restent, ce qui n'est pas négligeable, l'intérêt formel (en tant que système notationnel) et philosophique ; la théorie des catégories s'efforce de dégager tous les mécanismes "iconiques" (au sens de Peirce!) de la générativité en déduction mathématique (théorie des diagrammes de flèches). La combinatoire, et la théorie des graphes traitent de questions assez mal liées entre elles, de conceptualité mal définie. Reste l'algèbre, qui n'est discrète, je pense, qu'en apparence. La théorie des groupes finis est une théorie à générativité très fortement liée ; elle a connu récemment (avec la classification des groupes finis simples) de spectaculaires progrès. Un des aspects les plus intéressants de ces mathématiques discrètes concerne la théorie de la récursivité, et par suite la décidabilité de certains problèmes. Ainsi on a pu établir que reconnaître si deux groupes de présentation finie sont ou non isomorphes n'était pas décidable. De même pour l'équation diophantienne la plus générale ... La théorie des automates, jusqu'ici exclusivement discrète, bénéficierait grandement d'un peu de spatialisation ...

Ayant ainsi achevé cette présentation rapide des Mathématiques discrètes, nous revenons au courant central au point où nous l'avions laissé (historiquement, au début du 18° siècle). Ici le courant se divise grosso modo en deux branches ; l'une regarde vers les espaces, et est donc de nature essentiellement topologique. L'autre regarde les morphismes entre espaces, il est donc du ressort de l'Analyse. Après une période de divergence, ces deux courants rentrent en contact en des zones relativement disjointes : Algèbre linéaire et multilinéaire d'abord, puis Géométrie Algébrique et Analytique (moderno sensu) d'une part. D'autre part, en Géométrie Différentielle, groups de Lie et espaces homogènes associés avec leurs prolongements vers la théorie des représentations. Sur la branche espace, on a noté Topologie Générale (générativité nulle), Topologie Algébrique (homologie, homotopie), espaces fibrés, variétés différentiables, Topologie Différentielle ... La branche analyse porte un rameau récent, Analyse Différentielle ; mais la branche essentielle, plus ancienne traverse les fonctions analytiques, les fonctions spéciales, les espaces fonctionnels, l'analyse de Fourier, la théorie linéaire des Espaces Vectoriels Topologiques, des Équations aux Dérivées Partielles ... Un rameau plus récent issu de la branche espace est plus attiré par la description du monde réel : il porte les équations différentielles ordinaires, leur théorie qualitative et ses extensions (peu étendues) vers les Équations aux Dérivées partielles non linéaires et la théorie des chocs. Ce courant contient aussi la théorie des feuilletages, le calcul des variations, la théorie des structures et pseudo-groupes (au sens d'Ehresmann : les groupes infinis d'Elie Cartan). Un peu périphériquement, on peut rattacher à la branche Analyse la Théorie des Probabilités, née du besoin de considérer des modes "stochastiques" de propagation dans les espaces. A cette branche se rattache la Théorie Ergodique, elle-même adhérente aux Équations Différentielles ordinaires et aux feuilletages. En progressant encore plus vers les applications, l'optimisation donne naissance à deux disciplines encore plus développées : la Théorie du contrôle et la Théorie des jeux. Ces deux théories sont importantes par le point de vue neuf qu'elles introduisent en Dynamique, où l'homme lutte contre la nature – ou contre un partenaire ... les mathématiques qui y sont impliquées sont relativement neuves.

Nous n'avons pas mentionné l'Analyse Numérique, ni l'emploi des ordinateurs : il s'agit là de techniques ancillaires, bien qu'elles puissent parfois faire appel à de délicates questions d'Analyse.

Nous espérons dans ce portrait n'avoir rien oublié d'essentiel. Peut-être l'accent mis sur telle ou telle branche reflète-t-il des préjugés personnels, mais comment pourrait-on l'éviter? Sur le tableau, on a tracé en trait renforcé ce qu'on peut considérer comme une filiation conceptuelle essentielle (les rapports moins importants sont en pointillé, et bien entendu, pour ne pas surcharger la figure, on n'a pas fait mention de toutes les connexions existantes entre domaines). Le plus souvent la filiation essentielle s'est réalisée comme succession dans l'Histoire. La branche Analyse issue du Calcul correspondant grosso modo à la mathématique du XIX° siècle, la branche Espaces à celle du XX°. On voudrait suggérer ici un principe de récapitulation à la Haeckel-Müller ...

Appendice

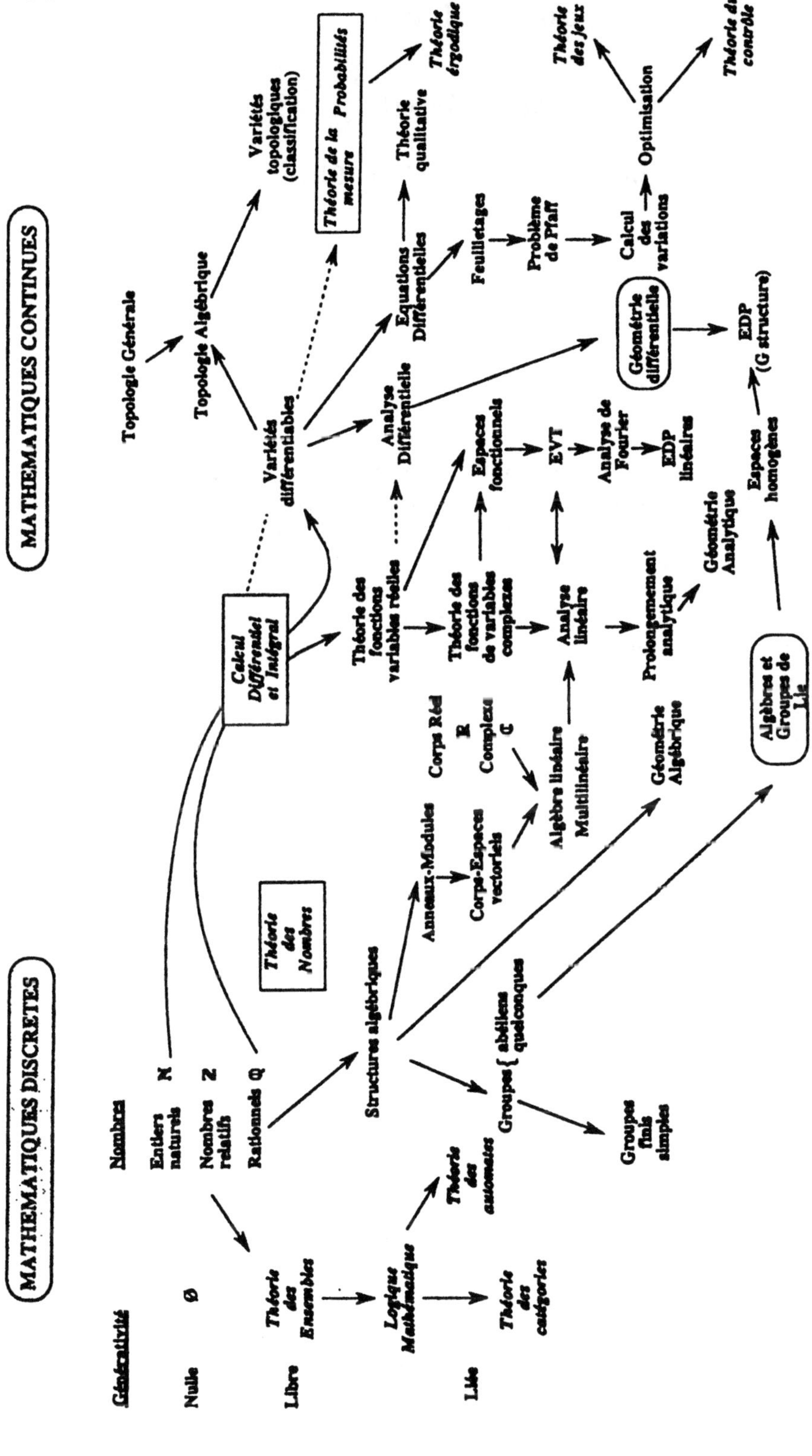
MATHEMATIQUES DISCRETES
MATHEMATIQUES CONTINUES
Généralités
Nulle
Libre
Liée
Ø
Théorie des Ensembles
Logique Mathématique
Théorie des catégories
Théorie des automates
Nombres
Entiers naturels N
Nombres relatifs Z
Rationnels Q
Théorie des Nombres
Structures algébriques
Groupes { abéliens quelconques
Groupes finis simples
Anneaux-Modules
Corps-Espaces vectoriels
Corps Réel R
Complexe C
Algèbre linéaire
Multilinéaire
Géométrie Algébrique
Algèbres et Groupes de Lie
Calcul Différentiel et Intégral
Théorie des fonctions variables réelles
Théorie des fonctions de variables complexes
Analyse linéaire
Prolongement analytique
Géométrie Analytique
Espaces homogènes
Espaces fonctionnels
EVT
Analyse de Fourier
EDP linéaires
Variétés différentiables
Analyse Différentielle
Topologie Générale
Topologie Algébrique
Variétés topologiques (classification)
Théorie de la mesure
Probabilités
Théorie ergodique
Equations Différentielles
Théorie qualitative
Feuilletages
Problème de Pfaff
Calcul des variations
Optimisation
Théorie des jeux
Théorie du contrôle
Géométrie différentielle
EDP (G structure)

Mathematical Progress as Synthesis of Intuition and Calculus

Klaus Volkert

Hochstr. 48, 6652 Bexbach (RFA)

In the following text the terms *calculus* and *intuition* are taken in a very broad sense. They are used as translations for the german words *Kalkül* and *Anschauung* .

1. The traditional conception of mathematics knows two sources of knowledge: intuition and calculus, or, otherwise taken, the senses - in particular the eyes - and the reason. Intuition leads to crude ideas, which can be true or false. The calculus on the other hand provides clearness and certainty: That, what is verified by calculation, is true (or demonstrated). Crude ideas (like the idea of a surface) are transformed into precise notions (like the notion of a differentiable manifold), which are linked by logical inferences. This process has been called *exactification* or (Carnap's term) *explication of a notion* (Begriffserklärung).

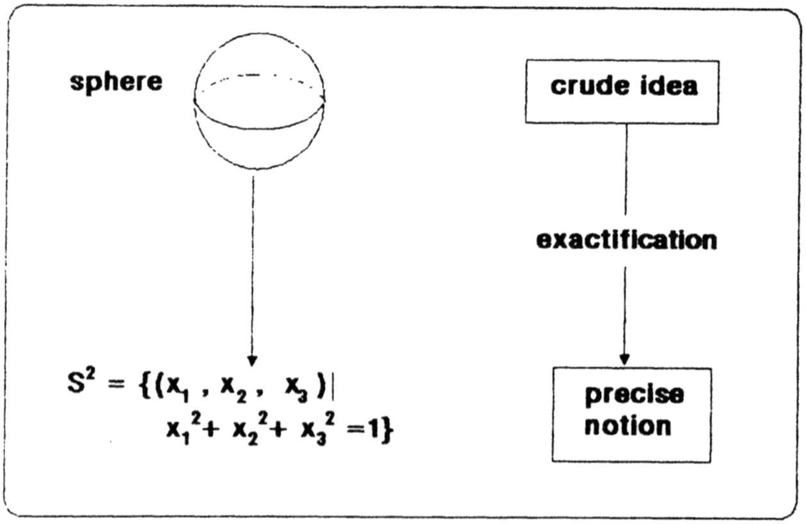

Fig. 1.

Intuition is part of the art of inventing (*ars inveniendi*), calculus belongs to the art of justification (*ars judicandi*). The doctrine, which is generally accepted in the historiography of mathematics says, that this strict separation between intuition and calculus was an achievement of the 19th century. It was the core of the *new mathematical rigour*; in the context of analysis it was obtained by the so-called *arithmetization* (F. Klein).

I want to treat here the relation between intuition and calculus in analyzing three examples taken from the history of mathematics in the 19th century. My thesis is :

> Mathematics in its full historical content cannot be understood by such a reductionist program. Mathematical progress is always a synthesis out of intuition and calculus : Calculus without intuition is empty, that is without content, intuition without calculus is blind. (cf. the *Critic of Pure Reason* by I. Kant)

It is difficult to choose examples; the choice taken can always be criticized. I hope that I have found typical examples, which are interesting.

2. In the year 1799 a very young german mathematician published *A new demonstration of the theorem that it is always possible to factorize every rational and entier function of one variable into factors, which are real and of first or second degree* (Gauss, 1799). In this work Carl-Friedrich Gauss gave a nearly complete demonstration of the fundamental theorem of algebra. In using polar coordinates – Gauss doesn't explain their geometrical signification (he is *purely analytical*) – the problem of the existence of a zero of the equation (or of the polynomical function) is reduced carefully to the existence of a point of intersection of two kinds of curves :

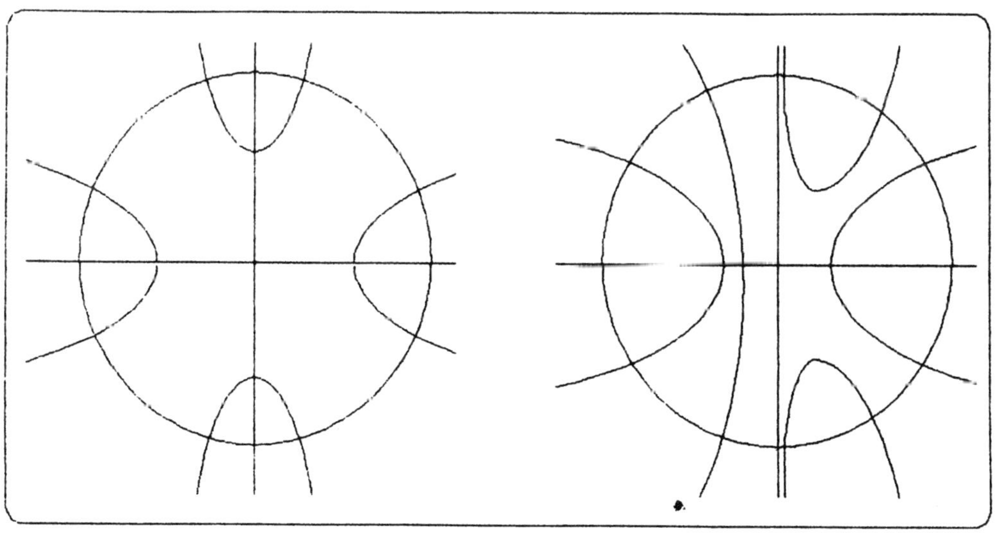

Fig. 2.

But the existence of this point is derived by a simple intuitional consideration which is illustrated by the picture Fig. 3.

This fact – that is the existence of the point of intersection – is treated by Gauss as something absolutely clear; he says nothing about it (whereas his paper is otherwise very explicit). Sixteen years later Gauss wrote in the preface to his second demonstration of the fundamental theorem (Gauss, 1815) : *My first demonstration depended partly on geometrical considerations...* . Here the term *geometrical* is synonymous with *intuitional*; it is opposed to *analytical*. Since all mathematics should be treated analytically, geometry

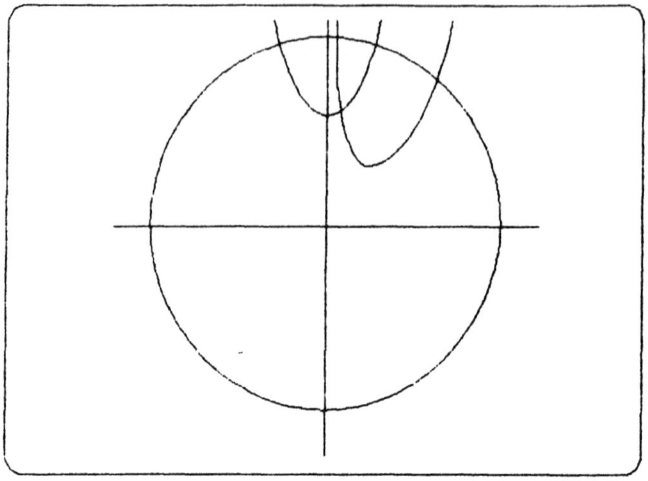

Fig. 3.

is inferior to algebra. The goal is to eliminate every reference to geometry or intuition. That's the credo of the ***mathematical positivism*** (my term). This positivism can be characterized as the attempt to reduce all intuition, which is necessary in mathematics, to the mere perception of simple signs (as letters or ciphers).

The demonstration given by Gauss is an example for a mecanism in the development of mathematics which might by described as an ***advancement of intuition in relation to calculus*** : A fact, which is intuitively absolutely clear (the intersection of the curves in our example), cannot be demonstrated analytically at a certain historical stage. Many of the famous mathematical conjectures are of this type (but certainly not all). The idea in our example, which wasn't ***exactified***, was the continuity of the plane. What one needs is a theorem like the Jordan curve-theorem (announced for the first time by C. Jordan in his famous *Cours d'analyse* (1887) and demonstrated by O. Veblen in 1905).

The problem of continuity in real analysis was already solved in the 19th century : The first step of this solution was the analytical definition of ***continuity of a function*** (Bolzano(1817), Cauchy (1821)), the second step being the construction of the real number system as a complete field (construction based on the rational numbers; Cantor/Méray, Weierstrass, Dedekind (all published around 1870)), or the axiomatic characterisation of the real numbers as a complete ordered field (Hilbert, 1900).

The mathematical progress was realized in this example as a synthesis : That what is intuitively obvious was vindicated later by the calculus. This method is considered as standard in the development of mathematics.

3. But there was (and still is) an inverse mecanism : that is there are situations, in which *the calculus is more advanced than the intuition.* A lot of examples for this mecanism are found in the history of number – the most famous being the introduction of the *impossible numbers* by G. Cardano (1545 – the term *impossible* was coined later; Cardano himself spoke of *very sophisticated quantities*). In the context of geometry one can cite the invention of the points at infinity by G. Desargues which led to the notion of the projective plane. The full intuitive content of this notion was not clarified until the second half of the 19th century by F. Klein and L. Schläfli.

A similar situation occured in the geometry of the 19th century : The so-called non-Euclidean geometry started by the formal deduction of the consequences of one of the negations of the axiom of the parallels (that is the fifth postulate of book one in Euclid's *Elements*). Lobatchevski, one of the inventors of this geometry (besides J. Bolyai and C.F. Gauss) wrote (1856) in his work *Pangéométrie* (this was one of his names for the new geometry; he also spoke of the *imaginary geometry*) :

> "Just one look at the equations, which formulate the relations between the sides and the angles of a rectilinear triangle suffices to demonstrate that the pangeometry at this moment turns out to be an analytical method, which replaces the analytical methods of the ordinary geometry." (Lobatschefskij, 1902, p. 75)

The pangeometry was - *in statu nascendi* – a mere logical- deductive theory without (or with a very restricted) intuitive content. It marks the beginning of epistemological rupture : The geometrical axioms don't describe any longer intuitive facts (cf. the Kantian doctrine of the Euclidian axioms as propositions which are synthetic *a priori* and his doctrine of space as a pure form of intuition). Towards the end of the 19th century the geometrical axioms became propositions which are established more or less arbitrarily and which are justified by their consistency (that is exactly the development from M. Pasch to D. Hilbert). The one intuitive geometry has given way for a multiplicity of *imaginary* geometries. The *reality of intuition* has been replaced by *logical possibility.*

But a theory without intuitive content wasn't considered as a satisfactory mathematical theory in the 19th century. One hesitated to accept the new geometry during a long time. A decisive step towards its acceptance was the discovery of the intuitive models for non-Euclidean geometry (by E. Beltrami, A. Cayley and F. Klein, H. Poincaré). These models were not only a means to demonstrate the relative consistency of the new geometries, but also showed their intuitive possibility. Apart from that they are also a very useful tool in demonstrations (because demonstrations are often guided by intuition – cf. part 2).

The models of non-Euclidean geometry were subject to several interesting debates in the philosophy of mathematics, beginning with H. von Helmholtz' *The origin and meaning of geometrical axioms*(1870). In the 20th century there was a discussion, today almost forgotten, between the German philosophers H. Reichenbach and O. Becker. Becker defended the thesis of an epistemological priority of the Euclidean geometry : He thought that this geometry is priviledged by intuition. That was the content of his *Habilitationsschrift* defended at Freiburg im Breisgau (under the direction of E. Husserl). Its title was *Contribution to the pheneomenological foundation of geometry* (1923). To Reichenbach, who was a logical empiricist, intuition had no relevance at all to mathematics. The only thing, that counts, is logic. To prove this statement Reichenbach cites an example (in the tradition of H. von Helmholtz and H. Hahn, who had proclamed in 1933 the *crisis of intuition* by similar arguments) : This is a movie, in which the live in a non-Euclidean world is shown (such a movie had been discussed by K. Gerhards in this work *Non-Euclidean cinomatography* (1932)). In this movie we live in a hyperbolic world : If we are moving towards the horizon, the things before us begin to become bigger, whereas things behind us begin to shrink. Therefore – so Reichenbach – it is possible to make the non-Euclidean world accessible to intuition. There is no intuitive priority of Euclidean geometry at all. But, replied Becker, if one is watching this movie, one is always aware of the fact, that one is watching a movie. The non-Euclidean world remains even in this movie an artificial and imaginary world.

The problem, which is involved here, is to discern model and reality, intuition and imagination, or - in traditional German terminology – *Anschauung* and *Einbildungskraft*.

> "Imagination (Einbildungskraft) is the faculty to represent an object in the intuition without its presence." (I. Kant. *Critic of Pure Reason* , B 151 - my translation)

The distinction between intuition and imagination, between dream and reality is basic for epistemology. To modern mathematics this is not so, because modern mathematics doesn't make a difference between possibility and reality.

Until the 19th century mathematical existence was assured by intuitive models and not by the consistency of the theory deduced from axioms. This was an intuitive criterion of existence, not a logico-deductive one.

(The difference between *impossibility* – that is non-consistency – and *possibility* – that is consistency – was discussed very clearly by Gauss in his paper of 1799. But the older mathematicians weren't satisfied with mere *possibility* ; they wanted *reality* . This type of *reality* occurs in Gaussens *Selbstanzeige* (1831) concerning the complex numbers.)

4. One discipline of geometry, which started towards the end of the 19th century, seems to me of a special interest for the problems discussed here. I am speaking of topology, defined as the project to grasp mathematically the basic facts about spatial intuition. Poincaré has spoken about that several times :

> "Analysis situs [his term for topology] is the science which tells us the qualities of geometrical figures not only in the ordinary space, but also in the space with more than three dimensions. The analysis situs in three dimensions is to us a knowledge nearly intuitive."(Poincaré, 1921, p. 100)

Here we meet the advancement of intuition over calculus in its clearest form : The difference between a sphere and a torus is absolutely evident – but how to express this difference in a calculus? The intuitive idea of a continuous deformation (or more generally of homeomorphism) was formalized in a process which lasted some decades (the idea of homeomorphism was used by Poincaré in its classical sense derived from analysis; in its general meaning, homeomorphism depends on the idea of a topological space, which was defined for the first time by F. Hausdorff in 1914). Poincaré invented one of the most important topological invariants : the fundamental group. He investigated and calculated systematically Betti-numbers and torsion-coefficients. To Poincaré the basic problem was the *classification problem* : The invariants are they strong enough to reflect well our intuitions, that is, to discriminate surfaces which are not intuitively equivalent? Surfaces with same invariants are they homeomorphic? To answer these questions – the second one being a generalisation of the well known *Poincaré – conjecture* – Poincaré undertook a series of very subtle investigations. It is interesting to look at the relations between intuition and calculus in this process. There is a stage in which calculus superceeded intuition (a complete reversion of the starting situation!) : The counterexamples by which Poincaré falsified the different modifications of his *conjecture* were defined in a purely formal way (by generators of a group and relations between them). There's no hint at all to an intuitive content behind these threedimensional *surfaces* . The intuitive knowledge of Poincaré's citation given above is completed by the experience gained in calculations. This experience may lead to new intuitive knowledge.

5. I discussed briefly the interactions between calculus and intuition. There's a lot of questions which one may ask. Is it possible that calculations alter the intuition? Can one get some intuition by calculating? Intuition is it a product of habit? What is generally accepted is : Intuitions can guide proofs. One makes a picture and tries to describe it with formulae. Here something important must be kept in mind : The axioms, which lie at the base of our calculations, are choosen more or less deliberately by us. The calculus depends on man, whereas the intuitive reality seems to be (more) independent of us.

What happens if there arise contradictions between our intuitions and our calculations? That's the situation in which one likes to speak of *monsters* or *pathological cases.* A well known example for this is the emergence of the so-called pathological functions at the end of the 19th century (Volkert, 1987). Is it possible to alter the intuition in order to make the unintuitive more intuitive? Must one change the calculus (its axioms and/or its rules) to reestablish the destroyed harmony between calculus and intuition? This seems to be a very bizarre idea to a modern mathematician, but this process is well known in certain branches of logic (for example in modal–logic). Or is intuition totally irrelevant to mathematics? This answer, which may be supported by many modern mathematicians, is certainly not able to explain the history of mathematics. I want to conclude with a thesis, maybe a little bit provocative :

> Mathematics always needs both intuitions and calculations. But the relations between them aren't those of a preestablished harmony. This harmony must be regained from case to case. Intuition is the human dimension of mathematics which distinguishes our calculations from those of a computer.

References

Becker, Oskar 1923 *Beiträge zur phänomenologischen Begründung der Geometrie und ihrer physikalischen Anwendungen.* In : Jahrbuch für Philosophie und phänomenologische Forschung 6; 385-560.

Becker, Oskar 1930 *Die apriorische Struktur des Anschauungsraumes.* In : Philosophischer Anzeiger 4; 129-162.

Gauss Carl Friedrich 1799 *Demonstratio nova theorematis omnem functionem algebraicum rationalem integram unius variabilis in factores reales primi vel secundi gradus resolvi posse.* In.Works Volume III. - Cited following the German translation by E. Netto in :

C.F. Gauss. *Die vier Gauss'schen Beweise für die Zerlegung ganzer algebraischer Functionen* (Leipzig/Berlin, 1913).

Gauss, Carl Friedrich 1815 *Demonstratio nova altera theorematis omnem functionem algebraicum rationalem integram unius variablilis in factores reales primi vel secundi gradus resolvi posse.* In : Works Volume III. - Cited following the German translation by E. Netto in : C.F. Gauss. *Die vier Gauss'schen Beweise für die Zerlegung ganzer algebraischer Functionen* (Leipzig/Berlin, 1913).

Gerhards, K. 1932 *Nichteuklidische Kinemathographie.* In : Die Naturwissenschaften 20, 925 - 928.

Helmholtz, Hermann von 1870 *Über den Ursrung und die Bedeutung der geometrischen Axioms.* In : Vorträge und Reden. Volume II (Braunschweig, 1887).

Lobatschejskij, Nikolaj Iwanowitsch 1902 *Pangeometrie.* (Leipzig - translated and edited by H. Liebmann).

Poincaré, Jules Henri 1913 *Pourquoi l'espace a trois dimensions.* In : Dernières pensées. (Paris).

Poincaré. Jules Henri 1921 - *Analyse de ses travaux scientifiques.* In : Acta mathematica 38; 3 - 135.

Reichenbach, Hans 1977 *Die Philosophie der Raum - Zeit - Lehre* (Braunschweig/Wiesbaden - with notes by A. Kamlah).

Volkert, Klaus 1986 *Die Krise der Anschauung* (Göttingen).

Volkert, Klaus 1987 *Die Geschichte der pathologischen Funktionen. Ein Beitrag zur Entstehung der mathematischen Methodologie.* In : Archive for History of Exact Sciences 37; 193 - 232.

4

The Philosophical Problem of Space

What is space?

Hans Freudenthal

Rijksuniversiteit Utrecht; Vakgroep onderzoek wiskundeonderwijs en onderwijs computercentrum. Tiberdreef 4. 3561 GG Utrecht

What is space ? The question sounds like it belongs to philosophy but the answer will give this suspicion the lie, provided philosophy is understood in its traditional sense.

In fact, as early as the time of the ancient Greeks, space has been a philosopher's concern. Geometers did not even have a name for the intuitive substratum that was indispensable to their research. Anaximander called it l'ἄπειρον, which means *illimited*, both spatially and conceptually. For Parmenides it was the μὴ ὄν, the *non-being*, against which he warns the readers of his poem. For the atomists and their master Demokritos, it was the κενόν, the *void*, in which the atoms are whirling. It was this void that was rejected by the great majority of philosophers from Aristotle to Descartes and Leibniz. Aristotle, followed by the scholastics, denied any global spatiality and accepts only the *place*, τοπός or χώρα, the first immobile envelope of the bodies, their container, as it were. Lacking any container, infinite bodies are impossible, as there is no infinity of bodies (any infinity being a potential infinity). On the other hand, a void would lack any resistance against movement and thus allow for infinite speeds, which proves the impossibility of the void. (Let us notice that shortly before, Aristotle had argued against the void since it lacked any preferred direction and, hence, any opportunity for movement.)

It is not my task to display and analyse the brilliant web of paralogisms of which the Aristotelian world system consists, and which over the centuries captivated the greatest minds, before Galileo and even after his intervention. Is it an evasion or mere cheating, if the master of the peripatetics allows the geometers to postulate line segments and circles of any dimension? Never does he grant potentialities that cannot be actualised by principle. But how to actualise the potentially infinite geometric objects in a world whose "place" is a last finite sphere that envelops the celestial vault, the sphere of the immobile motor, who is God?

The first disciple of the scholastics who rejected the Aristotelian surrogate of space is Descartes. The way out that he invents is *extension*, attribute of the matter, as thought is the attribute of the soul – the interaction between these two "substances" shall become the big problem of rationalism. Infinite space is admitted, though not the void, because where there is no matter, there cannot be extension.

Rather than in philosophy, the proper idea of space, which is going to dominate science, arises in mysticism. In the *Liber XXIV philosophorum*, a work of the 12th century, once ascribed to Hermes Trismegistos, one finds the pronouncement *Deus est sphaera infinita cuius centrum est ubique, circumferentia vero nusquam*, God is the infinite sphere whose centre is everywhere yet whose surface is nowhere. This thesis will be quoted and maintained by a long series of mystics up to the 18th century. Any absolute and infinite existence is a privilege of God; so it cannot be attributed to space, unless space has been deified. Did the author of the *Commedia Divina* know this way out? Anyway he preferred another one: space compactified not by the sphere of the first motor but by a single point, God, the source of all forces. On the contrary, for the mystics of the infinite sphere, each point was the centre and source of force and thus a place for the *unio mystica*. Nobody will be astonished to meet the inventor of the air pump, Otto von Guericke, in this company. Indeed, empty space as the precondition of the world, created in it, cannot but be God the creator himself.

I paid attention to this mystic current, mainly because it included none other than Newton, who was influenced by the mystic literature, and who in his *Principia mathematica philosophiae naturalis* badly needed this absolute space (and time) and the *actio in distans* – ideas, which, albeit judged unreasonable, could be justified by God's omnipresence and eternity as soon as space was considered as God's sensorium, his nervous system, as it were.

Against this "heretical" deification of space Leibniz stood up, who, for that matter, had still to settle an account with the Newtonians, who had accused him of plagiarising the new calculus. Against Newton, Leibniz defended the relativity of space (and time) as a relational sytem of coexistence (and succession) of things and events. Although Leibniz was still Aristotelian enough to reject relations ontologically, he used them methodologically in a way that remained almost unique up to the 20th century.

After recalling his "principle of sufficient reason" whereby nothing happens unless there is a reason why it happens so rather than otherwise, he continues:

> "I say that if space were an absolute being, something would happen for which there would not be any sufficient reason, which is against our axiom. This is proved as follows. Space is something absolutely uniform, and without the things placed in it, a point in space does not differ by anything from any other. From this it follows that (supposed that space were a thing in itself other than the order of bodies among each other) there cannot possibly be a reason why God, while preserving the same situation of bodies among each other, should have put the bodies into space as he did and not otherwise, and why, for instance, he would not have put all things the opposite way, interchanging East and West. But as soon as space is nothing else than that order or relation, and without the bodies nothing but the opportunity to get them placed, both states, the existing one and its opposite, would in no way differ from each other. Their seeming difference is merely the consequence of our chimeric supposition about the reality of space in itself. But in fact both of them are the same, since they are absolutely indiscernible, and consequently there is no occasion to demand a reason for the preference of the one above the other."

(Let us mention that in order to prove the relativity of time, Leibniz uses a similar argument: if time were absolute, there would not be any sufficient reason why God should not have created the world one year earlier than he did.)

I have quoted this passage at length because of its historical impact: in my view it was the germ of Kant's philosophy of space as found in the Transcendental Esthetic, the first part of the *Critique of Pure Reason*. In one of Kant's dissertations of the precritical period I think I have discovered Leibniz's influence – a negative one, since the confrontation with Leibniz's ideas was as important a stage in Kant's development, as for Leibniz his struggle against Descartes had been.

Kant's argumentation runs as follows:It is true that there exists no absolute place in the world. Each place is determined only by its relation to other ones. Nevertheless there is one thing spatial that is absolute. A right hand is not completely described by the relative position of its parts; in fact these relations are exactly the same for a left hand, though there is an undeniable difference between right and left hands. Up to this point the argumentation is correct but now Kant makes an error: rather than the internal properties of the hand deciding the question whether it is right or left, Kant claims, it is the position of the right hand in space that distinguishes it from a left one. So space precedes the object, since otherwise there would not be any relation of the object to space.

Obviously the argument is wrong. The property of a hand of being right is not caused by its relation to space but by that to some other object, say, a left hand, or any other object that shows some asymmetry. Today we would say that the concept of a right hand is only meaningful in a space that has somehow been oriented, that is, by means of some non-symmetric figure.

A few decades later Gauss unmasked Kant's paralogism, but then it was too late. Kant evoked the same sophism in another work, in order now to support not absolute space, which was then rejected, but subjective space, called "ideal" in the Transcendental Esthetic: space (as well as time) as a pure intuition, or rather perception, (where "pure" means a priori with regard to experience, that is, independent of it).

The Transcendental Esthetic is only the first part of the *Critique of Pure Reason*. Rather than the *a priori* nature of space (and time), it is that of the synthetic judgments around which Kant' s work turns. Since the ancient Greeks geometry had been the model of a deductive science. That by pure reasoning apparently one could find factual truth about real space that is not purely logical, had been an enigma: how was it possible that with his eyes closed towards the reality, and by pure reasoning, man could pass true judgments about this same reality? This then is Kant's famous problem: how are synthetic judgments *a priori* possible? And the answer would be that it is the soul itself that imposes upon the phenomena these forms of pure intuition (or rather perception) that are space and time, in order to rediscover them in the phenomena as observed.

At least, this is the official interpretation of Kant's philosophy of space (and time), which, in fact, does not explain anything. In mathematics a proof is suspected as soon as a supposition that looks essential is not used at all. According to this interpretation the soul would be the place where the forms of space and time are imposed up on the phenomena. Yet even while accepting this formality of space and time, one must ask oneself where this essential ideality, the transcendental subjectivity of the forms, enters. What, if anything, would change, if rather than being immersed in one's soul these forms were imposed on one's epidermis, for instance, planted upon one's nose like spectacles? Without losing anything one could even place them close to or within the objects, or in the *Ding an sich*. The ideality of the forms does not suffice to solve Kant's problem. What matters is not that something *a priori* is imposed on the sensations but that the reasoning mind is able to grasp it. How can reason profit from the fact that, thanks to this theory, the distance where the forms are settled, is diminished, as it were, from a few metres to a few centimetres?

Kant's problem was the possibility of synthetic judgments *a priori*, and in order to solve it one has to explain the relations of space and time with the judgments, rather than with the perceptions. According to Kant, space and time are forms of the intuition (or perception), rather than of the understanding. To be sure, there are also forms of the judgments, the logical categories. But this yawning gulf has been neglected in the discussions on Kant's philosophy of space and time.

No doubt Kant was aware of the problem. For its solution one shall take a look at the Transcendental Analytic, a chapter of the "Critique" that has seldom been read, and, if read at all, hardly been understood. Meanwhile it is my sad duty to state that the discussion of Kant's philosophy of space and time was based on a badly understood theory, and even worse, that the problem was not even understood.

Let us recall what distinguishes fundamentally modern from ancient science – the most striking example being biology! After the MIddle Ages the ancient idea of global links between causes and effects yielded to an analytic spirit, which amomg others was technically supported by the in vention of the microscope. So the problem arose how the soul manages to synthetise the stimuli, analysed by the retina and the other sensual organs, in order to recover the global images of reality in perception. The 19th century physiologists thought they had found the answer in Kant's work: by means of the forms of pure intuition (or perception), which are space and time. Today one would rather speak about programmes, available to the brain. Then space and time would merely be two among thousands of others in the organisation of the nervous system, all of them preceding experience as Kant meant it, of which each individual should acquire the mastery by means of its experiences.

And what are the formal conditions of the judgments? Today we would speak about models, which in turn are a vast manifold, ever and ever more refined, from the models of common sense to those of the most modern sciences. Space and time are merely two examples among them, albeit basic to many others.

If one tried to reinterpret whatever Kant's intentions might have been, one would hope to discover something like a congruence between the forms of perception and the models of representation, which in my view is a vain hope. On the contrary, all clues suggest that, to say it this way, the internal logic of the brain differs fundamentally from its external logic such as is revealed in our conscience – I dare say, as fundamentally as that of computers differs from both of them.

The end of the story? The ontological question on the nature of space is reduced to one of scientific methodology!

La "lineale Ausdehnungslehre" (1844) de Hermann Günther Grassmann

Dominique Flament

Recherches Epistémologiques et Historiques sur les Sciences Exactes et les Institutions Scientifiques (R.E.H.S.E.I.S.), U.P.R. 318 du C.N.R.S., 49, rue Mirabeau F-75016 PARIS (France)

Notre exposé[1] est en deux parties ; la première apporte quelques éléments d'ordre général servant à situer l'œuvre de Grassmann par rapport à celles de certains de ses contemporains et permet de voir ce qu'il est advenu de sa *Lineale Ausdehnungslehre* au cours du XIXe siècle. La seconde partie, reprend à la fois l'introduction que fait Grassmann à son ouvrage et la théorie générale des formes. Une troisième partie, qui ne sera pas exposée ici, développerait dans toute sa technicité d'époque, et en respectant le "style" de l'auteur, la théorie de l'extension elle-même. Nous avons préféré, au risque de décevoir, la seconde partie à celle-ci, convaincu que l'on ne peut parvenir à une véritable compréhension de la théorie de l'extension sans ce nécessaire préalable voulu par l'auteur.

Première Partie

Nous n'évoquerons ici que l'ouvrage de 1844 du mathématicien allemand Hermann Günther Grassmann, dont le titre complet est "*Die LINEALE AUSDEHNUNGSLEHRE, eine neuer Zweig der Mathematik dargestellt und durch Anwendungen auf die übrigen Zweige der Mathematik, wie auch auf die Statik, Mechanik, die Lehre vom Magnetismus und die Krystallonomie erläutert.*" [10] ; il correspond à la première partie, à la seule ayant été publiée, d'une théorie plus vaste, l'*Ausdehnungslehre*.

Nous passerons également sur les nombreuses et multiples difficultés que rencontra Grassmann pour obtenir un poste digne de ses compétences, disons simplement que celles-ci lui permirent d'enseigner dans plusieurs écoles de Stettin [Szczecin], et dans une école technique de Berlin, aussi bien les mathématiques, la physique, le latin, l'allemand que la religion. Ce n'est qu'en 1852, il a alors 43 ans, qu'il pourra enfin avoir un poste définitif de professeur de mathématiques et de physique en succédant à son père au Lycée de Stettin.

L'accueil fait à sa *lineale Ausdehnungslehre* en 1844 est très défavorable ; August Ferdinand Möbius [27], par exemple, dont les travaux sont voisins d'une partie de ceux de Grassmann (notamment tous ceux qui concerne son *Calcul Barycentrique* (1827) [25], refusera d'écrire un Compte Rendu de l'ouvrage. Selon lui, il y a pléthore de concepts

---

[1] Cette conférence a été en partie préparée en collaboration avec le Dr. B. Bekemeier.

nouveaux, de mots nouveaux, et excès de langage philosophique. Grassmann est parfaitement conscient de ces obstacles, sa préface le prouve ; il y dit explicitement que son introduction de caractère philosophique peut très bien être écartée sans que cela puisse nuire à la compréhension de l'ouvrage. En fait cette déclaration n'est pas totalement vraie, nous reviendrons sur ce point.

A diverses reprises, d'abord par un *Résumé* [11] puis par des publications en physique [12] (voir aussi ses *Werke* [16]) et en mathématiques, vues comme autant d'applications, il tenta de faire connaître sa théorie générale qui, dit-il, renferme "la seule méthode conforme à la nature, selon laquelle [doit] progresser toute application de la mathématique à la nature, et selon laquelle on doit traiter la géométrie, si on veut qu'elle conduise à des résultats généraux et fructueux" ([10], p.VIII). Il s'agit là d'une théorie générale qui écarte les coordonnées, les jugeant trop arbitraires et étrangères au sujet développé ; la méthode usuelle n'offrait jusqu'à présent, toujours selon lui, qu'un calcul consistant en un développement mécanique de formules, un calcul qui ne pouvait rien procurer à l'esprit et, donc, le "tuait" ; c'est un calcul direct sur les objets qu'apporte Grassmann .

Une publication a retenu l'attention dès sa parution en 1846, sa *Geometrische Analyse* [13] remporta un prix. Mais elle ne suffit pas à établir la renommée scientifique de Grassmann, si on s'en tient à ce qui suit. Möbius, un des censeurs du concours, se senti obligé d'ajouter à l'écrit de Grassmann, pour sa publication, un article intitulé "La théorie de Grassmann sur les grandeurs de points et sur les formes de grandeurs dépendantes" [26] ; il trouvait que la dernière partie de l'Analyse géométrique était beaucoup trop difficile à lire . L'année suivante, en 1847, Grassmann se porte candidat pour un poste de professeur de mathématiques dans une Université prussienne ; dans le rapport demandé par le ministère à Ernest Eduard Kummer on lit que l'*Analyse géométrique* renferme "du matériel méritoire, mais présenté sous une forme insuffisante". Ce jugement sévère, qu'il serait bon d'analyser de plus près, priva Grassmann de ce poste.

Au cours des années 1850, il reprend ses travaux pour un temps délaissés sur l'*Ausdenungslehre*, sans doute faut-il voir là le résultat des encouragements de Möbius, mais plus encore faut-il y voir le résultat de l'activité de Grassmann à faire valoir sa priorité. Dans un article en français daté de 1855 [14] il se réfère aux récentes communications de 1853 de A.L. Cauchy sur les *Clefs algébriques* [2], "Au premier coup d'oeil, je reconnus que les principes qui y sont établis et les résultats qui en ont été tirés, étaient absolumment les mêmes que ceux que j'avais déjà publiés en 1844". Les points communs entre ces deux théories sont nombreux ; par exemple, les "unités relatives" ["Grundänderungen"] de Grassmann, "ces quantités qui sont employées pour en déduire d'autres au moyen d'une multiplication par des facteurs algébriques", correspondent aux "clefs algébriques" ; "les facteurs symboliques" de Cauchy correspondent aux "quantités extensives". Mais, observe Grassmann, Cauchy regarde les clefs algébriques simplement comme un moyen pour résoudre divers problèmes d'analyse et de mécanique qui, une fois ces problèmes résolus, disparaissent ; en revanche, précise-t-il, "...d'après les principes établis par moi, on est en état, à chaque pas du processus, d'attribuer une signification indépendante aux unités relatives et aux quantités qui en sont composées, quelle que soit d'ailleurs la marche que l'on suive" ([14], p.127).

Par conséquent, Grassmann n'écrira pas le second volume de son *Ausdehnungslehre* projetté depuis longtemps, il préfère plutôt retravailler sa *lineale ausdehnungslehre* dont il publiera en 1862 une nouvelle version "complète et présentée de façon rigoureuse". Cet ouvrage va connaître le même sort que le premier. Si on en croit Hermann Hankel [23], cette fois la présentation n'est pas décourageante à cause d'une rédaction philosophique, "qui cependant lui était tout-à-fait appropriée", ou à cause de la "forme inhabituelle des opérations avec des grandeurs qui n'ont pas le caractère des grandeurs habituelles de l'arithmétique" ; elle l'est paradoxalement à cause de la trop grande généralité qu'a donné Grassmann aux théorèmes en étendant à n dimensions, pour n quelconque, leur validité. Les mathématiciens furent sans doute plus satisfaits par la nouvelle forme de présentation et par le désir exprimé de Grassmann à vouloir situer son approche par rapport à celles d'autres contemporains sur le même sujet, mais ils continuèrent cependant à ne pas apprécier la terminologie particulière, pourtant très épurée, de cette "étrange" théorie.

Dans les années 1860, jusqu'en 1872 pour être plus exact, Grassmann semble délaisser ses recherches mathématiques, ce nouvel échec de 1862 paraissant avoir eu raison de sa résistance. C'est pourtant au cours de la même époque que quelques mathématiciens commenceront à s'intéresser publiquement de plus près à son œuvre. Parmi ceux-ci signalons Hermann Hankel (1867), Rudolf Friedrich Clebsch (1872) et Viktor Schlegel (1872, 1875) [34]. Rendons aussi justice à William Rowan Hamilton qui très tôt reconnu une certaine valeur au travail de Grassmann ; dans des lettres adressées à Augustus DeMorgan au début des années 1850, il écrit [20] :

> "a very original work .... will perhaps set up in rivalship with mine." (26 Octobre 1852) "I have recently been reading ... more than a hundred pages of Grassmann's Ausdehnungslehre, with great admiration and interest. Previously I had only the most slight and general knowledge of the book, and thought that it would require me to learn to smoke in order to read it. If I could hope to be put in rivalship with DesCartes on the one hand, and with Grassmann on the other, my scientific ambition would be fulfilled!" (31 Janvier 1853) "Grassmann is a great and most German genius..." (2 Février1853)

En 1878, vu l'intérêt sans cesse croissant de la communauté scientifique, l'éditeur de Grassmann accepta de rééditer la version originale de l'*Ausdehnungslehre* de 1844, devenue très rare parce que mise au pilon depuis longtemps. Dans la préface qu'il eut le temps d'écrire avant de mourir, Grassmann revient sur certaines de ses positions abandonnées en 1862, il insiste notamment sur le caractère essentiel de la présentation philosophique pour une pleine compréhension de sa théorie.

Dès 1888, Giussepe Peano, séduit par le texte de 1844, mais plus encore, par le *Résumé* [11] de 1845 beaucoup plus facile à lire, reprit avec ses recherches sur le *Calcolo geometrico* [30] les travaux de Grassmann ; pour ce faire il dispose du langage ensembliste qui faisait défaut à Grassmann, il peut donc donner une définition générale des espaces vectoriels (sur le corps des réels) voisine de celle que l'on connait aujourd'hui. Les travaux de Peano (voir aussi [31] et [32]) servirent beaucoup la cause de Grassmann en permettant une plus large diffusion et plus grande acceptation de ses idées ; mais, là encore, le succès ne sera pas de mise ; c'est la définition d'Arthur Cayley, donnée en 1845 [5] et faisant d'un *vecteur* de l'espace à *n dimensions* simplement un système de *n nombres réels ou imaginaires*, qui aura les faveurs des mathématiciens de 1850 à 1930 environ. Pour Cayley, il s'agissait là surtout d'un langage commode. "C'est seulement lorsque les développements de l'analyse fonctionnelle rendront impératif l'abandon de la définition d'un espace vectoriel par des coordonnées, que, par contre coup, s'imposera la définition directe" ([7], t.1, p.95). L'*algèbre extérieure* de Grassmann, très loin d'être explicite dans le texte de 1844, ne commencera à être vraiment connue que lorsque Henri Poincaré [33] et surtout Elie Cartan [4] souligneront son importance en géométrie différentielle. "C'est seulement après 1930, lorsque l'œuvre de Elie Cartan a commencé à être comprise, que celle de Grassmann a repris la place centrale qui lui revenait dans toutes les applications de l'algèbre linéaire et multilinéaire" ([7], t.1, p.111).

Lorsque l'on compare l'œuvre de Grassmann à celles de ses prédécesseurs et de ses contemporains, on constate qu'en 1844 elle reste très largement supérieure à toute autre sur un plan strictement mathématique. On a vu grossièrement comment situer les résultats de Möbius, Cayley ou Cauchy, disons des approches de Warren [35] et de Mourey [28] qu'elles parvenaient dès 1828 à des systèmes presqu'achevés en ce qui concernait l'inventaire des axiomes devant caractériser les opérations sur des objets destinés à écarter les "quantités impossibles ou imaginaires". Chez Mourey, reconnaissent E. Study et E. Cartan, la théorie est, contrairement à celle d'Argand ou de Français,"exposée dans toute sa rigueur... [et] met bien en évidence la nécessité de définir les opérations sur les nouveaux objets qu'on veut introduire dans le calcul" ; cependant, son "algèbre émanée de la géométrie ou cette géométrie généralisée et rendue algébrique", ce "nouveau système de Géométrie" ([29], Préface, p.VII) qu'il propose aux "Amis de l'évidence" reste très élémentaire et exprimé avec trop de notations redondantes et souvent inutilement compliquées. Excepté Gaspar Wessel [36], lui aussi resté longtemps inconnu, et qui pourtant énonçait en danois dès 1797 des notions voisines de celles de Hamilton et Grassmann, dans un espace de dimension au plus égal à trois, tous les prédécesseurs s'étant plus particulièrement attachés à la recherche d'une représentation géométrique des quantités imaginaires ou impossibles dans le but de leur donner une plus grande légitimité et droit de cité comme entités mathématiques à part entière, n'ont en fait que très peu en commun avec Grassmann. Bien sûr de nombreux éléments ou concepts ont déjà été énoncés par d'autres[2], notamment par Bellavitis, Chelini, Justus Grassmann (son père), Saint-Venant ou Hamilton, mais comme le fait à juste titre remarquer Peano tout et plus est chez Grassmann, "le mérite de Grassmann est d'avoir étudié un système très général et organique d'opérations qui comprend comme cas particulier tout ce qui précède" ([30], p.168]).

Hamilton ([21] et [22]) est le seul qui soit vraiment très proche de Grassmann, une grande partie des résultats leur est commune. Mais, en fait, les vues de Grassmann sont beaucoup plus profondes que celles d'Hamilton. Ce dernier est trop attaché à définir un système d'unités imaginaires de son quaternion et à faire de lui un objet universel, alors que Grassmann essaye, pour ne citer que cet aspect, de dégager la notion "abstraite" de "structure d'espace vectoriel" (sur $\mathbb{R}$ ou $\mathbb{C}$).

Nous avons préféré pour la traduction en français de la *lineale Ausdehnungslehre*, dont nous sommes en train d'achever les commentaires, l'édition de 1844 à celle de 1862.

L'édition de 1862 est certes plus "confortable" sur un plan strictement mathématique, de nombreux concepts sont beaucoup mieux exposés et certaines définitions ne sont plus uniquement implicites (par exemple, celles de "combinaison linéaire" de "grandeurs", d'"indépendance linéaire", de "base" d'un "domaine" (i.e., à peu de chose près, celle d'espace vectoriel ou de sous-espace vectoriel), de "dimension", etc. sont plus clairement mises en valeur. Mais cette édition écarte trop de notions jugées essentielles par l'auteur lui-même, ou les minimise. Ainsi, la place occupée par l'*eingewandte Produkt* est très réduite dans l'édition de 1862, alors que dans celle de 1844 elle couvre près du quart de l'ouvrage. L'édition de 1862 délaisse en grande partie un vocabulaire crée par Grassmann

[2] Voir [6] où ces différentes contributions sont analysées très en détail par J.J. Crowe.

poussé en celà, entre autre, par un souci immodéré de "délatiniser" la langue allemande : par exemple, l'*eingewandte Produkt*, le *Spateck*, l'*ausdehnungsgrösse*, l'*eingeordnet*, etc. deviendront ou redeviendront respectivement, le *regressiv Produkt*, le *Parallelogramm*, l'*extensive Grösse*, l'*incident*, etc., et il en est ainsi de nombreux autres mots artificiels créés par lui ; certains survivront à cette épuration, le mot *Spath* , par exemple, continuera à designer le *parallélépipède.*

Enfin, et ce ne sont pas les moindres raisons, l'édition de 1862 délaisse l'approche philosophique et écarte la théorie générale des formes, elle nous prive donc, d'une part, d'éléments qui sont d'une grande importance sur le plan historique et, d'autre part, d'explications mathématiques de très grande valeur.

Grassmann n'est pas à proprement parler un "philosophe" et, ce qui peut aussi expliquer l'originalité de son style, il est un mathématicien "autodidacte". Sa formation reflète l'influence considérable qu'exercèrent sur lui son père, Schleiermacher et, bien qu'à un degré beaucoup plus difficile à apprécier, le *milieu* ["*Zeitgeist*"] Stettinois encore marqué par l'"Ecole Combinatoire" fondée à la fin du 18e. siècle par Karl Friedrich Hindenburg. Dans la préface de son livre Grassmann reconnait à plusieurs reprises que l'influence de son père à été déterminante pour la réalisation de sa *Lineale Ausdehnungslehre*, certains des éléments les plus caractéristiques, philosophiques ou mathématiques, de son œuvre sont déjà dans la *Raumlehre* [17], la *Trigonometrie* [19] et dans l'article *Sur le concept et l'étendue de la pure théorie des nombres* de son père.[18]

Nous n'insisterons pas ici sur l'influence manifeste ou non qu'exerça la *Dialektik* (1839) de Schleiermacher sur la présentation du contenu de l'ouvrage, ni sur celles de Schelling et de la *Naturphilosophie Romantique* de l'Allemagne du début du 19e siècle ; deux articles, l'un de A.C.Lewis [24] et l'autre plus récent de Michael Otte [29], ont largement développé ces aspects. Achevons sur ce point, en rappelant qu'en 1845, dans une lettre adressée à Möbius (cit. Lewis, [24] ; p.108), le professeur de philosophie E.F. Apelt se référant à l'*Ausdehnungslehre* de Grassmann écrivait :

> " Il me semble qu'une fausse philosophie des mathématiques est à la base. Le caractère essentiel de la connaissance mathématique, l'intuitif, semble en être complètement exclu. Une théorie abstraite de l'extension comme celle que souhaite Grassmann, peut être uniquement développée à partir des concepts ; la source de la connaissance mathématique se trouve non dans les concepts, mais dans l'intuition."

Deuxième Partie

Venons en maintenant au texte proprement dit, et plus particulièrement à son *introduction* et à la *théorie générale des formes*, partie qui fut considérée comme l'une des principales difficultés ayant rebuté et détourné un grand nombre de lecteurs mathématiciens et physiciens.

A. Déduction du Concept de Mathématique Pure

Grassmann commence ce paragraphe en faisant état d'un premier partage :

> "La division la plus élevée de toutes les sciences est celle entre les sciences réelles et les sciences formelles. Les premières figurent l'être dans la pensée – un être lui-même indépendant de cette pensée – et leur vérité est donnée par la concordance de la pensée avec cet être ; les secondes .... ont pour objet ce qui est posé par la pensée elle-même, et leur vérité est donnée par la concordance entre eux des processus de pensée. Dans les sciences formelles la démonstration ... reste purement dans la combinaison des divers actes de pensée." ([10], p.XIX)

Pour qu'il en soit ainsi, que ce dernier point soit possible, il faudra que les sciences formelles ne partent pas des *axiomes* comme le font les "sciences réelles" ; ce sont les *définitions* qui devront constituer leurs bases ([10], p.XIX]). La principale différence entre ces deux types de science, est dans la spécificité de leurs objets. Dans les *sciences réelles*, les objets (ou l'"être") sont les choses de la réalité, les substances. La science est un rapport entre le sujet et l'objet ; entre la pensée et la réalité. Dans les *sciences formelles* il n'y a pas un tel être en tant qu'objet ; l'être est d'abord posé par la pensée elle-même. La science est un rapport entre la pensée et la pensée posée précédente. Vient ensuite un nouveau partage, cette fois au sein des sciences formelles ; celles-ci

> "examinent, soit les principes *généraux* de pensée, soit le *particulier* posé par la pensée ; la première est la dialectique (logique), la seconde la mathématique pure... La dialectique, pour sa recherche de l'unité en toute pensée, est une science philosophique ; en revanche, les mathématiques suivent une direction opposée en prenant pour particulier tout ce qui est pensé isolé." ([19], p.XX)

Par conséquent, conclut-il enfin,

> "La mathématique pure est la science de l'être *particulier* en tant que devenu [né] par la pensée. Nous appelons l'être particulier, pris dans ce sens, une forme-pensée ["Denkform"] ou tout simplement une *forme*. Ainsi, la mathématique pure est *théorie des formes* ["Formenlehre"]." ([10], p.XX)

C'est ce type de déclaration d'allure philosophique qui n'était guère fait pour "séduire" à cette époque la majorité des mathématiciens, et qui les faisait se sentir "mal à l'aise". La division entre les sciences formelles, compte tenu de la division qui la prècéde, apporte à Grassmann la possibilité d'insister, à l'instar de Hamilton ou Boole, sur le fait que l'expression *théorie des formes* doit être préférée à celle de *théorie des grandeurs*, cette dernière ne pouvant convenir à l'ensemble des mathématiques. Par exemple, elle ne convient pas à la "théorie des combinaisons", qui selon Grassmann est une "branche essentielle des mathématiques", et ne s'applique que de manière figuré à l'*arithmétique*. Ce partage des sciences formelles est aussi l'occasion pour lui de revoir la place accordée à la géométrie au sein des mathématiques. C'est une branche du savoir qu'"on a inclu à tort dans les mathématiques". Compte tenu de la première division, il est évident que la "géométrie, aussi bien que la mécanique, [renvoyant] à un être réel, ..., l'espace pour la géométrie", ne sauraient être considérées comme des sciences formelles. Il est clair, dit-il, que l'espace ne peut être engendré par la seule pensée ([10], p.XX). Ainsi, la géométrie doit elle être considérée comme une application de la théorie des formes aux intuitions fondamentales de l'espace.

B. Déduction du Concept de l'Ausdehnungslehre

On l'a vu, les sciences formelles ont pour objet ce qui est posé par la pensé elle-même, leur vérité est donné par la concordance des processus de pensée et la mathématique pure est la science de l'être *particulier* en tant que devenu [né] par la pensée. C'est sur ce dernier point que Grassmann porte son attention :

> "Tout ce qui est devenu par la pensée ... peut l'avoir été de deux manières : soit par un *acte* simple *générateur* ["Erzeugens"], soit par un *acte* double de pose ["Setzens"] et de liaison ["Verknüpfens"] ; le premier acte donne la *forme continue* ou la *grandeur* au sens étroit ; le second donne la *forme discrète* ou *forme-liaison.*" ([10], pp. XXI-XXII)

Mais Grassmann n'en reste pas là, un tel *contraste* ["Gegensatz"] entre le *discret* et le *continu* est, dit-il, comme tous les vrais contrastes, fluide, mouvant ["Fliessender"]. C'est-à-dire, que le discret peut aussi être regardé comme continu et, réciproquement, le continu comme discret. Soit encore,

> "Le discret est regardé comme continu si ce qui est lié est compris .... comme quelque chose de devenu et si l'acte de liaison est compris comme un moment du devenir. Et le continu est regardé comme discret si les moments particuliers du devenir sont compris comme de purs actes de liaison et si ce qui est ainsi lié est regardé comme quelque chose de donné pour la liaison." ([10, pp.XXII-XXIII)

Vient ensuite un nouveau contraste :

> "Chaque particulier devient tel par le concept du *distinct* ["Verschiedenen"], par lequel il est coordonné à un autre particulier, et par le concept de l'*égal*, par lequel il est sub-ordonné avec d'autres particuliers à un général commun. Nous pouvons appeler *forme algébrique* ce qui est devenu par l'égal et *forme combinatoire* ce qui est devenu par le distinct." ([10, p.XXIII)

Là encore le contraste entre l'*égal* et le *distinct* est fluide, mouvant. L'égal est distinct, par ce que l'un et l'autre qui lui sont égaux sont d'une quelconque manière séparés ("sans cette séparation ils ne seraient qu'Un, donc rien d'égal") ; le distinct est égal, parce que les deux sont "liés par l'activité se rapportant aux deux, donc sont des choses liées." C'est à partir du croisement ["Durchkreuzung"] de ces deux contrastes, *continu-discret* et *égal-distinct*, dont le premier se rapporte à la manière d'engendrer ["die Art der Erzeugung"] et le second aux éléments de la génération ["die Elemente der Erzeugung"], que résultent les quatre genres de formes et les branches correspondantes de la théorie des formes ([10], p. XXIII).

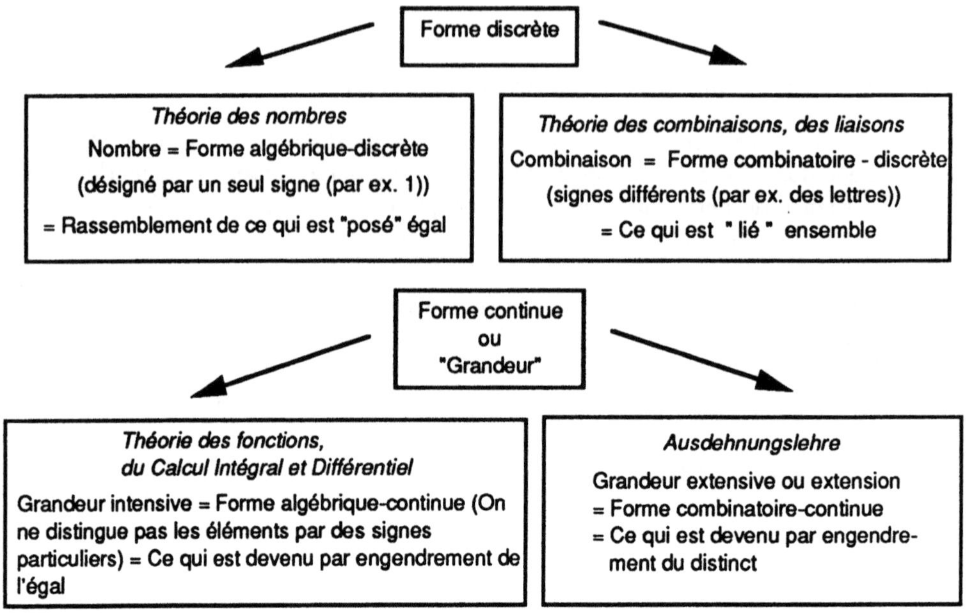

On peut encore représenter ceci plus simplement de la façon suivante :

LE CROISEMENT DES DEUX CONTRASTES

- CONTINU-DISCRET (se rapporte à la manière d'engendrer ; aspects "réels" des mathématiques)
- DISTINCT-EGAL (se rapporte aux éléments de la génération ; aspects "formels" des mathématiques)

CONSTITUE LE SYSTEME DES MATHEMATIQUES DE GRASSMANN :

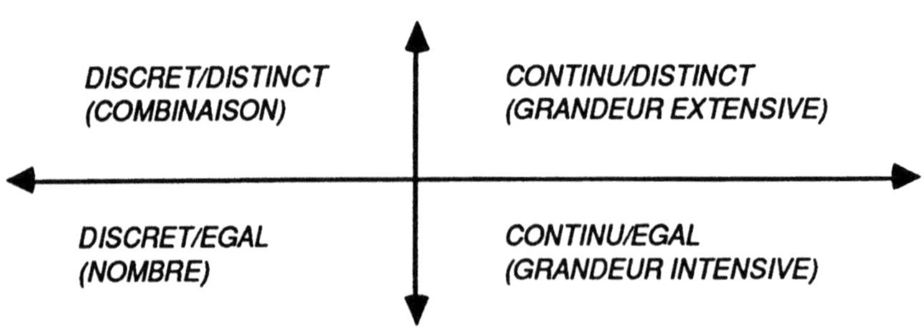

Dans le système des mathématiques de Grassmann on aura donc quatre branches :

I. DISCRET/DISTINCT ≡ THEORIE DES COMBINAISONS
II. DISCRET/EGAL ≡ THEORIE DES NOMBRES (Arithmétique)

III. CONTINU/EGAL ≡ THEORIE DES FONCTIONS (calcul intégral et différentiel)
IV. CONTINU/DISTINCT ≡ AUSDEHNUNGSLEHRE ("Théorie de l'extension")

C. Exposition du Concept de l'Ausdehnungslehre

Grassmann apporte ici des détails plus précis sur la "Théorie de l'extension", et retient donc le contraste "continu/distinct" :

> " Le devenir continu, analysé dans ses moments, paraît telle une génération continue ["stetiges Entstehen"] en fixant ce qui est déjà devenu. Dans la forme d'extension ["Ausdehnungsform"], ce qui est en train d'être engendré est chaque fois posé comme une chose distincte ; si maintenant nous ne fixons pas à chaque fois ce qui est devenu, nous parvenons au concept de *changement continu* ["stetigen Aenderung"]. Nous appelons élément générateur ["Erzeugende Element"] ce qui subit ce changement, et, quelque soit l'état que prend l'élément générateur lors de son changement, il est [appelé] élément de la forme continue." ([10], p. XXVII)

Pour rendre plus clair ce qu'il vient de dire, Grassmann fait appel à la *théorie de l'espace*. Dans celle-ci, le *point* est l'*élément*, le *changement de lieu*, ou le *déplacement*, se présente comme son *changement continu*, et, enfin, ce sont les *différentes positions* du point dans l'espace qui figurent ses *différents états*.

> " Le distinct doit se développer selon une loi pour que l'engendré ["Erzeugniss"] soit déterminé. Pour la forme simple, cette loi doit être la même pour tous les moments du devenir. La forme d'extension *simple* ["Die einfache Ausdehnungsform"] est, par conséquent, la forme qui résulte d'un changement de l'élément générateur qui suit la même loi ; nous appelons *système* ou *domaine* ["Gebiet"] l'ensemble de tous les éléments qui peuvent être engendrés suivant la même loi." ([10], p.XXVII)

De nouveau Grassmann fait appel à la géométrie pour illustrer son propos. Dans la *théorie de l'espace*, c'est l'"égalité de la direction qui embrasse les changements individuels" ; par conséquent, le "strecke" [ligne dirigée, ligne de grandeur et de direction, "vecteur" au sens d'Hamilton] et la *droite* ("infinie") correspondent respectivement à l'*extension simple* et à tout le *système*. Grassmann en vient alors à considérer deux et plus *lois de changement*, ce qu'il exprime de la façon suivante :

> " Si on applique deux lois différentes de changement, alors l'ensemble des éléments qui peuvent être engendrés forme un système du deuxième échelon. Les lois de changement, par lesquelles les éléments de ce système peuvent résulter les uns des autres, dépendent des deux premières ; si une troisième loi indépendante est ajoutée, alors on parvient à un système du troisième échelon, et ainsi de suite." ([10], p.XXVIII)

Encore une fois, la *théorie de l'espace* lui vient en aide, et calque exactement ce qui vient juste d'être dit :

> " Dans ce cas tous les éléments d'un plan sont engendrés à partir d'un seul élément par deux directions, l'élément générateur progressant à volonté dans les deux directions... et la totalité des points (éléments) ainsi engendrés font un [plan]. Le plan est alors un système du deuxième échelon ; une infinité de directions, qui dépendent des deux premières, y est contenue. Si on ajoute une troisième direction indépendante, alors tout l'espace infini est engendré (comme système du troisième échelon) au moyen de ces trois directions." ([10], p.XXVIII)

et il insiste à nouveau, "on ne peut pas aller au-delà de trois directions indépendantes (lois de changement), alors que dans l'Ausdehnungslehre pur" ([10], ibid.), par opposition à l'appliqué ou au concret qui désigne la géométrie, "le nombre de directions peut croître à l'infini." ([10], ibid.)

On a là un des principaux avantages de la séparation que fait Grassmann entre le développement théorique de l'*Ausdehnungslehre* et son application au "monde physique" : dans la première, le nombre de lois de changement indépendantes n'est pas restreint. On notera, au passage, et ceci se produit souvent, qu'il utilise dans le contexte de sa science abstraite le mot "direction", propre à la géométrie, à la place du mot "loi de changement".

D. Forme de Présentation

La forme de présentation constitue le dernier aspect qui retient particulièrement l'attention de Grassmann dans son introduction. Elle permet par la suite de comprendre pourquoi un lecteur, non averti, qui aurait sauté l'*Introduction*, comme cela est suggéré dans la *Préface*, rencontre de nombreuses difficultés avec la théorie de l'extension *linéale* et, notamment, est dérouté par la façon dont elle est exposée. Grassmann commence par un contraste entre la méthode philosophique et la méthode mathématique :

> " L'essence de la méthode philosophique est sa progression par contrastes pour arriver au particulier à partir du général ; en revanche, la méthode mathématique, progresse des concepts les plus simples aux concepts les plus composés et ainsi parvient, par la liaison du particulier, à des concepts nouveaux plus généraux (...). Comme les mathématiques et la philosophie sont des sciences au sens le plus strict, les deux méthodes doivent avoir quelque chose en commun, ce qui les fait être scientifiques. Mais attachons maintenant la qualité scientifique à une méthode de traitement si d'une part elle conduit nécessairement le lecteur à la reconnaissance de chaque vérité individuelle, et, d'autre part, le met en position de pouvoir embrasser à chaque chaque pas du développement l'orientation prise par sa progression." ([10],pp. XXIX-XXX])

Tout le monde admettra en effet le caractère indispensable de la rigueur scientifique ; en revanche, la "vue d'ensemble ... n'est pas encore suffisamment observée par la plupart des mathématiciens." Il ne saurait être ici question de faire nôtres les démonstrations, trop nombreuses, qui placent le lecteur dans l'impossibilité de savoir oú elles conduisent et l'obligent "à suivre aveuglément et au hasard pendant un certain temps chaque étape" du raisonnement. Ainsi, la méthode dont se fait le défenseur Grassmann aura cette double particularité : permettre au lecteur d'être, à tout moment du développement mathématique suivi, en mesure de savoir où il est, où il va, et d'anticiper ("pressentir") les étapes à venir. Cette "vue d'ensemble" correspond à une approche que nous qualifierons de didactique ou pédagogique.

Grassmann revient à plusieurs reprises dans son ouvrage sur cette méthode ; ce faisant nous pouvons observer les différents rôles qui sont distribués à la géométrie. On parle, d'abord de la relation analogique entre une science pure, l'*Ausdehnungslehre*, et son *application*, ou *forme concrète*, la géométrie. Les constructions géométriques sont ensuite introduites pour servir l'"intuition". Enfin, la géométrie est considérée pour elle-même dans la partie "applications" que l'on trouve à la fin de chaque chapitre du livre.

Avec cette nouvelle approche, que l'on trouve déjà mise à profit par des mathématiciens français, dit-il, "le lecteur se sent libre" et "n'est pas à l'étroit en des formes qu'il [doit] suivre avec servilité parce qu'il ne les maîtrise pas."

Le pédagogue Grassmann se sent chargé d'une mission dans son propre pays, où souvent les mathématiciens ont abandonné cette façon de faire en la désavouant.

Aperçu de la Théorie Générale des Formes

On peut faire précéder une partie générale à la séparation en quatre branches de la mathématique pure. Cette partie présentera les lois générales, c'est-à-dire les lois communes à toutes les branches. Cette partie, pour la distinguer de la "théorie des formes" qui désigne la mathématique pure, sera appelée "théorie générale des formes". Elle est présentée de la manière suivante par Grassmann :

> " Nous comprenons par théorie générale des formes la série de vérités qui, de la même manière se rapportent à toutes les branches des mathématiques et qui ne supposent donc que les concepts généraux d'égalité, de diversité, de liaison et de séparation. Ainsi, la théorie générale des formes devrait précéder toutes les branches spéciales des mathématiques ; mais comme cette branche générale en tant que telle n'existe pas encore, ... il ne nous reste alors qu'à la développer." ([10], p. 1)

Ces mots annoncaient ce qui doit être vu comme la partie la plus difficile pour un lecteur de l'époque, principalement à cause de la richesse de son contenu, de son degré très élevé d'abstraction et de la nouveauté des concepts. Grassmann part de formes originellement sans contenu et les relie au moyen de certaines liaisons ["Verknüpfungen"].

Ses premières explications sont pour l'égalité, la relation la plus fondamentale pour lui ; sa principale préoccupation est de sauver la simplicité de ce concept. A première vue, compte tenu de la "complexité", de la "nature", des "objets" à égaler ou du sens donné à l'égalité, il semblerait qu'il soit devenu nécessaire d'établir plusieurs concepts d'égalité. Dire par exemple que deux "vecteurs" ["Strecke"] sont égaux s'ils sont de même longueur, ou de même direction, n'est pas la même égalité que celle qui définie l'égalité de deux "vecteurs" s'ils sont à la fois de même longueur et de même direction. D'autres objets à comparer exigeraient encore de nouvelles relations d'égalité à définir. Pour Grassmann, le fait que ces relations soient liées à la nature des choses comparées suffit à lui seul à prouver que ces relations n'appartiennent pas au concept d'égalité. L'égalité reste, ce qui change ce sont les relations entre les objets considérés. Le concept d'égalité conserve sa simplicité et sa généralité, si il est déterminé comme suit :

> " Égal est ce dont on peut toujours dire la même chose, ou, plus généralement, égal est ce qui peut être mutuellement substitué dans chaque jugement [i.e. proposition]. Il

est évident, qu'on a à la fois dit ici que si deux formes sont égales à une troisième, elles le sont entre elles, et que ce qui est engendré de la même manière à partir de ce qui est égal est encore égal." ([10], p.2)

Avant d'aller plus loin, disons que cette partie, contrairement à ce que peuvent laisser entendre les mots introductifs et malgré le rôle qu'on souhaite lui voir jouer, n'est pas une description visant l'élaboration d'un système d'"axiomes". Ce que cherche Grassmann n'est pas un ensemble d' "énoncés non démontrés" (cf. [24]) ; en fait dans cette partie sont symbolisés des principes de liaison, exprimés au moyen des concepts généraux d'égalité et différence, de liaison et séparation.

Soient deux grandeurs ou "formes" ; on a vu que Grassmann préférait la seconde expression dont il disait qu'elle convenait mieux, bien que sa signification trop étendue eut pu lui laisser préférer l'expression "forme-pensée" ; mais il n'en fit rien car, ajoutait-il, la "forme", "dans sa signification pure, faisant abstraction de tout contenu réel, n'est rien d'autre que la forme-pensée". Les deux formes considérées, liées entre elles, sont appelées *membres* de la liaison ; la forme présentée par la liaison est son ***résultat***. Comme signe général de liaison est choisit $\frown$. Si donc on part d'un *membre antécédent a* et d'un *membre conséquent b*, le *résultat* de la liaison sera la nouvelle forme

$$(a \frown b). \tag{1}$$

Grassmann représente cette forme entre parenthèses pour accentuer le fait qu'on a affaire ici à quelque chose qui fait un ; c-à-d. que la forme résultante ne doit pas être vue en tenant compte de ses membres. Il s'agit d'une forme qui à son tour peut être liée à une autre forme pour en donner une nouvelle, et ainsi de suite. Par simplification et par commodité Grassmann supprime les parenthèses lorsque celles-ci renferment en totalité une expression ou lorsqu'elles sont situées à la suite d'un autre signe d'ouverture de parenthése ; on écrira par exemple :

$$(a \frown b \frown c). \tag{2}$$

au lieu de

$$((a \frown b) \frown c). \tag{3}$$

L'espèce particulière de liaison est maintenant déterminée par ce qui est fixé comme résultat par la liaison ; c-à-d. les conditions et l'étendue pour lesquelles le résultat reste inchangé. Les seules modifications que l'on peut faire, sans toucher aux formes elles-même qui sont liées, sont le *changement de parenthèses* et la *réorganisation des membres*.

Grassmann précisait que cette liaison était d'une nature telle que les deux équations suivantes pouvaient être données :

$$a \frown b = b \frown a \tag{4}$$

$$(a \frown b) \frown c = a \frown (b \frown c) = a \frown b \frown c. \tag{5}$$

La première désigne la *commutativité* ["Vertauschbarkeit"], la seconde l'*associativité* ["Vereinbarkeit"]. Il montrait ensuite que si ces équations étaient posées pour un nombre quelconque de formes liées, alors on pouvait toujours ré-arranger les membres et omettre les parenthèses ; ce qu'il exprimait par les deux théorèmes suivants :

" Si une liaison est d'une espèce telle que pour trois membres la façon suivant laquelle les parenthèses sont disposées ne cause pas une réelle différence, alors ceci est vrai pour un nombre quelconque de membres." ([10], p. 3)

et

" Si une liaison est d'une espèce telle qu'on a le droit, sans modifier le résultat, de disposer à volonté les parenthèses pour trois membres et de modifier l'ordre pour deux membres, alors la position des parenthèses et l'ordre pour un nombre quelconque de membres sont aussi indifférents pour le résultat." ([10], p.4)

Une telle liaison est dite ***synthétique*** ; elle est *simple* si elle vérifie les conditions précédentes, c-à-d. si elle est à la fois *associative* et *commutative.* Il est impossible, dit-il, d'avoir une détermination plus détaillée de la liaison, à moins de revenir à la nature des formes liées et donc d'aborder une des branches de la mathématique pure.

A partir d'une telle liaison synthétique, Grassmann déduisait ses ***liaisons analytiques.*** Le procédé analytique étant la "recherche d'un membre de la liaison, la liaison et son autre membre étant donnés." ([10], p. 5)

La liaison analytique, désignée par $\smile$, est telle que si la forme qui résulte est synthétiquement liée à l'une des deux formes prises comme membres originaires, il en résulte l'autre ; c-à-d. si a et b sont ces deux formes de la liaison analytique, et que nous lions le résultat, $(a \smile b)$, synthétiquement à b, on a par définition :

$$(a \smile b) \frown b = a. \tag{6}$$

Il montra alors que les équations suivantes sont vérifiées :

$$a \smile b \smile c = a \smile c \smile b = a \smile (b \frown c). \tag{7}$$

et

$$a \smile (b \smile c) = a \smile b \frown c. \tag{8}$$

Les résultats précédents sont réunis dans le théorème suivant :

" Si la liaison synthétique est simple, alors l'ordre dans lequel on lie synthétiquement ou analytiquement n'a pas d'importance pour le résultat ; aussi a-t-on le droit de mettre ou d'omettre les parenthèses après un signe synthétique si celles-ci ne renferment que des membres synthétiques ; cependant, après un signe analytique, on a toujours le droit de mettre ou d'omettre les parenthèses si seulement on inverse ici les signes entre parenthèses, c-à-d. si on change un signe synthétique en un signe analytique, et inversement." ([10], p.6)

Il faut voir dans cet énoncé le résultat le plus général que l'on peut obtenir, vues les hypothèses faites ; c'est-à-dire, que si l'on veut aller plus loin on cesse d'être général, on cesse de définir quelque chose de commun à toutes les branches spéciales des mathématiques, on définit quelque chose de particulier.

Mais les hypothèses considérées ne suffisent pas pour omettre ou non les parenthèses lorsque celles-ci sont précédées d'un signe synthétique et renferment un signe analytique. L'hypothèse que doit rajouter Grassmann est l'unicité du résultat de la liaison analytique.

Le lecteur de l'époque commençait à y voir enfin plus clair, lorsque Grassmann précisait que :

> " nous appelons alors (si cette unicité à lieu de façon générale) addition la liaison synthétique simple, et soustraction la liaison analytique correspondante." ([10],p.7)

Le procédé analytique permet également d'introduire de nouvelles formes, la "forme indifférente" (l'identité) et la "forme analytique". La première est obtenue en liant analytiquement une forme quelconque à elle-même ou deux formes égales ; si a est cette forme on la note $a \smile a$ et on remarque qu'elle est indépendante de la valeur a (d'où son nom). Cette nouvelle forme est représentée par le signe $\sim$, lorsqu'elle est reliée analytiquement à une autre forme, a , elle donne $\sim\smile a$; une forme désignée par $(\smile a)$ est appelée "forme analytique pure".

Si la liaison synthétique était l'addition, cette forme s'appelerait alors "la forme négative" ; la forme indifférente quant à elle s'appelerait "zéro". D'après ce qui précède, on observe directement qu'on a :

$$(a \frown \sim) = (a \smile \sim) = a \tag{9}$$

$$\frown (\smile a) = \smile a \tag{10}$$

$$\smile (\smile a) = \frown a. \tag{11}$$

En note ([10], p.8, (*)), Grassmann faisait remarquer que la multiplication arithmétique n'était pas une addition parce que la régularité (l'unicité) ne s'appliquait pas à zéro. Il observait également que l'*Ausdehnungslehre* comptait un grand nombre de liaisons analytiques qui ne sont pas régulières. Dans la même note enfin, il précisait que les principes une fois établis pour les grandeurs positives n'ont pas besoin de l'être à nouveau pour les grandeurs négatives.

Jusqu'ici Grassmann ne s'était occupé que d'une liaison synthétique ; il en vient ensuite à considérer la relation entre deux types différents de liaison synthétique. Dans ce but, il considère que le concept de l'une des liaisons doit être déterminé par l'autre. Et, à l'instar des hypothèses formulées dans le cas d'une seule liaison, il observe que la détermination conceptuelle dépend de la manière dont une expression, qui renferme deux espèces de liaison, peut être modifiée sans altérer le résultat final. Soient $\frown$ et $\stackrel{\frown}{\frown}$ les signes des deux liaisons, et considérons l'expression

$$(a \frown b) \stackrel{\frown}{\frown} c. \tag{12}$$

Si la deuxième liaison doit se rapporter d'une manière égale aux deux membres de la première, alors, écrit-il, la "modification la plus simple se présente comme étant celle qui soumet à la deuxième liaison chaque membre de la première et qui permet ensuite de prendre ces résultats particuliers comme membres de la première liaison." ([10],p.10) Si une telle modification laisse inchangé le résultat final, on a alors :

$$(a \frown b) \stackrel{\frown}{\frown} c = (a \stackrel{\frown}{\frown} c) \frown (b \stackrel{\frown}{\frown} c). \tag{13}$$

Cette "distributivité" caractérisera la seconde liaison, celle-ci étant appelée par lui la "liaison d'échelon plus élevé suivant" ["nächst höherer Stufe"]. Si la première liaison synthétique est simple et si sa liaison analytique correspondante est unique, la seconde liaison, qui satisfait la relation (13), est alors appelée "multiplication". Ce faisant, on constate que, pour lui, toute opération distributive est une multiplication. Bien sûr, à ce second type de liaison synthétique correspondait une liaison analytique ; appelée division si on considérait la liaison synthétique multiplicative. La fin de son exposé est plus facile à suivre pour le lecteur de l'époque, dans la mesure où Grassmann décide de revenir aux signes usuels d'addition, de soustraction, de multiplication et de division. C'est ainsi qu'il en revient à écrire les relations

$$(a+b)c = ac + bc \tag{14}$$

$$c(a+b) = ca + cb, \tag{15}$$

relations qui établissent le concept de multiplication. Il est plus facile ensuite pour lui d'établir la loi du partage du dividende, généralement vérifiée, que le résultat de la division soit unique ou non :

$$\frac{a \pm b}{c} = \frac{a}{c} \pm \frac{b}{c}. \tag{16}$$

Deux sortes de division doivent être considérées, la multiplication n'étant pas en général supposée commutative, selon que l'on recherche l'antécédent ou le conséquent de la liaison multiplicative.

Tout ce qui précède donne une idée des bases philosophiques que Grassmann fit siennes pour construire son système. Quelle que soit la valeur que l'on puisse accorder à sa philosophie, on ne peut pas l'écarter purement et simplement pour se suffir d'un texte mathématique vide de toute référence "extérieure" ; cette philosophie permit à Grassmann de se sentir à l'aise avec des notions nouvelles, comme celles de multiplications non commutatives ou d'espaces de dimension quelconque.

A cet égard, on peut le comparer au célèbre Hamilton lorsque ce dernier essaya de faire de l'algèbre la science du temps pur ; une approche qui, bien que non essentielle au développement des mathématiques et facile à écarter, lui permit la découverte d'une justification "purement algébrique" des "quantités imaginaires" en faisant d'elles des "couples algébriques", et qui le mit sur la voie des quaternions.

Le goût de Grassmann pour le général et l'abstrait l'a conduit à de nombreuses idées de valeur. La théorie générale des formes illustre bien cet aspect ; on l'a vu, Grassmann est parvenu à une compréhension très fine des concepts de commutativité, d'associativité et de distributivité, parmi d'autres plus abstraits encore dont nous n'avons pas parlé.

En 1844, Grassmann est un des très rares mathématiciens à parvenir à une telle compréhension ; des prédécesseurs tels que George Peacock ou encore Martin Ohm ["Essai sur un système achevé et conséquent des mathématiques" (1822)] n'ont jamais atteint une telle généralité.

Bibliographie

[1] J. R. Argand, *Essai sur une manière de représenter les quantités imaginaires dans les constructions géométriques* (1806). (2e.éd., Gauthier-Villars, Paris, 1874 ; nouveau tirage de la 2e. éd., Paris, A. Blanchard, 1971).

[2] A. L. Cauchy, *Œuvres complètes*, 27 vol. (2 séries), Gauthier-Villars, Paris, 1882-1974. (En particulier, voir : le vol. **XII** (1e. Série,1899), 439-445, le vol. **XII** (1e. série, 1900), 12-30 et 46-63, et le vol. **XIV** (2e. série, 1938), 417-466.

[3] A. L. Cauchy, Sur les clefs algébriques, *Comptes Rendus de l'Académie des Sciences*, **36** (1853), 70-75, 129-136 et 161-169.

[4] E. Cartan, *Œuvres complètes*, 6 vol., Gauthier-Villars, Paris, 195

[5] A. Cayley, *Collected Mathematical Papers*, 13 vol., Cambridge Univ. Press, N. Y. , 1889-1898 (En particulier, vol.1, 325).

[6] M. J. Crowe, *A History of Vector Analysis*, Univ. of Notre Dame Press, London, 1967.

[7] J. Dieudonné, (s.d.) *Abrégé d'histoire des mathématiques*, 1700-1900, 2 vol., Hermann, Paris, 1978 (En particulier les chap. II et IV).

[8] J. F. Français, Nouveaux principes de géométrie de position, et interprétation géométrique des symboles imaginaires, *Annales de Mathématiques* (Gergonne), **IV** (1813-1814), 61- 71 (voir aussi [1], 1971, 63-74).

[9] G. G. Granger, *Essai d'une philosophie du style*, Armand Colin, Paris, 1968 (En particulier le chap. IV, *Naissance du style vectoriel*).

[10] H. G. Grassmann, *Die lineale ausdehnungslehre, eine neuer Zweig der Mathematik, dargestellt und durch Anwendungen auf die übrigen Zweige der Mathematik, wie auch auf die Statik, Mechanik, die Lehre vom Magnetismus und die Krystallonomie erläutert.* Verlag von Otto Wigand, Leipzig, 1844 (Réed., Otto Wigand, Leipzig, 1878 ; voir également, [16], vol. **I**, 2-319).

[11] G. H. Grassmann, Kurze Uebersicht über das Wesen der Ausdehnungslehre, in *Archiv der Mathematik und Physik* (Grunert) **6** (1845), 337-350 (voir aussi [16], vol. **I**, 297-312).

[12] H. G. Grassmann, Neue Theorie der Elektrodynamik, *Annalen der Physik und Chemie* (Poggendorff) **64** (1845), 1-18 (voir aussi, [16], vol. **II**, 147-160).

[13] H.G. Grassmann, *Die Geometrische Analyse geknüpft an die von Leibniz erfundene Geometrische Charakteristik.* Leipzig, 1847 (voir aussi [16], vol. **I**, 322-399).

[14] H. G. Grassmann, Sur les différents genres de multiplication, *Journal für die reine und angewandte Mathematik* (Crelle), **44** (1855), 123-141.

[15] H.G. Grassmann, *Die Ausdehnungslehre. Vollständig und in strender Form bearbeitet*, Berlin, 1862 (voir aussi, [16], vol. **II**, 1-382).

[16] H. G. Grassmann, *Gesammelte mathematische und physikalische Werke*, 3 vol., Leipzig, 1894-1911.

[17] J. G. Grassmann, *Raumlehre, Ebene raumlehre Grössenlehre*, Berlin, 1824.

[18] J. G. Grassmann, Über den Begriff und Umfang der reinen Zahlenlehre, Programmschrift, Stettin, 1827.

[19] J. G. Grassmann, *Lehrbuch der ebenen und sphärischen Trigonometrie*, Berlin, 1835.

[20] R.P. Graves, *Life of Sir William Rowan Hamilton*, 3 vol., Dublin, 1882-1889.

[21] W. R. Hamilton, *Lectures on Quaternions*, Dublin, 1853.

[22] W. R. Hamilton, *Elements of Quaternions*, Dublin, 1866.

[23] H. Hankel, *Theorie der complexen Zahlensysteme*, Leipzig, 1867.

[24] A. C. H. Lewis,Grassmann's 1844 Ausdehnungslehre and Schleiermacher's Dialektik, *Annals of Science* **34** (1977), 103-162.

[25] A. F. Möbius, *Der Barycentrische Calcul* Leipzig, Verlag von Johann Ambrosius Barth, 1827 (Réed. Georg Olms, Hildesheim-New York, 1976).

[26] A. F. Möbius, Die Grassmann'sche Lehre von Punktgrössen und den davon abhängigen Grössenformen, [27], vol. I, 613-633.

[27] A. F. Möbius, *Gesammelte Werke*, 4 vol., Leipzig, 1885-1887.

[28] C. V. Mourey, *La vraie théorie des quantités négatives et des quantités prétendues imaginaires dédiée aux amis de l'évidence*, 1828 (2ᵉ. éd., Paris, Bachelier, 1861).

[29] M. Otte, The Ideas of Hermann Grassmann in the Context of the Mathematical and Philosophical Tradition since Leibniz, *Historia Mathematica* **16** (1989), 1-35.

[30] G. Peano, *Calcolo geometrico secondo l'Ausdehnungslehre di Grassmann, preceduto dalle operazioni della logica deduttiva*, Torino, 1888.

[31] G. Peano, Gli elementi di calcolo geometrico, *Opere Scelte*, vol. **III**, opuscolo n. **30** (1891), 41-69.
[32] G. Peano, Saggio di calcolo geometrico, *Opere Scelte*, vol. **III**, opuscolo n. **90** (1895-1896), 167-186.
[33] H. Poincaré, *Œuvres* , 11 vol., Gauthier-Villars, Paris, 1916-1956.
[34] V. Schlegel, *System der Raumlehre* , deux parties, Leipzig, 1872 et 1875.
[35] J. Warren, *Treatise on the geometrical interpretation of the square roots of negative quantities*, Cambridge, 1828.
[36] C. Wessel, *Om Direktionens analytiske betegning, et forsg anvendt formemmelig til plane og sphaeriske polygoners oplsning* (Nye Samling af det Kongelige Danske Videnskabernes Selskabs Skrifter, *Femte Del* Kjobenhavn, 1799 (Trad. française de Zeuthen, *Essai sur la représentation analytique de la direction* , H. Valentiner et T. N. Thiele, Copenhagen, 1897).

La Capture de l'Extension comme Dialectique Géométrique : Dimension et Puissance selon l'Ausdehnung de Grassmann (1844)

Gilles Châtelet

75, rue Rochechouart 75009 Paris (France)

Nous commentons la première édition (1844) de la Théorie de l'Extension de Grassmann. Rappelons que lintroduction fut très peu appréciée par les mathématiciens de son époque (considérée comme "inutile", "métaphysique", "obscure") et contrastait vivement avec l'hégémonie grandissante du "style" axiomatique. Ses découvertes mathématiques ne furent vraiment comprises que vers 1880 (Burali-Forti, Peano) et il a fallu attendre des articles contemporains (Rota et d'autres) pour rendre justice au produit "impliqué" (symétrique du produit "extérieur") et découvrir ses utilisations possibles en combinatoire.

Il est évident que Grassmann possédait une culture philosophique et faisait sienne cette idée de l'Idéalisme allemand que la science ne se réduit nullement à l'adoption de règles opératoirement fécondes et à la production de vérités simplement "déduites". Un texte mathématique doit aussi mettre le lecteur en "état" de se faire "une vue d'ensemble". Cette vue d'ensemble n'est pas une contemplation générale qui "survole" les "détails" mais se nourrit de la spécificité du singulier pour atteindre ce plan où naissent de manière *contemporaine* l'intuitif et le discursif.

Ce type d'exigences rapproche Grassmann de celles de philosophies de la Nature (Schelling–Hegel) qui refusaient la séparation accomplie de l'objet et du sujet et désiraient penser un savoir *inobjectif* de l'étant et qui pourtant n'est pas rien.

Comme l'intuition intellectuelle, la méthode scientifique exigée par Grassmann introduit à une connaissance qui ne laisse pas intact le dualisme objectif-subjectif, mais au contraire, se risque à la création même de l'objet, à affirmer l'indentité fondamentale du produit et de la productivité, identité d'autant mieux pensée, que le *mode d'articulation qui les distingue est plus précis.* On ne s'étonnera pas si les notions de dimension et d'orientation menacent toujours subrepticement la neutralité de l'observateur face à son objet. On ne peut les penser à la façon dont on s'assure d'une chose.

C'est tout le mérite de Grassmann d'avoir compris que le lecteur devait être capable de reconnaître dans le mode d'expression des formes du saisir de l'extension des degrés homologues aux étapes franchies par son entendement (l'Espace apparait ainsi comme un Entendement visible).

Nous montrons au chapitre III comment Grassmann par sa notion d'engendrement saisit profondément la dimension et quel parti il tire de l'orientation. Il peut ainsi se dégager des contraintes du spatial (les dimensions y sont figées et n'apprécient que les intervalles-résidus entre les choses) et c'est pourquoi il parvient à saisir un geste qui déploie les "formations d'extension d'ordre n" sans le secours du "réalisme" géométrique.

Dans la trace d'Oresme, Grassmann utilise toute la puissance du diagramme (base, hauteur, amplitude) et montre qu'une dimension posée en appelle une autre permettant donc de continuer des "systèmes" de dimension arbitraire.

Philosophie de la Nature et Géométrie

Cette philosophie a mauvaise presse. Il semble que les historiens de la philosophie acceptent de retenir de Hegel, Schelling et Fichte la partie politique ou esthétique de leur oeuvre, mais renoncent à "sauver" les philosophies de la Nature de l'idéalisme allemand. Le début du XIXème siècle est pourtant en Allemagne une époque d'échanges prolongés entre scientifiques et philosophes. Schelling avait la "passion du positif" et Hegel ne voyait pas dans la positivité un ennemi irréductible de la spéculation. Certains travaux au Etats-Unis, en Allemagne et en Angleterre insistent sur cet aspect de la philosophie de la Nature et sur l'importance qui y est accordée à la métaphysique comme "matrice d'idées" ; ces analyses, cependant, en restent généralement au niveau de la sociologie. L'idéalisme allemand a pourtant tenté de repenser l'articulation entre sciences exactes et spéculation philosophiques sans se limiter à la problématique de la philosophie de la connaissance et en situant l'horizon du pensable bien au-delà des conditions de possibilité de l'objet. Les systèmes ainsi développés s'inscrivent dans la tradition des grands visionnaires de la Renaissance. Un certain type de dialectique géométrique est à l'oeuvre explicitement dans les travaux de Steiner, Kummer ou Grassmann.

Rappelons que Schelling disait que la Nature était l'Entendement visible. Grassmann, Hamilton, Kummer pressentaient bien qu'il existait une relations profonde entre le procès d'individuation du sujet connaissant et le procès d'articulation des formes de saisie de l'espace.

Ceci nous amène à étudier le concept d'articulation et à remarquer que le début du XIXème siècle, apprécié comme l'âge d'or des grands systèmes philosophiques, comme celui de l'émergence de la notion de structure, est peut-être aussi celui où le centre change de statut : le point à l'infini de l'Age clasique cède la place aux points neutres d'indifférence qui marquent les degrés d'équilibre et d'où jaillissent continument les polarités. L'existence d'un tel centre d'indifférence dans les diagrammes électriques et magnétiques fascine physiciens et philosophes spéculatifs. Ce point neutre permet de dépasser les oppositions et montre bien que ce dépassement se paie de l'émergence d'une singularité. Les procès d'individuation ne sauraient être épuisés par des différences quantitatives et supposent un lien où l'entendement fonctionnel s'affaisse, où une incertitude apparait.

Les philosophes classiques avaient beaucoup spéculé sur l'infini car les "formes indéterminées" qu'il secrétait, permettaient d'abolir les différences quantitatives. De tout manière, l'infini était pour les mathématiciens du début du XIXème une "singularité" bien inoffensive et apprivoisée par la perspective qui "ramenait" facilement les pôles à distance finie !

Le "singulier" menaçant est celui qui révèle (souvent cruellement) la naïveté du "d'une part et d'autre part" et qui se manifeste en mathématiques par l'effondrement d'un certain mode "canonique" de correspondance : les points de ramification, où se proposent plusieurs saisies fonctionnelles. Ces saisies sont toujours fondées sur l'articulation "évidente" d'une ordonnée et d'une abcisse (paramètre) "générique" et oblige à opter pour telle ou telle détermination.

Le centre, comme point à l'infini, permettait certes de dominer les différences quantitatives qui séparent les être finis, mais en dominant *extérieurement* les particularités : le dispositif de porte-à-faux qui le détecte permet d'envelopper tout un faisceau de virtualités sans les confondre : *la symétrie est redécouverte comme productrice de différences.*

I. La Capture de l'Extension comme Dialectique Géométrique

> "Une ontologie qui passe sous silence la Nature s'enferme dans l'incorporel et donne, pour cette raison même, une image fantastique de l'homme, de l'esprit, de l'histoire"[1]

Merleau-Ponty aurait pu adresser le même avertissement à toute philosophie (et peut être à toute science !) qui ne connaît que cet Espace étendu, toujours donné en face de nous, ce récipient contenant bon gré, mal gré, des particularités individuées par *un repère universel* : le monstrueux point-origine de Descartes et Newton. Que reste-t-il de vivant dans cet écoulement, qui tombe perpétuellement en dehors de soi, de cette persévérance dans l'épuisement et l'indéfinité ? Les particularités subjuguées par l'Absolu dont elles sont détachées à jamais, ne peuvent qu'entrer dans une compétition féroce pour les places.

L'espace "espacé", étendu n'est qu'un produit définitif, un agrégat transparent de parties absolument extérieures, rigoureusement actuelles, clairement liées. Aucun *envoloppement*, aucune *articulation* ne sont accordés à ces coquilles vides. Elles s'entrechoquent impatientes : l'Espace étendu est bien "cet être en dehors de soi absolu qui est tout aussi bien purement et simplement ininterrompu, un perpétuel être autre qui est identique à soi"[2].

Car ces êtres exsangues, non conformes à leur concept, excroissances permanentes toujours orphelines du geste qui les pose, ne connaissent que la cruauté des lois de l'impact et de la rareté[3]. Toutes les positions s'aigrissent en pétitions et l'extériorité homogène censée assurer la co-existence paisible des choses, exaspère les contrariétés.

L'espace étendu comme *forme affirmée des choses en tant que séparées* n'entretient qu'une guerre perfide des natures mortes. Guerre d'ailleurs sans enjeu : toutes les places y sont fondamentalement équivalentes !

L'espace étendu ne peut tenir aucune promesse : les possibles qu'il distribue sans pudeur ne sont que des *possibilités prises en tant que possibilités*, sans espoir de rédemption par actualisation prochaine (ce sont des possibilités actualisée en tant que possible !). L'espace étendu se donne sans réserve parce qu'il est terminé... fini...

Pourtant Platon avait bien vu que le réceptable participe de l'intelligible mais d'une manière fort obscure. *L'espace est ce dans quoi il se trame quelque chose et qui n'est pas nécessairement le chaos.*

S'il veut avec le poète s'écrier : "l'espace était splendide !", le philosophe doit se tenir aux avant-postes de l'obscur, assumer de plein fouet la verticalité de l'Etre, le jaillissement

---

[1] Cf. [14], p.91.
[2] Cf. [9], p. 171.
[3] Cf. [16], p. 207.

des dimensions, *il doit se laisser hanter* par l'espace. La capture de l'extension, entreprise par Grassmann et d'autres grands mathématiciens du XIXème ne peut se satisfaire "d'une connaissance qui, comme celle de la géométrie ordinaire est tout à fait abstraite et privée de vie"[4].

Il s'agit d'appréhender toute la positivité du sortir de soi, de cet acte permanent d'auto-production, sans l'idolâtrer et sans oublier que le positif véritable est "celui qui se souvient d'avoir été posé" (Hegel). La chose concrète est ..... Elle n'est plus l'obstination têtue d'une *pétition d'existence.*

Cette passion du positif saisit l'instant où l'Espace étendu frissonne enfin des virtualités qui l'habitent, et nous invite à *éprouver* la dimension comme *invention* d'une articulation. Elle nous conduit pour ainsi dire par la main pour réapprendre le mouvement qui sépare et lie à la fois.

C'est bien l'articulation qui permet de se situer résolument en dehors de toute opposition et donc de dominer toute opposition. Car il s'agit plus d'épargner les vieux dualismes : sujet-objet, forme-contenu, etc... ni de se laisser submerger par la confusion de quelque "bouillie primordiale".

Une articulation convenable permet certes d'intégrer positivement toutes les forces prisonnières des contrastes, mais elle s'accompagne toujours de la naissance d'une singularité. C'est précisément ce qui la distingue du vulgaire *recollement* ou du douteux *agrégat.* Si l'articulation libère *un nouveau mode de l'homogène*, c'est toujours en introduisant une torsion.

Si avec Schelling, le philosophe-géomètre veut prendre résolument "le point de vue de la productivité dans le fini", il doit se faire forgeron et ne plus dégrader la particularité en débris.

Il saura alors apprécier dans un simple fragment l'adresse et la continuité d'un geste. Sa science se montre alors capable de réactiver une productivité jamais éteinte dans son produit et revèle que le "même esprit est à l'oeuvre dans les productions de la Nature et les créations de la liberté".

C'est bien ce qui rend remarquable la théorie de l'Extension de Grassmann. Elle constitue une véritable pédagogie des formes du saisir de l'Espace.
Pour l'apprenti géomètre qui "sait penser librement, l'assimilation des connaissances devient alors une vraie reproduction" à condition de l'initier au sujet de manière "scientifique". La géométrie présente un caractère "scientifique" (il faut comprendre ici "systématique"), comme la Philosophie, si elle conduit le lecteur à la nécessité d'admettre chaque vérité singulière (la déduction) surtout si elle le met dans une situation où il peut avoir une "vue d'ensemble"[5]. Cette vue d'ensemble n'est pas une contemplation distante de dillettante ; elle participe à l'action : c'est une *intuition intellectuelle* au sens où l'entend la philosophie de la Nature. Elle nous transporte dans cette zone privilégiée où intuition et discursivité nouent une vivante unité. Elle n'est ni a priori, ni a posteriori : elle est *contemporaine* de ce qu'elle saisit. Elle prend chaque être à son niveau d'unité sans le décomposer en éléments, ni en le *déposant* dans un "fond de réalité" plus vaste.

[4] Cf. [18], p. 226

[5] Grassmann dans son introduction de l'Ausdehnunglehre ([7]) ne se satisfait pas des démonstrations obtenues "après avoir subi aveuglément et au hasard pendant un certain temps chaque pas". Une telle démonstration ne laisse peut être rien à désirer du côté de la rigueur mais elle n'est pas scientifique : il lui manque la deuxième exigence : la vue d'ensemble (Übersichlichkeit).

Cette démarche scientifique ne se réduit nullement à l'adoption d'une panoplie de règles (fût-elle "opératoirement féconde"). Grâce à elle, le lecteur doit être capable de reconnaître dans le mode d'exposition des formes des degrés homologues aux étapes de *son* propre cheminement. La capture de l'extension ne concerne pas seulement la *connaissance du sujet, elle se nourrit de l'individuation de la connaissance du sujet*[6]. On ne construit pas l'Espace en assemblant des formes comme les pièces d'un mécano (on tiendrait alors pour acquises les articulations !). Il ne s'agit pas d'ailleurs de construire l'Espace mais de se laisser envoûter par un rythme : celui qui noue et tisse des homogènes gorgés de tensions.

Bien plus qu'un "ordre en progrès"[7], la capture de l'extension est une individuation en progrès : celle du lecteur qui peut alors *pressentir* "directement la vérité suivante", il prend l'habitude d'esquisser des mouvements singuliers de l'Espace comme Devenir : pour connaître, il faut d'abord se pénétrer du rythme de l'apprendre. C'est en ce sens que la capture de l'extension rend progressivement l'Esprit à lui même. Elle fait découvrir que, comme la Nature, l'Espace est l'Entendement visible[8].

Comme l'intuition intellectuelle, la méthode scientifique exigée par Grassmann introduit à une connaissance qui ne laisse pas intact le dualisme objectif-subjectif, mais au contraire, se risque à la création même de l'objet, à affirmer l'identité fondamentale du produit et de la productivité, identité d'autant mieux pensée, que le mode d'articulation qui les distingue est plus précis. On ne s'étonnera pas si les notions de dimension et d'orientation jouent un rôle aussi crucial dans la capture de l'extension. Elles menacent toujours subrepticement la neutralité de l'observateur face à son objet. On ne peut les penser à la façon dont on s'assure d'une chose. Elles suggèrent l'existence d'un savoir *inobjectif* de l'étant et qui pourtant n'est pas rien.

La dimension n'est ni une propriété, ni un attribut de l'objet accompli. Elle concerne à la fois l'objet et son mode de production. Je peux bien sûr, connaître les mensurations de telle ou telle chose, en itérant un certain geste (toujours le même !), transporter un étalon unité arbitraire (déjà choisi). Cette unité commode *s'applique* sur l'objet mais ne *m'implique* pas dans son procès de connaissance ou de production.

Une pensée de la dimension ne saurait se satisfaire d'un point de vue quantitatif aussi trivial : la dimension n'est pas seulement un "nombre", c'est un degré, un échelon qu'il a fallu gravir. Elle participe d'un évènement de pensée : celui où l'espace est apprécié par le savant comme déploiement prochain d'une phase, comme libérant un relief, comme possibilité de polariser une grandeur[9].

Par la dimension, l'intervalle qui n'était que le résidu d'espace entre deux choses, se charge de tensions et se révèle être la condition positive de la naissance d'une structure. La dimension sanctionne l'accès à un degré inattendu de l'Etre : elle prend la mesure d'un Devenir :

L'orientation ne ménage pas non plus l'Etendue morte : en articulant un "côte à côte" avec "un haut et un bas", elle révèle que ce qui a été *juxtaposé*, l'a été dans un certain ordre. En manifestant une pré-férence, elle éveille une puissance de scission permanente

---

[6] Cf. [20], p. 30.

[7] Cf. [8], Préface.

[8] "La Nature n'est que l'organisme visible de celui qui existe invisiblement dans notre entendement, elle ne peut agir autrement qu'avec régularité et finalité" ([19], p. 163).

[9] Cf.[17]

au sein de ce qui a été posé : elle peut se réactiver sans s'affaiblir : elle peut donc être partie- prenante d'un engendrement dialectique. Dialectique est ici à reconnaître comme "l'intelligence qui sépare et en raison même de cette séparation, introduit dans les choses un ordre et leur impose une forme". La dialectique n'abolit pas la différence, elle l'éclaire et l'approfondit. Engendrer (Erzeugen) doit être ici associé à la capacité de *manifester le ressort de quelque chose*, (sich in Zeug legen : faire tout ce qu'on peut).

Pour faire ressortir l'étoffe du spatial, je dois le saisir comme incitation à s'amplifier (erweitern). L'espace ne s'exténue plus à écouler un même être-autre. Il se resaisit par la vitesse et l'orientation, et celles-ci ne se dissipent pas dans une pure expansion. L'espace peut être saisi d'emblée comme existence au sens de "procès hors de soi, auto-manifestation et accession soi-même en cette auto-manifestation"[10].

II Déduire ou Articuler

L'introduction de Grassmann prétend "déduire" le concept de Mathématique pure puis celui de l'Ausdehnunglehre.

Il ne s'agit d'ailleurs pas d'une déduction d'énoncés à partir d'un corps d'axiomes, mais de mettre en oeuvre une discipline de discernement qui sépare des degrés d'articulation. Pour Grassmann, unir n'est pas coaguler mais être capable de mobiliser une puissance de distinction issue du flux même du Devenir. ("flüßig gewordene").

Grassmann distingue d'abord (ce qui est classique depuis Aristote) entre les Sciences "réelles" et les Sciences "formelles". Toutes deux se donnent immédiatement comme résultant d'un faire-face entre l'être représenté (produit par la pensée) mais, pour les premières l'être représenté existe par lui même alors que pour les secondes il est posé par la pensée.

Le caractère problématique du faire-face induit par les sciences "réelles" saute aux yeux ! Pour les sciences "formelles", l'épreuve est encore plus pénible : cette fois, la pensée réflexive se déchire elle-même !

Elle doit négocier avec "son" formalisme qui se dresse maintenant devant elle comme menaçant. En effet, même si la preuve est administrée par une série normée d'actes de pensées, le caractère "scientifique" au sens de la "mise en état" fait cruellement défaut[11].

C'est pourquoi par réflexivité, les sciences formelles, à peine posées, éclatent en une Dialectique vouée "à la recherche de l'unité" et une Mathématique commise à l'étude du particulier (à l'individu en tant que partie). On pourrait croire, dès lors, que Grassmann reste prisonnier des dualismes dénoncés avec véhémence par la Philosophie de la Nature.

Nous verrons au contraire, que toute son introduction et son style d'exposition (pas assez "axiomatique" et "obscur" selon certains)[12] impliquent un effort inoui pour saisir au vol le "comprendre du comprendre", pour accéder à une plus grande intelligibilité de la géométrie par la notion d'engendrement, mobilisant ainsi des forces lovées au sein même de l'Espace.

---

[10]Cf. [10], p. 90.

[11]Grassmann "In den Stand gesetzt wird".

[12]"... Grassmann est sans doute le plus philosophe : c'est aussi incontestablement le plus obscur" ([6], p. 89).

A ce brutal face à face du réflexif, qui entraîne le philosophe-géomètre dans une confrontation stérile avec la rebellion de ses propres objets, l'Idéalisme allemand et Grassmann substituent la continuité d'un procès, d'une involution intensive dont chaque spire enveloppe un degré supérieur d'épanouissement des formes, l'invention d'une articulation frôlant avec plus de subtilité l'origine géométrique de la pensée. "Le lien le plus intime ne se dessine qu'à la faveur d'un épanouissement progressif, chaque degré de séparation des forces est marqué par l'apparition d'un nouvel être..."[13].

On devine facilement que la dialectique du continu et du discret comme génération et procès de distinction de *phases* (ce qui ne présuppose aucune successivité) joue un rôle crucial dans la déduction dialectique de l'Ausdehnunglehre. Continuité et discrétion se font face comme des contraires. Il s'agit d'assumer radicalement cette hostilité pour mettre à jour le procès qui les engendre et les distingue.

La permanence triviale du "stetig" nous livre immédiatement la forme continue[14]. Le Devenir se donne alors comme débit avec la monotonie d'une constante diversité. Au "stetig" fait face la promptitude zèlée du Discret, de l'acte qui pose et lie. Le travail dialectique s'efforce de saisir la position et la liaison, non comme des actes, mais comme les moments d'une processualité continue (au sens du "flüßig gewordene"). Si je pose des termes a, b que je lie *ensuite*, je n'obtiens qu'une *abstraction* de la relation. Pour le penser concrètement (ou ici ce qui revient au même dialectiquement), je ne dois plus opposer des opérateurs à des termes, mais penser comme contemporains la liaison et la position du deuxième terme : ce dernier ne préexiste pas à la liaison. Leibniz avait déjà bien vu que l'élément différentiel dx d'une variable x ne se juxtapose pas à x comme le ferait une excroissance, mais articule une extension déjà disponible et une intensité sur le point de s'étaler (cf. figure 1) :

Fig. 1.

C'est donc uniquement la scission abstraite du flux du Devenir en un avant et un après qui légitime l'opposition Continu-Discret. Ce qui est posé et lié – le Discret – peut être compris comme continu (comme devenu) et ce qui "s'écoule continuement peut être séparé par l'entendement en restaurant la disjonction entre la liaison et la position. Dans ce derniers cas, ce que Grassmann appelle les moments singuliers du Devenir[15], sont transis en purs actes de liaison opérant sur des termes données.

La confrontation du Discret et du Continu doit donc être pensée en termes de degré et de fluidité plus ou moins grand (Grassmann dit : "le contraste entre continu et discret

[13]Cf. [18], p. 251.

[14] Stetig : incessant, assidu (stets) : en adverbe : toujours.

[15]"Einzelne Momente des Werderns".

est *fluide*"[16]. Ceci nous mène au coeur de la processualité dialectique : la question de l'invention d'une *articulation*.

L'articulation ne prétend pas concilier deux contrastes A et -A, elle contourne leur affrontement. Elle invente un nouveau mode de fluidité en déployant un éventail de degrés dont -1 et 1 sont les branches extrêmes : (ils ne sont plus opposés mais *s'écartent* de 180°). Ce qui se tenait en face-à-face hargneux est maintenant pensé comme les deux volets d'un diptique. L'articulation préfigure la forme et mobilise l'Espace. Ni "matérielle", ni "formelle", elle est *manuelle* : elle palpe et attendrit. (Schelling parle de "la tendresse de l'articulation"). Elle permet de séparer sans détacher, de tendre sans rompre. Elle exalte le conflit de termes sans subordination : les polarisations se substituent aux oppositions. Les chocs, les impulsions ne séduisent que les pensées fascinées par l'actualité et les individus accomplis. Articuler c'est toujours se ménager un nouvel enveloppement, découvrir un matériau plus *ductile* que celui des côtés[17].

Une articulation ne relie pas deux contenus ou deux segments disjoints qui lui préexistent : elle saisit l'émergence même de ses côtés à partir d'un point d'indifférence. Dans sa fourche, l'articulation retient le produit et la productivité. Elle participe toujours de la libération d'une dimension. En esquissant dans l'air tout un *spectre* de virtualités, elle manifeste que c'est *toujours un geste qui multiplie le savoir*. Elle ne classifie pas, elle discerne des degrés et éveille des rythmes. L'articulation s'esquisse dans le ruissellement du Devenir : sans elle, la constance se dégrade en subsistance (le "stetig"). Il y a bien une *patience mobile* de l'articulation qui sait se maintenir hors des éternités qui n'en finissent pas et des prothèses qui cessent trop vite.

Comme les contrastes Continu-Discret, l'Egal et le Distinct résultent d'une activité de polarisation : c'est ainsi que peuvent être discernées les Formes Algébriques "devenues par l'Egal" et les Formes Combinatoires "devenues par le Distinct". Il s'agit de trouver l'articulation qui permet de passer continument de l'Egal au Distinct. L'écriture suggère la solution. Lorsque j'écris $A_1 = A_2$, l'expression $A_1 = A_2$ indique bien que, de A_1, on peut dire "la même chose" que A_2 ("l'égal est ce qui peut être mutuellement substitué dans chaque jugement"[18], mais l'indexation rappelle que A_1 et A_2 sont "quand même" différents : à la lecture je suis obligé de séparer A_1 et A_2. $A_1 = A_2$ peut se lire A_1 "est" A_2, A_1 *fonde* A_2, ou mieux, A_2 expose A_1 ou A_2 se dégage de A_1. Une égalité parfaite s'engloutit alors immédiatemet dans l'Identité $A = A$. On ne peut donc oublier que l'égalité posée est toujours issue d'une procès d'égalisation et d'évanouissement des inégalités, d'un *dispositif de porte-à-faux*.[19].

Les formes Egal et Distinct émergent avec l'ouverture d'un compas où A_2 expose progressivement A_1. Comme l'égalité parfaite, la diversité radicale n'est qu'une abstraction : si A_1 est donné différent de A_2, cette séparation terminée a certainement mobilisé une activité qui les liait en quelque manière.

Le concept du Distinct par "lequel un particulier est coordonné à une autre particulier" et le concept d'Egal par "lequel il est subordonné avec d'autres particuliers à

[16]Cf. [7].

[17]Pour la question des "ductus" cf. [11], p. 307, 309.

[18]Cf. [7], chp. concernant l'égalité.

[19]Ceci expliquerait pourquoi les "mesures par zéro" en physique sont les plus précises et les plus fondamentales. Elles s'immiscent dans le mouvement même d'engendrement de l'objet.

un genre qui leur est commun", s'articulent donc par des degrés de congruence que le philosophe-géomètre se doit de discerner.

Remarque

L'entrelacement illustre l'intrication entre position et liaison (cf. Figure 2).

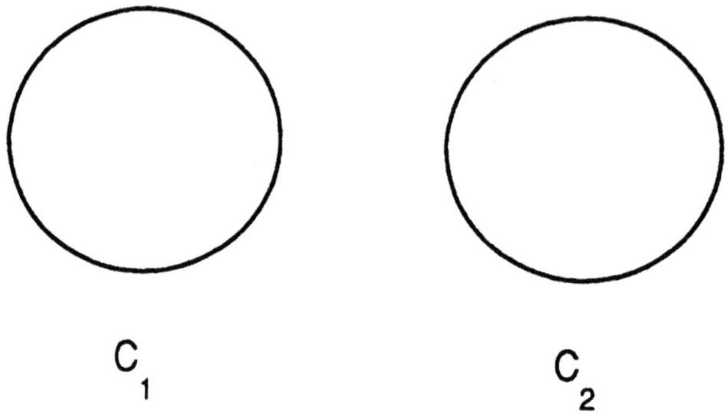

Fig. 2.

Je peux dire des deux cercles de la figure 2 qu'il sont "simplement l'un à côté de l'autre", l'acte de pensée qui les lie reste totalement disjoint de l'acte qui les pose. Je peux tracer l'un ($C1$) ***puis*** l'autre indépendamment : l'indexation est totalement arbitraire et simplement là pour rappeler qu'ils sont "deux à être pensés ensemble" et qu'ils reposent *sur* un fond indifférent.

Ici, la juxtaposition ignore toute prédisposition. On peut y voir le diagramme même de l'addition avec sa synthèse totalement réversible, sa commutativité (Versetzbarkeit), sanctionnant une séparation accomplie et l'effacement radical de la main qui a tracé.

L'entrelacs de la figure 3 se manifeste de toute autre manière.

Ici la connexion se noue. Il n'est plus question maintenant de dessiner les deux cercles séparément : je dois impérativement lever la main pour capter les dessus et les dessous de cette entité rebelle à toute immersion dans le "plat". Je ne peux prétendre avoir posé d'abord et lié ensuite (ou l'inverse!), il s'agit bien du diagramme d'une relation vivante, nullement extérieure à ses termes. Quelque chose a été tissé et, pour séparer, je dois briser.

Cette cohérence qui semble nous narguer n'est pas celle de la solidité déposée dans certains volumes, elle résiste comme une intrication capable d'affirmer le relief par elle-même.

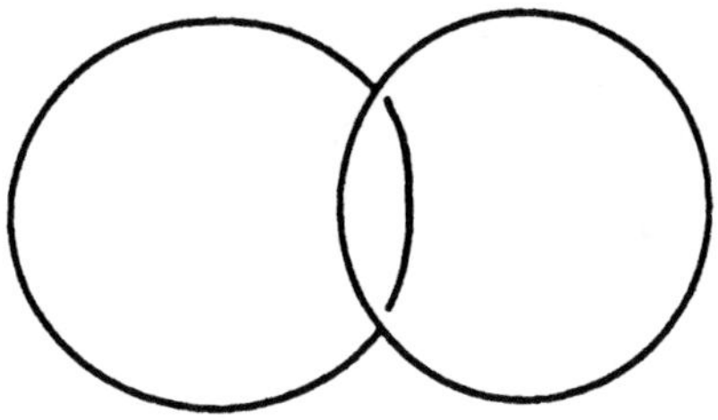

Fig. 3.

III Le Produit Géométrique Multiplie les Formes sans Corrompre l'Unité

> **"Nous comprenons par théorie générale des formes la série des vérités qui, de la même manière, se rapporte à toutes les branches des mathématiques et qui ne suppose donc que les concepts généraux de l'égalité, de la diversité, de la liaison et de la séparation".**

Les contrastes et les articulations précédentes invitent, en effet, à dessiner un losange (figure 4) où l'Egal et le Distinct, le Continu et le Discret s'affrontent aux sommets opposés.

Nous pouvons maintenant lier les sommets adjacents, mettant ainsi en évidence la Forme Algébrique discrète : le Nombre ("le rassemblement de ce qui est posé comme égal"), la Forme Combinatoire discrète : la Combinaison ("le rassemblement de ce qui est posé comme distinct"), la Forme Algébrique continue ("la théorie des Fonctions") et enfin la Forme Combinatoire continue (la théorie de l'Extension).

Ainsi sont distingués les quatre pôles des "sciences formelles". Ceux-ci pourtant ne doivent nullement être pensés comme *disjoints* mais *articulés.*

La dialectique intensif-extensif, appliquée au Nombre, permettra de mieux saisir la théorie de l'Extension comme mobilisation de l'Espace par l'orientation qui *prédispose* avant toute extension. C'est-à-dire qu'ils permettent de rassembler le plural dans une détermination simple. Le degré ne tente pas d'épuiser une multitude mais sanctionne l'achèvement dans un ordre de perfection.

A l'opposé, la quatification extensive comptabilise les agrégats, en collationnant des unités. Elle évalue le multiple par décomposition : la dispersion lui est essentielle.

La formule "évidente"

$$n = \sum 1 \tag{1}$$

illustre la définition de Grassmann "le nombre est le rassemblement de ce qui est posé comme égal". (1) montre que n est le pôle vertical-intensif qui capte par la simplicité

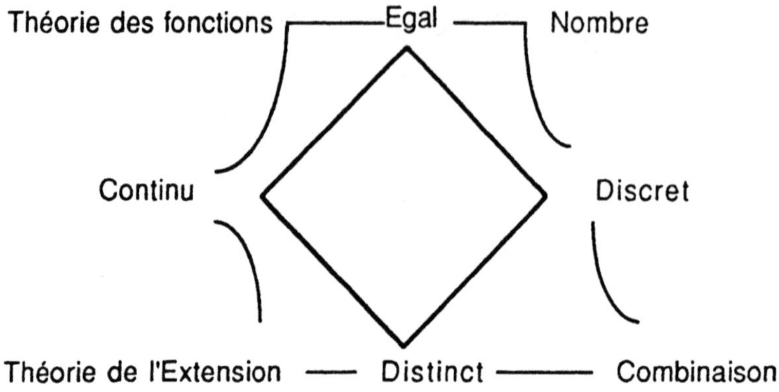

Fig. 4.

d'un acte la répétition monotone de la même unité 1[20]. Le signe = est l'articulation de la décomposition discrète qui expose le nombre à la fois comme même "échelon" et donc comme degré et comme 1 répété n fois (respectivement comme Zahl et Anzahl). Le diagramme de la figure 5 rend flagrant l'articulation de la fragmentation et de l'unification.

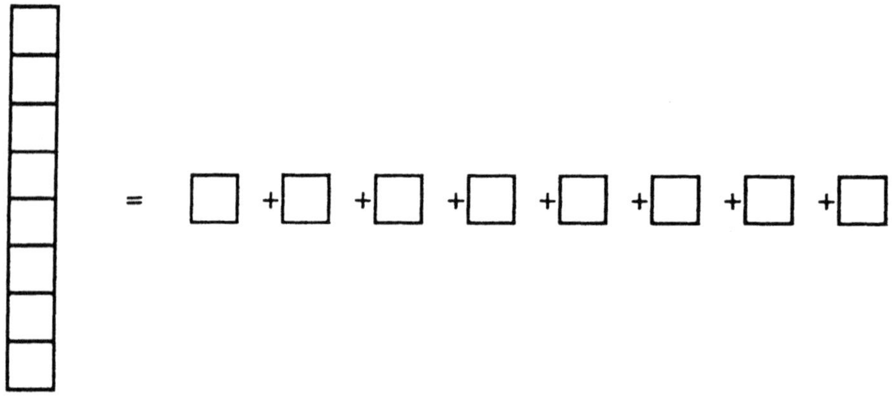

Fig. 5.

Ce type de diagramme tire toujours sa richesse de la tension vertical-horizontal et de la dimension qu'elle convoque subrepticement.

La formule

$$1 + 2 + \ldots + n = \frac{n(n+1)}{2}$$

[20]Cf. à ce sujet [14], p. 161.

"me saute au yeux" si j'organise un dispositif qui permet de *sommer des degrés* (cf. figure 6).

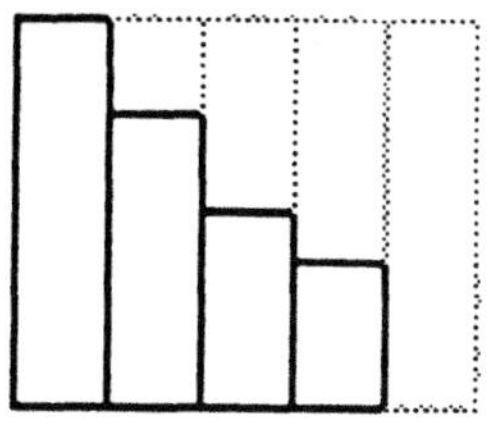

Fig. 6.

En complétant par des pointillés je fais apparaître un rectangle, "une *nouvelle dimension*". Cette dimension étant acquise, la formule (2) suit par division par deux.

Retenons que la conjugaison horizontal-vertical permet de mobiliser la pure dispersion de l'auseinander-seinende (le "étant l'un-à-côté de l'autre simplement extérieur") en la redressant suivant différentes hauteurs tout en déployant le vertical.

La polarisation intensif-extensif se retouve également dans la formule $E = \{a_1, \ldots \ldots, a_n\}$. Le symbole E rassemble dans l'unité d'un acte ce que le membre de droite exhibe comme agrégat d'éléments différents. Une fois de plus le signe = permet d'exposer une pluralité en multitude. L'égalité apparait comme la forme du "faire" et du "défaire". L'articulation = autorise un jeu de sommation-décomposition dont "E" et $a_1, \ldots, a_n$ ne sont que les moments abstraits.

Si maintenant je considère la décomposition en unités égales comme *Devenir*, j'obtiens la formule fondamentale du Calcul Différentiel et Intégral (Stokes)

$$L = \int_0^L 1dx \tag{3}$$

qui exprime à la fois comme grandeur intensive (membre de gauche) et comme grandeur extensive (membre de droite). Rappelons que les degrés sont distincts mais ne sont pas disjoints puisque la simplicité est essentielle à la quantité intensive. La quantification s'effectue de l'extérieur par un dispositif de développement et d'enveloppement. ("Base et Roulante").

Un tel dispositif me permet de passe de tel degré (d'arc ou de température) à tel autre degré : ce peut être une colonne de mercure qui se dilate ou un cercle qui roule sur une droite. La figure ci-dessous montre bien que l'intensif est associé à une *charnière* : le point de vitesse nulle qui articule la longueur L_1 sur le cercle et la longueur L_2 sur la droite [fig. 7] 0

Ces dispositifs permettent de développer les degrés en somme d'unités. Grassmann dit très bien que la grandeur intensive "ne se *défait* pas"[21] *en* elle-même. Elle *s'étale* seulement sur le mode de l'accroissement.

[21] "Hingegen als Beispiel der intensiven grösse etwa einen mit bestimmter kraft begabten Punkt, under hier die Elemente nicht sich entaüssern". "entaüssern" "se défaire" dans soi, s'affaisser.

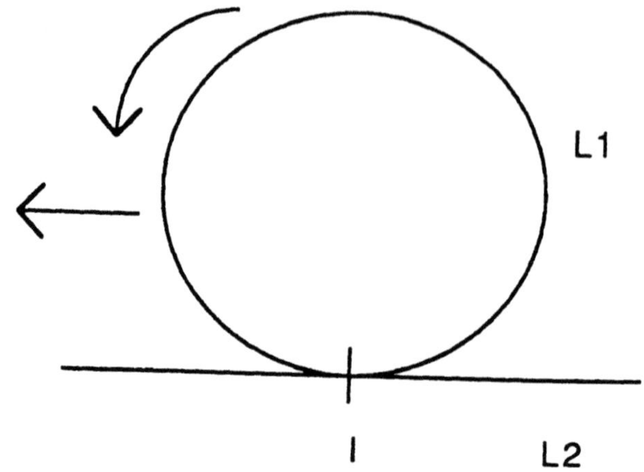

Fig. 7.

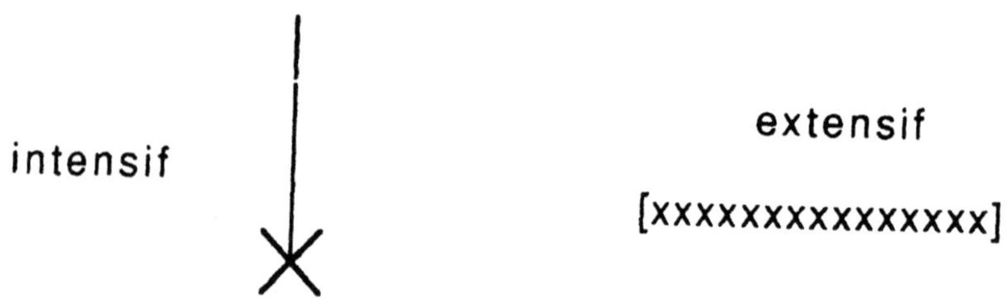

Fig. 8.

Les deux diagrammes de la figure 8 opposent bien l'intensif et l'extensif la force appliquée en un point sans aucune dispersion, extériorise une pure verticalité et contraste avec le segment extensif, essentiellement dispersé.

La formule (1) peut être comprise comme une sommation- décomposition mais nous allons tenter de mieux la voir comme exposition d'une vitesse suivant une longueur. Les cinématiciens-philosophes du Moyen-Age avaient bien su visualiser par diagrammes le transport d'un sujet plus ou moins affecté par la vélocité[22]. Ainsi, se trouvaient articulées, les figures abstraites de la Force et du Segment. L'intensif n'est plus prisonnier de sa

La force concrète (au sens dialectique) étale son intériorité. La force abstraite n'est qu'une pure extérioration (Aüsserung) qui semble "attachée" à la matière par un point. La force concrète produit le travail $W = FL$ ce concept permet *d'envelopper* l'extérioration.

[22]Cf. [3].

propre pétulence. Promu en capacité d'étalement, il peut enfin "se défaire" comme une longueur. Du coup, la longueur du segment ne se réduit plus à une distance morte : elle évalue l'achèvement d'un parcours. La trajectoire est bien *ce qui met en œuvre l'Espace.* La quantité extensive prise "seule", se contente en quelque sorte de prendre acte de places déjà données, d'adopter indifféremment divers "états"[23] alors que maintenant elle quantifie l'élément dans son jaillissement.

L'extensif est redressé, l'intensif est discipliné par la forme mouvement unité.

Le père de Grassmann "par analogie avec le produit arithmétique" avait déjà défini le produit géométrique : "le rectangle lui même est le vrai produit géométrique si l'on prend la multiplication dans son sens le plus pur et le plus général....
La multiplication est seulement une construction de degré plus élevé... " En géométrie, le point est l'élément producteur et la ligne émerge de lui par une construction... Le rectangle émerge si nous traitons la droite de la *même* manière que nous traitons le point. La situation est identique en arithmétique. Dans ce cas, l'unité est l'élément producteur originaire. L'unité est vue simplement comme *donnée* ... Si maintenant le nombre est pris comme nouvelle base pour compter (en le prenant comme nouvelle unité)... la multiplication apparaît bien comme la production d'un nombre d'ordre plus élevé, *un nombre dont l'unité est déjà un nombre.* Un rectangle est le produit *géométrique* de sa base et sa hauteur et de produit se comporte comme un produit arithmétique "..." ou encore "en géométrie le point est l'élément, la synthèse est le mouvement d'un point dans quelque direction"...[24]

H. Grassmann poussera plus loin en comprenant que le rectangle n'est pas une "abstraction" qui flotte dans l'Espace et qui "soutient" l'intuition. Il se délivre de la pesanteur du réalisme géométrique en saisissant la dimension comme capacité d'exposition et l'orientation comme articulation avant toute dissipation dans la grandeur numérique (longueurs, aires, volumes, etc ...).

Contrairement à l'évaluation de l'aire du rectangle, qui traite base et hauteur sur le même pied, Grassmann, dans la trace d'Oresme exige que la distinction des pôles intensif et extensif, soit manifeste dans la règle de formation du produit.

Je peux relier les diagrammes du segment dispersé et de la force appliquée (cf. figure 9) en exhibant la formation d'extension $I_1 \wedge I_2$ produit *dans cet ordre* par I_1 *puis* I_2 comme l'indique la figure 10.

J'anime chaque point de I_2 (dont les points sont essentiellement dispersés) par le pur transport conduit selon I_1. Le premier terme intensif se présente comme celui qui prendra la *mesure de l'engendrement*, le deuxième terme comme la *base de l'engendrement.*

La formation d'extension $I_2 \wedge I_1$ permute les rôles de I_2 et I_1.

Son diagramme est représenté la figure 11.

H. Grassmann définit l'addition des formations et montre ensuite que :

$$I \wedge (I' + I") = I \wedge I' + I \wedge , I"$$
$$(I' + I") \wedge I = I' \wedge I + I" \wedge I$$

[23] La mobilité permet de poursuivre constamment un même changement fondamental ([7], ch. 13). "Il est important de retenir que tous les éléments ainsi engendrés ne doivent pas être conçus comme déjà donnés auparavant. Ils doivent être saisis dans leur jaillissement originaire (Ursprüngliche Erzeugung").

[24] Cité par MJ. Crowe dans [18], p. 58-59.

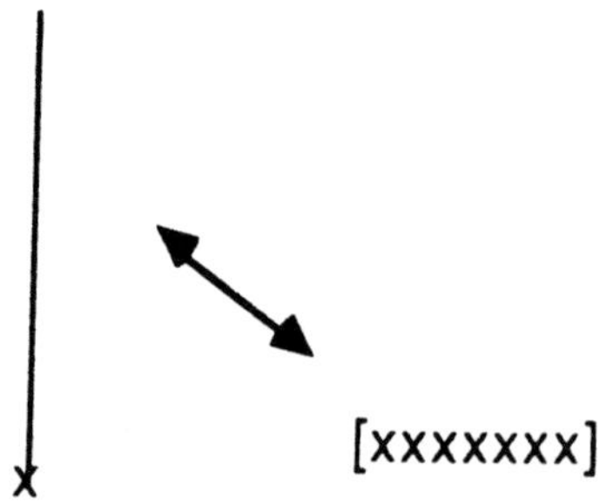

Fig. 9.

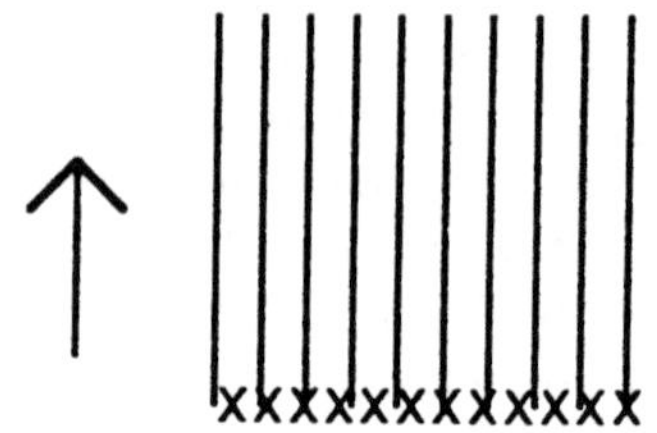

Fig. 10.

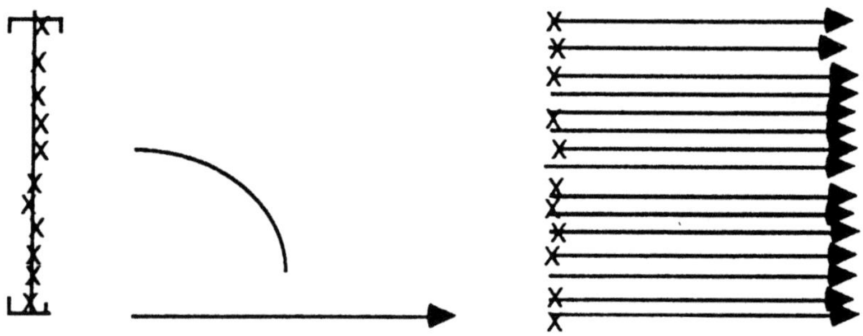

Fig. 11.

Ces propriétés ne le différencient pas d'un produit de grandeurs numériques contrairement à la relation :

$$X_1 \wedge X_2 = -X_2 \wedge X_1 \qquad (4)$$

qui *dévoile le rôle fondamental de l'orientation.*

Le produit de Grassmann est bien une véritable multiplication : une opération productrice de pluralité. L'addition se contente de pouvoir "penser ensemble" une multitude d'entités en les juxtaposant, comme on place des jetons sur un fond indifférent, "l'un à côté de l'autre". Elle ne résoud que "formellement" l'éparpillement et les "sommes" qu'elle fabrique risquent toujours de basculer dans la fragmentation : pour l'addition, le multiple est toujours issu d'une décomposition dégradée de l'unité.

La multiplication distingue un multiplicande et un multiplicateur. $N = P_1P_2$ ne s'obtient pas en juxtaposant P_1 et P_2 mais en considérant déjà l'un des membres comme une unité "N est le produit de P_1 *par* P_2". La multiplication numérique est bien sûr "commutative" : mais la synthèse privilégie toujours l'un des membres. C'est comme dirait Grassmann une liaison de 2ème degré : elle est compatible avec les décompositions éventuelles (distributivité par rapport à l'addition). ***Elle permet enfin de penser une multiplicité, qui, loin de corrompre l'unité, avance dans la diversité pour mieux en manifester le déploiement.***

Ici l'unité du plural est héritée de la même mobilité (celle de la hauteur) qui emporte tous les éléments de la base. Le produit géométrique renforce d'ailleurs la distinction entre l'agent (ce qui mesure) et le patient (ce qui nourrit l'engendrement) en exigeant de préciser soigneusement l'ordre de formation des produits : a.b s'obtient par déplacement des points du *premier* segment (la base) sous l'action du *deuxième* (la hauteur). Ce produit n'est pas disponible comme une chose étendue, là, dans l'Espace ; ce n'est pas une figure de la Raumlehre. L'ordre des symboles, qui conduit la main, manifeste l'intervention géométrique du "par". La juxtaposition, se contente de "penser ensemble" deux éléments qui, restent étrangers l'un à l'autre : "A+B" n'est pas vraiment un "mot". Ici la main s'implique dans le produit ; elle coordonne une successivité et un côte à côte : elle apporte la nouvelle dimension en incarnant une orientation : grâce au produit géométrique, le plan s'empare d'une mobilité qui semblait réservée au vecteurs[25]. Cette mobilité anime le parallélogramme $ABCD$ ou le triangle ABC : ce qui me permet de choisir, d'énoncer a puis b ou b puis a, me permet aussi d'opter entre deux virtualités de circulation : c'est ainsi que le "spath" étendu $\diamond$[26] gagne de l'épaisseur par scission dans soi.

Lorsque je parle de $ABCD$ (ou lorsque j'écris le *mot* $ABCD$) je le conçois déjà comme un parcours orienté. Les deux circuits qui s'esquissent en pointillé dans le seul énoncé du produit géométrique ne s'ajoute pas à la figure comme une excroissance (une géométrie plus "sophistiquée") mais sont reconnus par J et H. Grassmann comme ce qui permet d'éveiller l'étendue par dédoublement. Ce sont bien deux manières de "tourner autour", qui enveloppent réellement le domaine alors que $\{A, B, C, D\}$ n'est qu'un simple agrégat de lettres, mais $ABCD$, *dans cet ordre* est déjà un mot : c'est bien l'enjeu du non-commutatif : saisir ce qui ne peut être "pensé ensemble", ce qui se donne irréductiblement comme un procés.

C'est peut-être cette manière d'articuler une base, une hauteur, et un volume par l'orientation qui fait qu'une ***dimension posée en appelle une autre***. Une "ligne" peut être parcourue dans un sens ou dans l'autre à une vitesse v mais pour saisir dans l'intuition ces deux mouvements, je dois ménager une profondeur : v et $-v$ se réfléchissent l'un dans l'autre par réflexion (ils ne sont pas "l'un à côté de l'autre") et pour sortir de ce piège je dois inventer une nouvelle dimension en les repliant l'un sur l'autre.

Ce geste de pliage s'esquisse avant toute immersion dans le Spatial (ici avant tout parcours accompli sur la droite). Cette manière de saisir "n'a suivant le concept aucune limite"[27] et nous pouvons dès lors parvenir à des systèmes de dimension arbitraire (cf. figure 12).

Il suffit de supposer une "base" E_n d'ordre n déjà construite, une autre "hauteur" et une "enveloppe orientée" d'ordre $n+1$ E_{n+1} (un volume associé à la donnée d'un mot de $n+i$ lettres). Nous obtenons des formules du type

[25]Cf. en particulier l'introduction des "lectures on quaternions" de R. Hamilton.

[26]Pour Grassmann : spath : parallélépipède

[27]Cf. [7], ch. 16

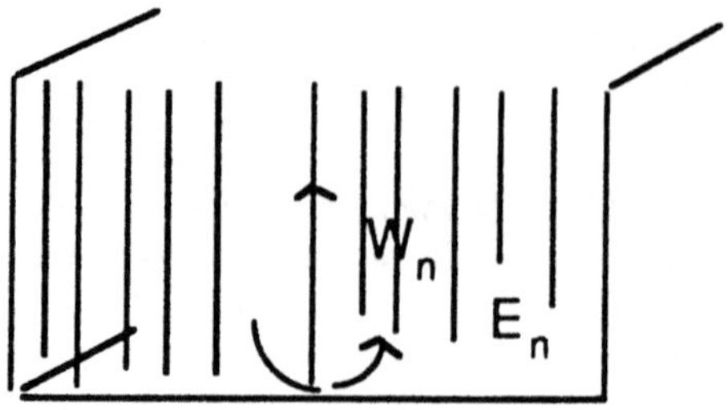

Fig. 12.

$$E_n = W_n.E_{n-1}$$
$$E_n = W_n.W_{n-1}.W_1$$
$$A.B = (-1)^{rs}\, B.A$$

(A et B sont des formationsd'extension d'ordre r et s)

Nous obtenons ainsi des formations d'extension d'ordre $n+1$. Celles-ci, par réflexion de miroir et dilatation conduisent aux *systèmes* linéaires d'ordre$n+1$.
Le produit extérieur permet d'itérer l'articulation fondamentale que réclame l'intuition d'un circuit à partir des mouvements de deux mobiles sur une droite ou de l'espace à 3 dimensions requis pour le rabattement d'un plan orienté sur un autre.

Cette itération associée à l'orientation et à l'idée que l'Espace n'est pas la cœxistence de figures transies dans l'étendue mais qu'il prolifère comme des mots, enrichit la notion même de dimension. Celle-ci n'est pas seulement un rang qui marque de l'extérieur tel ou tel système "son degré d'indépendance", je peux opérer *sur* elle ; ce n'est plus un "ordinal abstrait".

Je peux certes définir la dimension $n+1$ à partir de la dimension n par *adjonction* formelle mais je ne peux prétendre l'avoir vraiment produite, l'adjonction me permet simplement de "penser ensemble" un vecteur W_{n+1} et un système $E_n = \{W_1 \ldots, W_n\}$ auquel il n'appartient pas alors que le produit extérieur ne sanctionne positivement que les formations d'extension nouvelles : il s'annule pour toute adjonction triviale[28]. De plus, l'amplitude orientée $W_1 \ldots\ldots W_{n+1}$ *enveloppe* le système des vecteurs $\{W_1, W_n, W_{n+1}\}$, elle ne se contente pas de *réunir* E_n et W_{n+1}. L'intuition de l'enveloppement précède la construction effective de la formation d'extension d'ordre n+1 et, comme les degrés d'arc appellent une rectification qui permettent de les sommer (cf. : figure 3), le produit extérieur me permet d'étaler ces degrés d'indépendance.

Le produit régressif ou impliquant (eingewandte Produkt)[29].
Le produit extérieur "explique" l'expansion spatiale : il construit la forme d'extension qui saisit et enveloppe au plus près l'étendue de la réunion issue des engendrements X et Y.

[28] Cf. [7], ch. 31
[29] Cf. [7], ch. 125.

La pénétration de Grassmann lui perme de sentir que cette priorité naïve de l'expansion fondée sur les degrés d'indépendance, devait être équilibrée par une démarche symétrique : une multiplication qui exposerait la *contraction* spatiale et répondrait à la question : quelle serait la forme qui envelopperait ce qui est commun à deux systèmes A et B ? Grassmann définit donc le produit régressif qui enveloppe l'intersection comme le produit extérieur enveloppait l'union. Si par exemple :

$$A = (e_1 \ldots e_n)(e_{n+1} \ldots e_k)$$
$$B = e_1 \ldots e_n . e_{n+1}$$

Le produit régressif de A et B se note $A.B = e_1. \quad .e_n$.

Ce produit régressif permet de *descendre* d'un système de dimension n jusqu'à 1 la dimension O comme le produit extérieur avait permis d'épuiser E en "partant" de la dimension O (l'unité de dilatation = 1). En démasquant le caractère unilatéral et "naturel" de cette expansion, le produit régressif fait éclater les limites d'un constructivisme, cramponné à la finitude de ses itérations et foncièrement incapable de concevoir, les horizons (ici l'enveloppe orientée E) toujours donnés sur le mode de l'implicite et du coposé.

Grassmann a génialement senti la nécessité d'équilibrer cette situation : *E peut être vu comme un horizon à partir de 1 et 1 comme un horizon à partir de E* et ceci par une symétrie qui n'échange plus les figures de l'Espace mais concerne la dynamique même de la pensée.[30]

IV L'Ambiguïté où Germe l'Opération

Apprécié comme l'âge d'or des grands systèmes philosophiques, comme celui de l'émergence de la notion de structure, le début du XIXème est peut être aussi celui où le centre change de statut : le point à l'infini de l'Age classique cède la place aux points neutres d'indifférence qui marquent les degrés d'équilibre et d'où jaillissent continûment les polarités. L'existence d'un tel centre d'indifférence dans les diagrammes électriques et magnétiques fascine physiciens et philosophes spéculatifs[31] comme les pôles Nord et Sud d'un aimant semblent issus d'un point neutre, il est tentant de penser que les différences *reçues*, les grosses inégalités quantitatives ne peuvent amorcer les procès d'individuation. Leur hétérogénéité arrogante ne produit aucune dimension : les formes nouvelles ne surgissent que lorsque s'esquisse *une manière de congruence* et cette congruence semble bien associée aux inflexions de la main qui trace une courbe, à l'ambiguité du choix des racines d'une équation, à ces points étranges où les charges s'équilibrent, bref à ces lieux où l'entendement vacille (cf. Figure 13). *Aux centres d'indifférence, est atteinte la plus haute incertitude et exigeant donc la décision la plus irréversible.*

Les philosophes classiques avaient beaucoup spéculé sur l'infini car les "formes indéterminées" qu'il secrétait, permettaient d'abolir les différences quantitatives. De toute manière, l'infini était pour les mathématiciens du début du XIXème une "singularité"

[30] Pour différents exposés de la théorie de Grassmann voir [15], [2] et aussi l'article remarquable de Barnabei-Brini-Rota [1].

[31] Cf. [5]. Schelling et d'autres se sont beaucoup inspiré de cet auteur ("Toute qualité est polaire et électrique" (cf. le beau commentaire de Marquet dans [12] p. 115-120, et aussi [17].

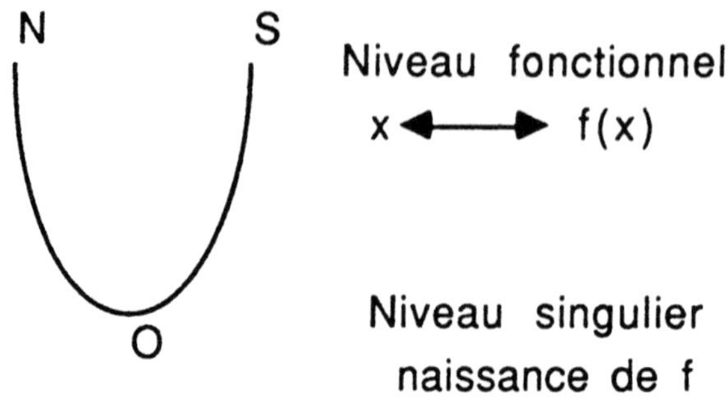

Fig. 13.

bien inoffensive apprivoisée par la perspective qui "ramène" facilement les pôles à distance finie !.

Le "singulier" menaçant est celui qui révèle (souvent cruellement) la naïveté du "d'une part et d'autre part" et qui se manifeste en mathématiques par l'effondrement d'un certain mode "canonique" de correspondance. Les points de ramification, où se proposent plusieurs paramétrages ($Z_i = y_i(t_i)$), rendent donc possibles plusieurs saisies fonctionnelles. Ces saisies sont toujours fondées sur l'articulation "évidente" d'une ordonnée et d'une abcisse (paramètre) "générique". Cette nécessité d'opter pour telle ou telle détermination qui *accompagne toujours la rencontre d'un centre d'indifférence*, trouble les géomètres : il devient évident que ces points promettent un foisonnement de correspondances bien plus vertigineux (et bien plus difficiles à contrôler !) – que les "formes indéterminées".

Le centre, comme point à l'infini, permettait certes de dominer les différences quantitatives qui séparent les êtres finis, mais en subjuguant *extérieurement* les particularités. Le centre d'indifférence nous installe au cœur de l'individuation : le dispositif de porte-à-faux qui le détecte permet d'envelopper tout un faisceau de virtualités sans les confondre : *la symétrie est redécouverte comme productrice de différences.*

Le problème de la mesure des angles fournit un exemple des pouvoirs de ces centres : je crois pouvoir saisir facilement l'écart qui sépare les "côtés" d'un angle aigu (cf. figure 14) : il semble être maîtrisé par le simple concept de *distance.*

Si maintenant, je désire me rendre présent le tour de main qui l'a fait apparaître, (ce qui revient à se donner un paramétrage continu) l'existence de *deux* manières de tourner indique bien que l'angle entre deux droites n'est pas disponible comme le serait la distance entre deux points d'une ligne. Un angle se dérobe aux saisies instantanées : il se n'expose que progressivement à partir d'un choix. Un point à l'infini n'implique aucun dilemne pour le géomètre il peut facilement être ramené à distance finie par une transformation "convenable". Une "position" se "donne clairement" alors que l'angle se démasque très vite comme procès, comme "choix continu d'une détermination" et ce choix conduit à une réelle perplexité avec les "angles plats" ! Je ne puis évidemment disposer de ces angles [π ou $-\pi$] comme de grandeurs numériques ou distances franchies

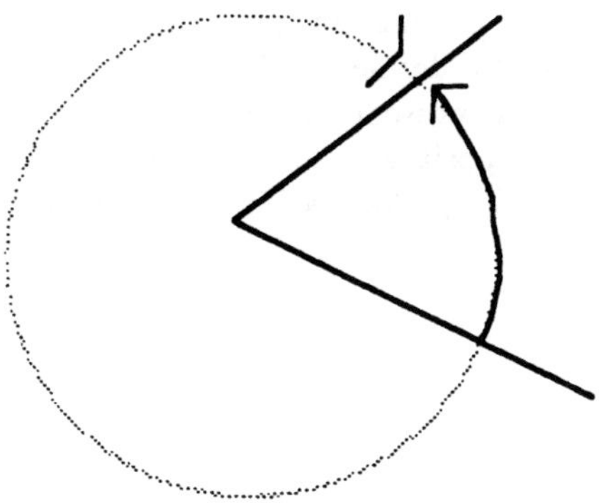

Fig. 14.

d'un coup d'œil. π et $-\pi$ *sont des exposants* : ils *s'impliquent dans leur propre mesure* pour ainsi dire ("l'exposant est ce par quoi l'étant prend sur lui même sa mesure") : je dois faire revivre le mouvement qui leur a donné naissance en butant dès le début sur une indétermination.

La simple donnée d'une saisie fonctionnelle

$$y = f(x) \tag{1}$$

suppose trop vite acquise la séparation tranchée entre une position et ce qui a permis cette position. Une correspondance établie du type de (1) naît toujours d'une déchirure ("principe de naissance par phase stationnaire"). Bien avant la fonction, l'exposant capte une puissance d'opération sans "mise à plat". *Exposer n'est pas expliquer* mais ménager une niche continuum au singulier en enveloppant toutes les déterminations virtuelles issues d'un problème (c'est l'intérêt des "fonctions multiformes" ou les corps de Galois associés à une équation).

Dans la notation X^2, 2 ne fonctionne ni comme un $2^{\text{ème}}$, ni comme $1 + 1$, mais permet enfin de faire "interagir la grandeur avec elle-même" ... Une ambiguité s'immisce toujours où apparaissent les exposants : c'est le prix à payer pour nouer produit et productivité et *donner de la mémoire à la forme.* Cette mémoire n'est pas déposée dans la forme : elle n'est réveillée que par *l'intuition de l'intuition : l'ambiguité.* L'intuition géométrique de la *Raumlehre* appelle toujours au secours la certitude, l'inertie rassurante d'un spatial enrobant tranquillement les choses. Cette intuition de l'intuition (dont rêvait le jeune Hegel) entreprend de disloquer les fausses évidences, invite à traverser les paliers d'ambiguité maximale pour remobiliser le savoir déjà acquis. Dans l'exemple de la mesure des angles, c'est précisément l'existence de deux chemins symétriques qui montre que l'intuition créatrice se meut dans l'espace des chemins.

L'exemple qui suit, le dédoublement des réels positifs, illustre nettement *l'ambiguité associée au retournement dans soi de la grandeur.*

Les réels positifs étant donnés, je peux obtenir les négatifs par une symétrie de miroir et *ajouter* ensuite l'élément nul. Ce dernier se détache alors nettement comme une excroissance et ce qui est construit ne possède que la cohérence médiocre de l'agrégat. Mais je peux aussi penser les nombres positifs comme un des feuillets issus d'un pliage de la droite (cf. figure 15) : le point O apparaît alors comme point de pivotement et donc comme ***articulation***.

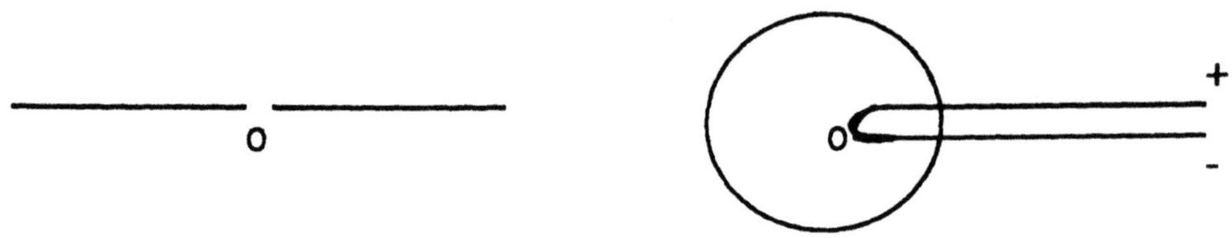

Fig. 15.

Ici encore l'ambiguité, l'intuition de "second degré", ne prétend pas "soutenir" un calcul (la simple considération des ***produits*** des opérations définie sur les nombres positifs ne me permettra jamais de les "dédoubler"...) mais suscite ***en pointillé*** une nouvelle manière de voir. Il y a bien deux manières d'étaler ou de replier les "côtés", autrement dit ***d'exposer*** les réels.

Il est clair que O n'est pas un simple pôle : aucun changement de point de vue ("collinéation projective") à distance finie ou infinie ne donne accès à la scission dans soi de la positivité la O est le ***point de ramification*** où se nouent ***deux*** saisies fonctionnelles ($y = \sqrt{x}$, $y = -\sqrt{x}$). Une nouvelle dimension se manifeste donc avec l'énoncé du *problème* :

$$(?)^2 = a$$

Celle-ci n'était nullement donnée dans la droite construite axiomatiquement. Les réels positifs doivent être saisis comme fendus ***en*** eux-mêmes ; ils acquièrent ainsi une épaisseur.

Nous touchons le point sensible de la Philosophie de la Nature : celui de l'intution des degrés d'articulation, ceux où l'Etre porte à faux, et où les différences quantitatives et choix "canoniques" de paramétrage s'avanouissent. L'ambiguïté accompagne ces seuils critiques car elle seule confère une ***posivité dynamique à la séparation.***

Deux quantités grossièrement inégales nous comblent de la certitude des dualismes établis. Victoire à la Pyrrhus ! On ne domine pas un antagonisme en jouant sur la contrariété de deux termes mais en réveillant l'affinité secrète qui les lie, en se glissant dans le mouvement par lequel la positivité se creuse une profondeur.

Si je ne donne pas deux solutions à l'équation $X^2 = a$ (*égales en "valeur absolue"*), la grandeur ne se mobilise pas ; elle s'affaisse dans soi et la productivité s'épuise dans le produit. De la même manière, si le produit de Grassmann ne reposait pas fondamentalement sur l'ambiguité d'un choix : ... d'abord *ce* terme puis l'autre, et donc afin ***d'épaissir*** *en deux feuillets* le rectangle de l'intuition naïve, il manquerait la capture de l'extension faute de révéler le centre neutre où tout va se décider.

Par l'ambiguïté de la racine-exposant, ce qui était posé dans l'unité d'un acte : appliquer une règle R, se déploie dans les continuité d'un geste. La positivité disponible, naïvement palpable, se fissure pour envelopper un spectre de solutions suivant la formule : $(?)^2 = R$. Cette formule, les Grecs l'avaient pressenti : ils savaient que "la droite à la puissance de son carré"[32].

Opposition, écartèlement, impossibilité, peuvent être les symptômes d'un degré d'intuition à franchir mais la dimension nouvelle ne surgit pas "à cause" de l'opposition ou de l'impossibilité. Ces dernières m'invitent simplement à mépriser la certitude des clichés "intuitifs" mais elles ne "mènent" pas à la solution : l'articulation est un saut : ni épuisement déductif, ni une induction "abstraite" à partir de caractères communs.

Car le centre d'indifférence, *l'exposant*, s'impose toujours comme *un degré de l'Etre*. Produit et productivité y sont indiscernables et les formes semblent avoir triomphé de l'ambiguïté. L'exposant-racine offre sa mémoire et sa profondeur. Sa conquête participe de ce "pressentiment" dont parle Grassmann et qui agit bien avant la mise à plat de tous les moments de la vérité nouvelle.

Enfin peut être pensé l'événement d'une séparation. Appréciée trop souvent comme *rupture*, d'ailleurs vite engloutie dans le fracas de disparités quantitatives, cette séparation des formes se joue maintenant dans l'infime recul dégagé par l'exposant : plus haute est la symétrie et plus irrévocable est la décision. Ce n'est pas le moindre des paradoxes de ce paradis étrange où les formes semblent procèder d'elles- mêmes sans blessure.

Remerciements

Messieurs Flament et Beckemeier m'ont permis d'accéder à leur traduction en cours de l'Ausdehnunglehre de 1844. Qu'ils en soient remerciés.

Références

[1] Barnabei-Brini-Rota On the Exterior Calculus of Invariant Theory (*Journal of Algebra* **96** 1985 (120-160).
[2] Burali-Forti : *Introduction à la géométrie différentielle suivant la méthode de H. n* (Paris-Gauthier-Villars 1897).
[3] G. Châtelet : La Toile, le Spectre, le Pendule (*Fundamenta Scientiae*) (10)2 1989.
[4] MJ. Crowe *History of Vectorial analysis* .
[5] Eschenmayer *Versuch die magnetischen Erscheinungen a priori abzuleiten* (1795).
[6] G.G. Granger : *Essai d'une philosophie du style.*
[7] H. Grassmann : *Die lineale Ausdehnunglehre* Verlag von Otto Wigand, Leipzig, 1844.
[8] W.R. Hamilton – *Lectures on Quaternions*, 1853.
[9] G.F. Hegel – *Science de la logique – logique de l'Être (I)* trad. Labarrière.
[10] M. Heidegger – *Schelling* – Trad. Courtine.
[11] W. Leibniz : Dynamica de Potentia (section I, III), Gerhardt.
[12] Marquet *Liberté et Existence.*

[32] Cf. [11], section III (2) p. 364.

[13] M. Merleau-Ponty *"La Prose du Monde"*.
[14] M. Merleau-Ponty – Le concept de la Nature dans *Résumés de cours*.
[15] G. Peano : Calcolo geometrico secondo l'Audsdehnung *Lettra di H. n* (Fratelli Bocca Turin 1888).
[16] J.P. Sartre – *Critique de la Raison dialectique.*
[17] Schelling – *Déduction générale du processus dynamique* (1800).
[18] Schelling – *Nature de la liberté humaine* (N L H) trad. Jankelevitch.
[19] Schelling – *Un système de la philosophie de la Nature* (trad. Jankelevitch).
[20] G. Simondon – *L'individuation psychique et collective* (Aubier).

Helmholtz and Poincaré's Considerations on the Genesis of Geometry

Gerhard Heinzmann

Université de Nancy II

What historical sources are at the bottom of Poincaré's distinction between the status of arithmetical and geometrical judgments? On the one hand, one may think of Kant, on the other hand, of Riemann, Helmholtz and Lie. Indeed, in geometry, Poincaré confines himself to metrics compatible with the group of rigid motions (Lie and Helmholtz) so that restricted hypotheses (Riemann) are opposed to Kant's *a priori* judgments. These hypotheses – which in reality are conventions – replace the *facts* in Helmholtz's approach and displease at he same time the upholders of Kantian orthodoxy and the defenders of empiricism. Roughly speaking, Helmholtz is, surely, an empiricist, but if one looks closer it will appear that the apparently empiric fact of the existence of freely movable rigid bodies is not systematically put out of question by Helmholtz. In this paper, I want to consider these *prima facie* atypical doubts in the context of Helmholtz's theory of sensual knowledge[1].

In 1866, Helmholtz formulates as the principal theorem of the empirical conception the following thesis : "Die Sinnesempfindungen sind für unser Bewußtsein Zeichen, deren Bedeutung verstehen zu lernen unserem Verstande überlassen ist"[2]. Helmholtz also proposes a dualism according to which objects are in themselves quite devoid of sense and only become a vehicle of knowledge if our mind is able to interpret them as signs. However, the semiotic interpretation of our sensations is by no means based on a realistic theory of the relation between the sign and the signified. On the contrary, the external world is understood, i.e. its elements are conceived as objects of our discourse, only if they are the result of an abstraction in termes of a formation of invariants. The relations between the sign and the signified "beschränkt sich darauf, daßdas gleiche Objekt, unter gleichen Umständen zur Entwirkung kommend, das gleiche Zeichen hervorruft"[3]. But in order to introduce the cause-effect-relation into the field of our sensations, we have to relinguish the dualistic attitude with regard to time : for, if we presuppose,as Helmholtz does, the simultaneity of the sign and the signified[4], the sense of simultaneity and succession must be given from the very beginning. This semiotic theory of sensations has an important consequence for the aim of this paper : if one accepts it, the two characteristic elements of the Kantian conception of external intuition, that is its

[1] For more details cf. Heinzmann 1991 (?), chap. 2.11–2.12.

[2] Helmholtz 1910, p. 433.

[3] Helmholtz 1978, p.18/19.

[4] Cf. Helmholtz 1869, p. 56.

singularity and the immediate awareness of its objets, will be disposed of. And of course, according to Helmholtz,a ***repraesentatio singularis*** is ***sub reptum*** always a ***repraesentatio discursiva***. Hence, the Kantian conception of intuition is clearly modified : intuition is no longer an independent and immediate source of knowledge, but is related to a certain use of language. We say that language is used in an intuitive manner, if it remains – in principle – in connection with sensual constructions. Knowledge is therefore intuitive, if we are able to imagine a semiotic interpretation of a sequence of sensations coordinated with it. Bearing this sort of linguistic turn in mind, we will better comprehend that Helmholtz himself considers his modification of the concept of intuition as the "most essential progress of the times we live in"[5].

He uses this progress in order to justify the intuitive character of "metamathematical" spaces. The construction of the amorphous space is no longer based on the analytic postulate of a quadratic measure (Riemann), but only on the hypothesis of a certain manifold and, especially, on a congruence-condition obtained through a psycho-physiological analysis : the free mobility of rigid bodies. I omit here the corrections put forward by Lie. Apparently we have thus linked up with the classical interpretation according to which Helmholtz is an empiristic long before the choice of a metric is relevant. But, to take this interpretation for granted would imply that Helmholtz lays himself open to the well-known circle mentioned by Poincaré, Schlick and Dingler[6] : on the one hand, the rigidity of a body serves as a condition to formulate a concept of congruence, on the other hand you must use the concept of congruence to control the empirical fact of the rigidity of a moving body. But, as Torretti has pointed out[7], the reproach of having committed such an error does not quite correspond with Helmholtz's consciousness of the problem in question. In fact, the latter begins his essay "Ueber die Tatsachen, die der Geometrie zugrunde liegen" with the following question :

> "Wieviel von den Sätzen der Geometrie hat objektiv gültigen Sinn? Wieviel ist im Gegenteil ... nur von der Form der Darstellung abhängig?"[8]

It may be claimed that rigidity possesses an objective sense. Indeed, this is not the case. For Helmhlotz adds :

> "Diese Frage ist meines Erachtens nicht so ganz einfach zu beantworten, da wir es in der Geometrie stets mit idealen Gebilden zu tun haben, deren körperliche Darstellung in der Wirklichkeit immer nur eine Annäherung an die Forderung des Begriffes ist, und wir darüber, ob ein Körper fest, ob seine Flächen eben, seine Kanten gerade sind, erst mittels derselben Sätze entscheiden, deren tatsächliche Richtigkeit durch die Prüfung zu erweisen wäre."[9]

Let us comply with Helmholtz's doubt. The notion of a rigid body would also be a transcendental notion with regard to the representation of factual objects and, further,

[5] Helmholtz 1878, p. 48.
[6] Cf. Poincaré, SH, p. 70/71, Schlick 1921, p. 30, note 31 and p. 33, note 40, Dingler 1925.
[7] Cf. Torretti 1978, pp. 167 sqq.
[8] Helmholtz 1868, p. 38.
[9] *Ibid.*

this transcendental status would be justified from the standpoint of espistemology and would no longer express a simple reminiscence of Kant. Yet, in opposition to Kant, the transcendental element would be related to a linguistic representation and not to a pure intuition or better, it would concern an idealized norm in view fo a production in correspondence with the aim of an action.

We are now prepared to establish a connexion with Poincaré's occasionalism. Although referring to Helmholtz and Lie, Poincaré, in his reflexions about the genesis of geometry, attempts to invert their program of foundations :

> "Pour Helmholtz et Lie la matière du groupe existait avant la forme ... Pour moi au contraire la forme existe avant la matière."[10]

The formal structure of the group is prior to its model, that is to the *fact* of the possibility of rigid movements in the infinitesimal.

What is the sense of this priority? In order to answer thc question we briefly have to examine Poincaré's genetic remarks : he interprets the localisation of an object in space in terms of a reflection upon the procedure of an action in order to reach this object, that it is in terms of a reflection upon sequences of muscular sensations. The starting-point of his reconstruction is the *natural* observation that a set of impressions can be modified according to distinct manners : there may be an involuntary change in our impressions which is not accompanied by muscular sensations and there can be a voluntary, produced by a driving action and accompanied by muscular sensations. The first case is called an *external change*, the second case is called an *internal change*. This fundamental observation suggests now a *conventional* classification of external changes: among external changes some can be compensated by an internal change, others cannot. The first are called *changes of position*, the second *change of state*. The relation 'x and y are compensated by a same internal change' is an equivalence relation in the set of changes of position and the corresponding equivalence classes are called *displacements*. The laws to which these displacements are subjccted form the objects of geometry.

Closing here my description I want to add some comments and to draw conclusions :

1) In the space called 'sensible' or 'representative' the changes of position are never realized exactly, because the starting-position would have to coincide with the final state. Hence, the compensations leading to the changes of position must be read as an order to act in view of an ideal norm. We act as if realizing the norm. However, what is realized is not the norm, but the order to realize this norm. In such a reconstruction experience plays a double part : it gives occasion to create the norm (its *ratio cognoscendi* and it is an occasion to test the norm in its role as an instrument for the conceptualisation of reality.

With regard to Helmholtz, Poincaré has accomplished a rationalistic turn. This is shown by the fact that the general concept of a group is no longer a form of our intuition, but a form of our reason[11]. This form, mediated by language, is the successor of Helmholtz's discoursive intuition.

Since Poincaré in his occasionalism emphasizes the active aspect of this form, one clearly sees that Poincaré's conception of geometry is neither analytical nor empirical ;

[10]Poincaré, FG, p. 60.

[11]Poincaré, SH, p. 93.

considering at the same time the norm leading to the compensative action as result and cause of experience, it seems to be more adequate to call Poincaré's approach an operative conventionalism – operative, because the norm is certainly conventional, but is nevertheless an idealization of an action.

2) Stressing on the one hand the relative starting point of the compensation-procedure and also the abolition of an absolutely rigid body, and on the other hand the sequential conception of the theory of compensation, Poincaré opens the condition of the possibility – seen already by Helmholtz – of a simultaneous dilatation of the external object and the internal movement, briefly, the possibility of non-Euclidean displacements :

> "la géométrie n'est que le résumé des lois suivant lesquelles se succèdent des images. Rien n'empêche alors d'imaginer une série de représentations, de tout point semblables à nos représentations ordinaires, mais se succédant d'après des lois différentes de celles auxquelles nous sommes accoutumés."[12]

Like Helmholtz, Poincaré is well aware that the sequence of sensations, that is their distinction *de numero*, presupposes a form of mind wich he calls 'time'. Time is a form on which depends our faculty of repetition, a faculty not only indispensable for the constitution of arithmetical objects, but also for the constitution of geometrical objects, namely for the metrical determination of the group of motions. Therefore, it will no longer be sufficient to refer to the Kantian forms of time and space in order to distinguish arithmetic from geometry. In the algebraic foundations based on the Erlangen-Program, the symmetry is moderated in favour of time :

> " C'est la répétition qui a donné à l'espace ses caractères essentiels ; or la répétition suppose le temps ; c'est assez dire que le temps est antérieur logiquement à l'espace."[13].

3) For Poincaré, the essence of the geometrical method is to be found in the transition from an *a priori*-continuum to a *measurable* quantity. The measure is the result of the introduction of a conventional law concerning a continuum considered as a computable quantity. Hence, the passage determining a new method is not the transition frome the discrete to the continuous, but a change in the perspective with regard to the quantity : from the computable one passes to the measurable.[14]

References

Dingler, Hugo, 1925, Über den Zirkel in der empirischen Begründung der Geometrie. *Kantstudien* **30**, pp. 310- 330.

Heinzmann, Gerhard, 1991 (?), *Zwischen Objektkonstruktion und Strukturanalyse. Überlegungen zur Philosophie der Mathematik von J.H. Poincaré und C.S. Peirce.* Forthcoming.

[12]Cf. Poincaré, SH, p. 88.

[13]Poincaré, SH, p.98.

[14]I'am indebted to Mrs. Prim for her help in translating the french version of this paper.

Helmholtz, Hermann von, 1910, *Handbuch der Physiologischen Optik*, 3. Auflage, Bd. III ([1]1856), (Voss), Hamburg and Leipzig.

Helmholtz, Hermann von, 1869, Über das Ziel und die Fortschritte der Naturwissenschaft. In Helmholtz, *Das Denken in der Naturwissenschaft*, (Wissenschaftliche Buchgesellschaft), Darmstadt 1968, pp. 31-61.

Helmholtz, Hermann von, 1868, Über die Tatsachen, die der Geometrie zugrunde liegen, in Helmholtz, *Schriften zur Erkenntnistheorie*, ed. Paul Hertz and Moritz Schlick, (Springer), Berlin 1921, pp. 38-55.

Helmholtz, Hermann von, 1878, *Die Tatsachen in der Wahrnehmung*, reprinted by Wissenschaftliche Buchgesellschaft, Darmstadt 1959.

Poincaré, Henri Jules, FG, Des fondements de la géométrie, ed. L. Rougier, (Chiron), Paris (s.d., (1921)), [1]1898/99.

Poincaré, Henri Jules, SH, *La science et l'hypothèse* ([1]1902), (Flammarion), Paris 1968.

Poincaré, Henri Jules, VS, *La valeur de la science* ([1]1905). (Flammarion), Paris 1970.

Schlick, Moritz (with Hertz, Paul), 1921, *Helmholtz, Schriften zur Erkenntnistheorie* (ed.), (Springer), Berlin.

Torretti, Roberto, 1978, *Philosophy of Geometry from Riemann to Poincaré* (Reidel), Dordrecht.

Le Continu Contre L'Espace

Jean-Michel Salanskis

Fondements des Sciences, UPR n°265 du C.N.R.S., 3, rue de l'Université, 67084 Strasbourg Cedex

Nous voudrions faire réfléchir sur la relation qu'entretient le continu avec l'espace. Pour la plupart d'entre nous, croyons nous, une certaine figure de l'identification du continu et de l'espace a force d'évidence : le continu, c'est $\mathbb{R}$, et par extension, n'importe lequel des $\mathbb{R}^n$, mais l'espace aussi, de son côté, se laisse incarner par $\mathbb{R}^3$, du moins en premier lieu. Bien entendu cette assimilation a son origine dans l'avènement de la géométrie de coordonnées, c'est-à-dire dans l'avancée décisive de la mathématique survenue au XVIIème siècle, à laquelle sont attachés notamment les noms de Fermat et de Descartes. Mais elle est aussi corroborée par le visage technique que prend le plus souvent pour nous le domaine de la physique mathématique aujourd'hui : celui de travaux dans le cadre de la théorie des variétés différentiables. Ce dernier concept, en effet, fait-il autre chose que pluraliser et généraliser une certaine idée de la complicité indéchirable entre espace et continu?

Rendons nous pourtant attentif à une voix qui parle dans un autre sens, même si c'est depuis une tout autre position de parole, *philosophique* : celle de Hegel. Il est de bonne méthode de ne pas décider par avance que de telles vues ne sont pas pertinentes.

Hegel : le Drame Conflictuel du Continu et de l'Espace

Hegel expose une *généalogie*, qui commence par l'espace : le continu, en liaison avec l'infini, vient après l'espace, et s'oppose à lui, nous allons le voir, comme *meilleur*. C'est que l'espace dans son rôle de commencement signifie purement et simplement *l'absence de pensée* pour celui qui raconte cette histoire. Voici en effet la définition donnée par Hegel de l'espace au §254 de l'Encyclopédie [Hegel 1830] :

> "La détermination première ou immédiate de la nature est l'*universalité* abstraite de son *être-en-dehors-d'elle* son état d'indifférence sans médiation est l'*espace*."[Hegel 1830, 142]

"première ou immédiate", cela nous dit le commencement de la généalogie. Quant au contenu, donc, l'espace est identifié comme l'être-en-dehors-de soi élevé à la puissance de l'universel, de l'englobant. Mais cet être-en-dehors-de-soi envahissant est de plus *indifférence sans médiation*. Dans cette formulation est implicite le jugement hégélien de l'espace : selon lui l'espace veut que ce qui est "en lui", se trouvant distribué selon la stricte extériorité du *côte-à-côte*, et selon elle seulement, soit dénué de tout rapport. La juxtaposition signifie en quelque sorte l'échec ou l'impossibilité de toute émergence de sens rassemblant les juxtaposés. Etre-avec-autre-chose-dans-et-selon-l'espace, c'est n'avoir pas un rapport qui fasse sens ou qui lance la pensée. Dans le passage qui suit, et qui présente aussi l'espace, dans la *Science de la Logique* [Hegel 1812, 1813, 1816], on trouvera l'explicitation du "jugement" de non-pensée que nous venons de dégager par commentaire :

> "D'un autre côté, l'ob-jet abstrait est encore l'*espace*, quelque chose de *non sensiblement sensible* ; l'*intuition* est élevée dans son abstraction, il est une *forme* de l'intuition, mais est encore intuition, un sensible, l'*extériorité réciproque* de la sensibilité elle-même ; sa pure *absence de concept.*"[Hegel 1816, 352]

On peut ajouter que le discours hégélien, identifiant l'espace comme "absence de concept", se prolonge en un discours sur les mathématiques. Cela est naturel, parce que la science mathématique se pose éminemment comme science de l'espace, et l'espace est, pour ce discours, si fondamentalement l'ennemi du concept que la mathématique se trouve emportée dans l'absence de concept par sa seule complicité avec l'objet espace. On sait que la fameuse Préface à la *Phénoménologie de l'esprit* [Hegel 1807] s'exprime avec une certaine violence au sujet des mathématiques :

> "La *matière* au sujet de laquelle la mathématique garantit un trésor consolant de vérités est l'*espace* et l'*Un*. Or l' espace est l'être-là dans lequel le concept inscrit ses différences comme dans un élément vide et mort, au sein duquel ces différences sont également sans mouvement et sans vie ... Dans un tel élément sans réalité effective il y a encore seulement un vrai sans réalité effective, fait de propositions rigides et mortes ;"[Hegel 1807, I, 38].

Où l'on voit comment le "défaut-de concept" de l'espace se transmet à la mathématique, dont sont dénoncées les "propositions rigides et mortes".

Mais, nous l'avons dit, la généalogie possède un second temps. L'espace ne s'en tient pas à l'espace, il se laisse concerner, envahir, capturer par le continu et l'infini, "après" le commencement : cette transfiguration de l'espace dans un "après-coup" relève bien sûr du *procès dialectique*, élément capital et dominateur du hégélianisme.

Ce second temps se marque par exemple comme tel dans le passage cité tout à l'heure du *Précis de l'Encyclopédie des Sciences Philosophiques*, consacré à une définition de l'espace : le fragment dont nous avons déjà fait état est imédiatement suivi de la considération suivante, qui le complète :

> "Il n'est pas admissible de parler de points spatiaux comme s'ils constituaient l'élément positif de l'espace, puisqu'à cause de son indifférenciation, il n'est que la possibilité et non la *position* de l'état de séparation et de négation et que pour cette raison, il est continu ; le point, l'*être pour soi*, est par suite au contraire, la *négation* de l'espace posé en lui. C'est aussi ce qui résoud la question de l'infinité de l'espace (§100, *Rem.*)."[Hegel 1830, 142-143].

Ni les "points" spatiaux, ni l'infinité de l'espace ne sont là dans le premier moment de l'espace qu'a posé la définition hégélienne. Pour qu'ils comparaissent, il faut que le point vaille comme ***négation*** de l'espace, et cela il le fait en tant que ***continu*** et qu'***être-pour-soi***. Une négation, chez Hegel, est une temporalisation, la négation de l'espace nous place "après" l'espace. En quelque sorte, l'idée de Hegel est que l'espace est d'abord un pur divers-en-extériorité, en lequel les points ont un statut aristotélicien de simple possibilité ; ***ensuite***, le point s'affirme dans une autonomie identitaire, comme quelque chose qui "enveloppe" son voisinage, et dans ce mouvement se constitue en même temps une totalité infinie et continue. La référence à l'Etre-pour-soi nous indique par ailleurs que ce mouvement est un retour du ***qualitatif*** : l'Etre-pour-soi est dans [Hegel 1812] un des noms de l'infini qualitatif, le "bon infini". Dans le même texte, on peut d'ailleurs aussi trouver l'assertion complémentaire que le mauvais infini, le quantitatif, est complice de l'être-là spatial : c'est donc ici du premier moment de la spatialité, le seul auquel soit lié le mot espace chez Hegel, qu'il s'agit.

En résumé, le continu et l'infini sont regardés comme des déterminations qui reviennent à l'espace, mais au prix de sa destruction (par l'instance de l'Etre-pour-soi, infini qualitatif) comme ce qu' il ne peut néanmoins pas cesser d'être : l'objet de l'appropriation de la mathématique géométrique. Ceci veut dire encore que la positivité conceptuelle de l'espace ne peut que rester étrangère, pour Hegel, à ce que la considération mathématique saisit et développe. Nous en trouvons plusieurs témoignages dans les écrits de Hegel, comme cet autre passage de la préface de [Hegel 1807] :

> "En effet, ce que la mathématique considère, c'est seulement la grandeur, la différence inessentielle. Ce qui scinde l'espace en dimensions, et détermine le lien entre elles, c'est le concept ; mais la mathématique fait abstraction de cela ; elle ne considère pas, par exemple, la relation de la ligne à la surface, et quand elle compare le diamètre du cercle à la périphérie, elle se heurte à leur incommensurabilité, – une relation vraiment conceptuelle, un infini qui échappe à la détermination mathématique."[Hegel 1807, I, 39].

On pourrait encore invoquer la grande remarque sur le calcul infinitésimal, dans [Hegel 1812], qui ne traite que de l'infini du calcul, et pas directement du continu, mais qui aboutit exactement à la même sorte de verdict, au même type de partage entre mathématique et philosophie.

Tout à l'heure, nous décrivions le premier moment de l'espace hégélien un peu comme un moment aristotélicien, le second moment, en revanche, celui de l'émergence concommittante du point, de l'infini et du continu, un peu comme un moment "cantorien". Le texte nous y autorisait, mais il faut bien voir que ce second moment, devant échapper tout à la fois à l'espace lui-même et aux mathématiques qui l'étudient, doit être en fait un moment purement *conceptuel*, pour Hegel. Nous ne devons donc pas être surpris que le discours de Hegel dénonce par avance l'idée d'une *ensemblisation* de l'espace et du continu. Poincaré, prouvant qu'il ne voyait pas dans l'ensemblisation la constitution d'un véritable continu, écrivait :

> "Le continu que nous offre la nature et qui est en quelque sorte une unité est-il semblable au continu mathématique, tel que l'ont défini les plus récents géomètres, et qui n'est plus qu'une multiplicité d'éléments, en nombre infini, mais extérieurs les uns aux autres et pour ainsi dire logiquement discrets. [Poincaré 1913, 187]"

Le passage suivant de Hegel, pris dans [Hegel 1812] nous semble lui faire écho :

> "La *continuité* est donc rapport égal à soi-même, simple, qui n'est interrompu par aucune limite et exclusion ; non pas pourtant l'unité immédiate, mais l'unité des Un étant-pour-soi. Est donc contenue là l'*extériorité-réciproque* de la *multiplicité*, mais en même temps comme une [multiplicité] non différenciée, *ininterrompue*. La multiplicité est posée, dans la continuité, telle qu'elle est en soi ; en effet les multiples sont chacun ce que sont les autres, chacun égal à l'autre, et la multiplicité, par conséquent, égalité simple, dépourvue-de-différence. La continuité est ce moment de l'*égalité-à-soi-même* de l'être-en-extériorité-réciproque."[Hegel 1812, 168].

Il est indiqué fort nettement, en effet, que le continu est du côté de l'égal et de l'un, et que par conséquent son "rapport " au principe d'extériorité universelle de l'espace ne peut être qu'un rapport d'antagonisme, ainsi qu'il était d'ailleurs déjà dit dans notre deuxième citation du §254 de [Hegel 1830]. Si le continu "existe" néanmoins dans cet antagonisme, à en croire ce passage, c'est par ce que le "moment de l'égalité-à-soi- même" est un moment conceptuel, philosophique, et non pas un moment géométrique ou mathématique. On peut parier que Hegel aurait jugé l'unité de $\mathbb{R}$ une simple unité *immédiate*, confirmant l'extériorité spatiale.

Supposons donc que nous ayons entendu le message hégélien. Supposons de plus qu'il problématise pour nous l'évidence du lien continu/espace, dont nous évoquions tout à l'heure la forte installation en nous. Nous pouvons poser deux questions :

1) y a-t-il un précédent philosophique où nous puissions trouver au contraire un soutien pour notre sentiment de la complicité du continu et de l'espace?
2) le développement contemporain de la mathématique parle-t-il dans le sens du conflit du continu et de l'espace, ou dans le sens de leur solidarité?

Kant : l'Intuition comme Rapport Herméneutique à l'Espace ; la Hiérarchie Herméneutique Espace/Continu/Infini ; le Thème de la Localité

Pour aborder d'abord la première question, nous croyons qu'un précédent du type cherché nous est offert par la philosophie critique et transcendantale de Kant. Comme ce n'est pas ici le lieu d'argumenter de façon précise et approfondie à partir des textes, nous allons d'un seul et même mouvement présenter notre manière d'entendre le message "esthétique" de Kant, et les enseignements relatifs à l'espace et au continu, ainsi qu'à leur rapport, que nous puisons dans ce message.

Premièrement, et ceci correspond à une manière de s'inscrire dans un débat qui a été volontairement placé au centre de ce colloque, nous ne tenons pas la théorie kantienne de la "forme a priori de la sensibilité" *espace* pour une théorie qui institue la géométrie euclidienne comme *dogme* commandant de manière intemporelle à la science de la nature, et confinant une fois pour toutes le savoir mathématique dans des bornes intangibles.

Pour nous, la doctrine de l'intuition pure, notamment spatiale, exprime la troisième détermination de la *finitude* du sujet de la connaissance :

- la première détermination consiste en ce que le sujet ne constitue pas au plan "intellectuel" ce qu'il a à connaître (comme Dieu) ; pour lui penser et connaître ne coïncident pas, le réel de sa connaissance doit *s'annoncer à lui*, doit lui venir de l'extérieur, comme une adresse ou une provocation, que la connaissance viendra recouvrir ;
- la deuxième détermination consiste en ce que cet "externe" qui s'annonce ne s'annonce pas comme *singulier* ; ce qui "entre" est un *divers* de la sensibilité, une multiplicité que Kant nous demande de nous représenter d'abord comme inorganisée, sans structure, sans unité ; pour dramatiser au maximum cette détermination, disons qu'elle signifie qu'il n'y a pas d'abord et d'emblée des *objets*, dont on pourrait décrire les propriétés dans un langage logique du premier ordre ;
- la troisième détermination, donc, consiste en ce que *pour nous*, le divers est inexorablement pris dans un cadre spatio-temporel (le *forme* de forme a priori ne veut rien dire de morphologique, il désigne tout à la fois le cadre et son caractère incontournable).

La lecture de la doctrine esthétique kantienne comme dogmatisme comprend que le "cadre" est donné avec une information, quelque chose comme un "mode d'emploi", qui serait le texte de la géométrie euclidienne. Par dessus le marché, elle interprète que la légitimité de ce texte réside en ce qu'il émane du sujet comme une de ses composantes innées et nécessaires (l'esthétique transcendantale serait une pièce de la psychologie cognitive kantienne).

Nous comprenons au contraire que le cadre est mis en scène par Kant comme quelque chose qui excède la capacité de pensée articulée et conséquente du sujet, tant et si bien qu'émane de lui comme une "question", à laquelle le sujet ne peut échapper, bien loin que lui revienne une certitude intime sur laquelle il se guide. Le point 3) de la première rédaction de l'exposition transcendantale de l'espace, pour nous, exprime que la géométrie est une *interprétation* de l'espace, en tant que celui-ci est infinitaire et dépasse le pouvoir représentatif finitaire de l'entendement : le caractère "régulateur" de l'espace ne gît pas dans le fait qu'il y a un discours fixe et immuable de l'espace, mais dans ceci que le discours de l'espace, la géométrie, fait toujours système, en sorte que la nécessité de chaque énoncé sur l'espace renvoie à celle du système, et non à la contingence de l'information empirique. Kant dit que nous n'affirmons pas la tridimensionnalité de l'espace "parce que nous n'avons jusqu'ici rencontré que des espaces tridimensionnels" : il ne dit pas sur le mode positif que cet énoncé nous est dicté par l'intuition pure. Cette formulation négative doit être comprise, selon nous, comme l'expression de ce que, pour lui, tout jugement sur l'espace est toujours pris dans la nécessité d'une géométrie qui précède toute expérience, et que c'est en cela que consiste la régulation apriorique de la connaissance par l'intuition pure de l'espace.

Nous nommons *herméneutique* ce rapport à l'espace : l'espace fait question, le sujet de la connaissance est originairement à son égard dans une situation qui est à la fois de *désaisissement* et de *familiarité* ; le désaisissement signifie que le sujet ne possède ni ne maîtrise l'espace, que l'espace est inépuisable et infinitaire ; la familiarité signifie que le sujet a toujours préjugé de la cohérence, de la logique de l'espace, qu'il a toujours anticipé l'essence des rapports spatiaux, de la structure spatiale. Ce "projet" de l'espace s'inscrit dans une géométrie, aujourd'hui dans une *théorie géométrique*. L'espace est toujours modèle d'une théorie géométrique. Mais les théories sont plusieurs, dans la succession et la simultanéité. Comme l'espace est en excès sur le pouvoir représentatif, la question de l'espace persiste à faire question par-delà les théories qui lui répondent, et de nouvelles théories voient le jour, qui reprennent l'impensé ou l'implicite des précédentes, en "revenant" à ce qu'elles posent comme le plus intime de la question. Comme ce jeu, qui doit être appelé *herméneutique de l'espace*, est joué librement au long des siècles, il arrive aussi que le "projet de l'espace" se diversifie de façon contradictoire, et donne lieu à plusieurs versions du "mystère de l'espace", irréductibles et concurrentes à une époque donnée. En première approximation, c'est le cas aujourd'hui. Mais nous aurons à en discuter un tout petit peu plus précisément.

Le deuxième élément essentiel que nous trouvons chez Kant, c'est l'idée d'une structure de double renvoi qui fonde une *hiérarchie herméneutique* des trois termes infini-continu-espace. En effet, les points numérotés 4) et 5) de l'exposition métaphysique de l'espace[1] disent que le "préjugé spatial", le fond herméneutique de notre rapport à la question de l'espace, contient à titre de noyau thématique le caractère global, insécable de l'espace, et solidairement son caractère chargé d'infini. L'espace est quelque chose en quoi l'infini est déposé, distribué : c'est d'ailleurs où réside l'indice de son caractère d'intuition plutôt que de contenu conceptuel. En ce sens la question de l'espace renvoie à la question de l'infini comme à une question plus fondamentale. Mais les développements ultérieurs de Kant dans la *Critique de la Raison Pure* ([Kant 1787]) nous indiquent qu'entre l'espace et l'infini s'interpose le continu : cet infini se déposant ou se distribuant, c'est aussi, c'est d'abord le caractère *continu* de l'espace, nominalement identifié par Kant à la *divisibilité à l'infini* dans les *anticipations de la perception* : d'après cette identification, donc, le continu est lui-même "fondé" sur l'infini. Or c'est comme continu que l'infini charge l'espace, ou encore c'est le caractère continu qui "contraint" à l'infini l'intuition de l'espace. On aurait donc la hiérarchie suivante : la question de l'espace renvoie à celle du continu (comme à son préalable, comme à une question plus fondamentale), la question du continu renvoie à celle de l'infini (de la même manière). Selon cette hiérarchie, notons le immédiatement, le continu n'est pas "contre l'espace", au contraire, il est un constituant essentiel du sens de l'espace (l'espace ne se comprend que par le continu). Il est donc attesté que cette autre perspective philosophique contrecarre le point de vue que nous trouvions chez Hegel (historiquement, le rapport polémique fonctionne bien évidemment dans l'autre sens).

Par ailleurs, et sans entrer dans les détails, cette hiérarchie nous semble rencontrer le sentiment qu'on a naturellement, à partir d'une formation mathématique contemporaine (ensembliste) : le concept d'espace paraît, à travers la géométrie de coordonnée, et son héritière la géométrie différentielle moderne, construit "sur" le continu, et son investigation requérir en effet la connaissance du continu (des $\mathbb{R}^n$) comme préalable ; le concept de continu lui-même se cristallise dans la construction du $\mathbb{R}$ de Cantor-Dedekind, dont nous savons qu'elle comporte comme moment décisif l'admission d'un infini actuel des totalités ensemblistes, où il faut voir une grande figure de l'herméneutique de l'infini ; donc l'idée que la théorie du continu renvoie à la formalisation de l'infini est naturellement nôtre.

L'ultime enseignement qu'il est naturel d'aller chercher chez Kant concerne le rapport de l'herméneutique de l'espace à celle du continu : si Kant pense le "renvoi" dont nous avons parlé, de l'espace au continu, y-a-t-il chez lui une approche de ce qui dans la ques-

[1] Toujours dans sa première rédaction.

tion de l'espace excèderait la question du continu, de ce qui serait un noyau thématique propre à l'espace dans sa distinction d'avec le continu?

Le continu, dans sa cohésion et son intégralité, est selon Kant intuitionné *a priori* comme global. Mais le mot "global" est ici maladroit, surtout devant une assistance de mathématiciens, parce qu'il évoque son pendant mathématique, le local. Or il s'agit plutôt, dans l'exposition métaphysique de l'espace, d'une *intégralité* qui est pure cohésion justement, pur caractère continu, et ne s'oppose pas encore au local, n'est pas encore négativement polarisée à l'encontre de celui-ci.

Justement, notre thèse est qu'on peut lire chez Kant l'idée que la pensée spatiale n'est pleinement elle-même que lorsque s'ajoute à la pensée (essentiellement intuitive, rappelons le) de l'intégralité infinitaire déposée-distribuée celle du *local* : celle-ci fait apparaître par contraste une signification de globalité, laquelle, dans une inscription théorique, pourra s'attribuer à l'intégralité, mais sera plus générale et plus libre. Ce supplément intuitif propre à l'espace, nous le voyons comme ce qu'apporte le thème kantien de l'*intuition formelle*, évoqué à la fin de la déduction transcendantale des catégories, au §26 de [Kant 1787]. Donnons donc rapidement une idée de la lecture que nous proposons de ce passage célèbre et énigmatique.

Kant, dans ce passage, conclut la "déduction transcendantale des catégories de l'entendement" en expliquant que le divers sensible, en tant que "spatialisé-temporalisé" (pris dans les *formes pures de la sensibilité*) n'est pas seulement "jeté" en l'espace comme en un fond ou un cadre, il y a aussi un moment de l'intuition formelle, qui est le moment de l'*intuition unitive*, laquelle est encore "représentation de l'espace comme objet". Ce que vise Kant, c'est bien l'intuition a priori de l'être-ensemble dans l'espace plutôt que de l'espace comme cadre et fond, plus précisément de cet être-ensemble comme un "objet" que le fond spatial n'est jamais, puisqu'il est un cadre, un medium-norme, quelque chose d'enveloppant et pas d'ob-jeté. L'être-ensemble ob-jeté dans l'espace, c'est, traditionnellement, quelque chose comme la figure, la forme au sens morphologique, non kantien ; dans l'exemple de la maison que donne Kant juste après avoir évoqué l'intuition formelle, on voit que ce moment de l'intuition correspond au tracé du dessin de la maison[2]. Si l'on se place *a priori*, c'est-à-dire si l'on envisage un divers jeté dans l'espace absolument quelconque, le "contenu" de l'intuition formelle ne pourra plus être aucune forme particulière, mais seulement la qualité synthétique du divers au sein de l'englobant spatial, qualité qui fait valoir le divers comme uni en un lieu : le corrélat absolument général et apriorique de l'intuition formelle, même si Kant le voit sans doute plutôt comme la figure, c'est le *lieu*, mieux c'est la localité du lieu (le schème du lieu). Kant associe explicitement ce moment de l'intuition formelle à la *géométrie* : même si la géométrie est déjà évoquée dans l'exposition métaphysique de l'espace, même si l'espace est déjà l'entier infinitaire continu corrélat d'une théorie géométrique à ce stade de l'exposé de Kant, il semble que pour lui, l'identité de la géométrie ne soit pleine qu'avec l'intervention de cet intérêt pour la figure, ou plus radicalement pour la localité comme telle.

Le "plus" de la question de l'espace par rapport à la question du continu serait donc, en suivant Kant (c'est-à-dire aussi en le précédant), la question du local : dès lors que l'entier infinitaire continu, interrogé selon la localité, devient *espace* en un sens plus complet, l'intégral de l'exposition métaphysique, abritant le local, vaut comme *global* dans son contraste avec lui.

---

[2] Cf.§26 juste après la division, p. 139 dans notre édition.

Nous sommes naturellement conduits à nous demander ce qu'il en est aujourd'hui de ce dispositif : la question du local, en mathématiques, est-elle toujours sous la dépendance de la question du continu? le concept de figure est-il toujours ce qui recouvre le thème de la localité, et nomme l'intérêt prévalent de la géométrie?

Il nous semble à la vérité que, regardés à la distance et dans l'esprit convenables, les grands événements de l'histoire de la géométrie célébrés par ce colloque vont tous dans le sens d'une accentuation du "poids" de la thématique du local.

La Géométrie Moderne

Soit par exemple ce qu'on appelle le "point de vue Erlangen" en (ou sur la) géométrie. Réduit à l'adage que tout le monde connaît, il s'agit du point de vue qui comprend essentiellement la géométrie comme étude de substrats munis de groupes de transformation en lesquels réside toute l'identité de chaque géométrie considérée. Or, n'est-il pas clair que la donnée d'un groupe de transformation spécifie justement comment il est possible de "permuter" globalement les localités ultimes que sont les points du substrat? Si le groupe, dans la mesure où il agit transitivement, "témoigne" de l'homogénéité de l'espace-substrat, dans tous les cas, il atteste également l'écart entre l'intégralité de l'espace et chacun de ses points, écart qui autorise une pluralité de façons de bouleverser le substrat en conservant sa structure globale, mais en altérant de nombreux points (envoyés "ailleurs"). Dans ce dialogue entre l'intégralité de l'espace et la localité ponctuelle, nous ne retrouvons pas la localité de tout à l'heure, celle de l'intuition formelle, qu'on peut en première approche identifier comme celle de la figure. Mais le point de vue Erlangen a tout de même sa façon de la traduire : dans une géométrie donnée par un couple substrat-groupe, à toute figure sera associé un sous-groupe du groupe fondamental, le stabilisateur de la figure. Alors l'écart entre le local ultime ou local ponctuel et le local de la figure se trouvera en quelque sorte reflété ou traduit par l'écart entre le groupe fondamental et son sous-groupe (le quotient étant le groupe fondamental d'un espace dont les points seraient les localisations de la figure). C'est plutôt comme cela que comme une "algébrisation" de la géométrie que nous demanderions à l'histoire et à la philosophie des mathématiques de comprendre le point de vue Erlangen : comme un moment herméneutique à la faveur duquel est proposée une "version" de la *localité*, pièce essentielle de l'individuation de la géométrie. Cette version dramatise tout à la fois l'écart entre l'entier de l'espace et la localité ultime du point, et l'écart entre cette localité ponctuelle et la localité de la figure.

Le second événement présent à l'esprit de nous tous, réunis pour ce colloque, est sans conteste l'émergence du concept de variété différentiable, dont on saisira volontiers le premier moment, sinon l'accomplissement, dans l'article de 1854 de Riemann. Cependant, du point de vue qui est le nôtre, il convient plutôt de passer par dessus l'enchaînement historique voulant que la géométrie différentielle moderne se soit développée avant et indépendamment des concepts topologiques. Aujourd'hui, nous comprenons les variétés différentiables comme une catégorie, éminente et paradigmatique certes, mais une catégorie tout de même parmi les espaces topologiques. Et cette notion plus générale mérite d'être d'abord commentée, parce qu'elle n'est pas autre chose qu'une élaboration de la *localité* pour elle-même, une élaboration de cela même qui motive, disions-nous, une herméneutique de l'espace ayant son autonomie et son thème propre vis-à-vis de l'herméneutique du continu. "topologique", cela veut bien dire *qui a trait au lieu*, l'étymologie, en l'occurrence, manifeste le caractère non arbitraire de la terminologie. L'élément crucial du concept d'espace topologique est la mise en avant d'un concept axiomatique du lieu comme "sous-ensemble bien abritant", qui est *l'ouvert*. L'usage marque cette option, puisque "localement" veut dire presque toujours "dans quelque ouvert" (variante : dans quelque ouvert contenant ce qui est nommé "le point" dans le contexte). Notamment, dans sa contribution à ces journées, monsieur Gromov, qui discute des cas dans lesquels il est possible de passer du local au global, donne à local ce sens (si nous comprenons bien, son problème est plus précisément de déterminer dans quel cas la validité de quelque chose "universellement localement" entraîne sa validité "globalement").

Les axiomes des ouverts chargent d'un sens implicite le "lieu". Le lieu est quelque chose qui reste soi-même si on le "raffine" (par intersection) de manière *finie*, mais risque de se dénaturer si on se laisse aller à resserrer d'un seul coup *infiniment* la localisation. Le lieu donne matière à la définition du *voisinage*, et se redéfinit alors comme ce qui est voisinage de tous ses points, propriété que nous proposons de gloser comme propriété d'être *bien abritant* (c'est exactement le langage que tient Heidegger pour caractériser le lieu de façon pré-métrique dans [Heidegger 1954]). Mais ce lieu est essentiellement indéterminé, des sous-ensembles de taille absolument imprévisible ont la charge signifiante du lieu, notamment l'espace entier (toujours) ; la logique pure du lieu abandonne absolument tout contrôle quantitatif du lieu, s'en tenant à sa qualité d'*ouvert*. Si néanmoins on envisage tous les lieux contenant un point donné, alors la "synthèse" de ces lieux, sans donner naissance à un lieu infinitésimal homogène à chacun des lieux indéterminés pris en compte, suscite une *perspective infinitésimale*. L'impossibilité d'intersecter, et donc d'assigner l'infinitésimal *dans l'espace* induit classiquement une démarche détournée donnant naissance à des objets infinitésimaux : celle du passage aux "germes". Moyennant une complication de "type", c'est-à-dire l'invocation de classes d'équivalence d'entités elle-mêmes définies sur les lieux variables, on obtient des incarnants de l'infinitésimalité. Pour ce genre de raison de principe, l'infinitésimalité est plus universelle, *a priori*, que la localité indéterminée. Le concept topologique du local fait aussi apparaître un concept corrélatif du global : global se dit alors de ce qui est "recollement" de localités. Donc n'importe quelle réunion d'ouverts, ou plutôt n'importe quel objet en tant qu'on le fait valoir comme recouvert par une famille d'ouverts. Un cas particulier extrêment important est celui de ce global qui ne nécessite jamais vraiment une multiplicité infinie de points de vues locaux pour être couvert : si une telle multiplicité est donnée, on peut toujours la réduire à une multiplicité finie effectuant déjà le "recouvrement". Ces "globalités" *essentiellement* finitaires sont les parties dites *compactes* de l'espace. Mais ce sont en fait aussi des localités : elles seraient même quelque chose comme le répondant, dans le dispositif topologique, du local de l'intuition formelle kantienne, celui de la figure. La *figure* paradigmatique, pour Kant et pour la pensée mathématique courante, notamment géométrique, est en fait la figure se prêtant à une synthèse finie, comme le cercle ou la sphère. Or, ces "compacts" sont cela même, auquel on a retiré les propriétés de régularité, de teneur locale, qui font des cercles et sphères des objets beaucoup plus riches que des compacts généraux.

Les compacts, ainsi, sont des parties qui exemplifient la localité en même temps qu'un certain mode de la globalité. Corrélat du local, le global est lui aussi indéterminé. Seule sa limite supérieure a une individuation invariante : le maximum que puisse prendre en lui un lieu synthèse de lieux, c'est l'intégralité de l'espace. On aboutit ainsi au diagramme suivant, le diagramme de la localité topologique :

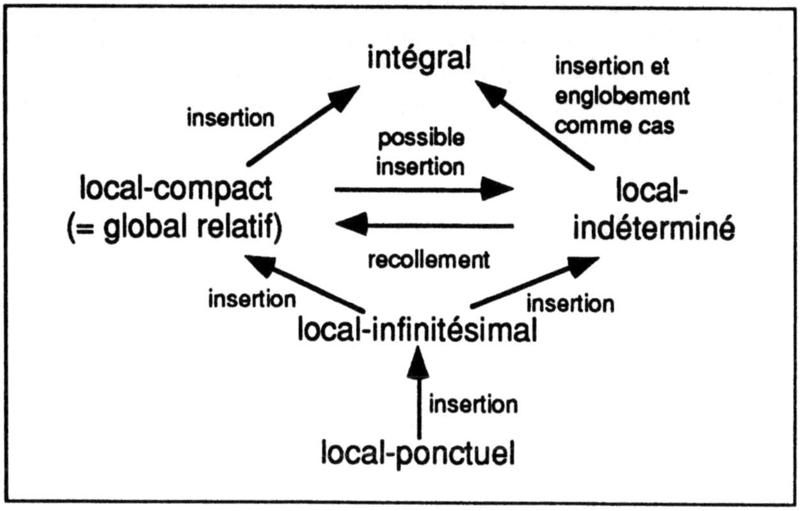

Mais nous l'avons dit, "en fait", cette pensée de la localité se développe à partir d'un approndissement généralisant de la pensée euclidienne de l'espace qui est la promotion du concept de variété différentiable. Ceci possède deux aspects et deux significations importantes pour nous :

1) le contenu topologique de la notion de localité ne fait pas, à lui tout seul, une localité géométrique ; être topologue n'est pas "déjà" être géomètre, même si, de facto, un topologue pur, qui ne s'intéresserait pas à ces espaces topologiques particuliers que sont les variétés différentiables, est chose parfaitement improbable : la condition supplémentaire, pour qu'on ait de la géométrie au sens strict, est, semble-t-il, une certaine concession au suffixe *métric* ; non pas nécessairement qu'on dispose d'une métrique globale ou locale, mais tout de même qu'il y ait une présence locale du nombre, que l'espace puisse être vu localement selon quelque chose comme le nombre, la mesure, le paramètre.

2) la géométrie différentiable, fondée de manière moderne sur le concept de variété différentiable, est le lieu d'un développement considérable de la pensée géométrique de l'espace où celle-ci reste liée au continu. Elle envisage le continu comme un "noyau" de la spatialité, tout en attestant le caractère définitoire, pour ce que la géométrie possède en propre, du thème de la localité : ce thème fonctionne, dans cette géométrie, conformément au diagramme général, avec des déterminations particulières liées au fait que les "espaces" sont localement homéomorphes à des ouverts de $\mathbb{R}^n$, et ce de manière différentiablement inter-rapportable. Notamment, le problème général de la substantialisation des objets infinitésimaux se trouve redoublé du problème du caractère *intrinsèque* des définitions, problème qui découle naturellement du concept d'*atlas* (premier problème de cette sorte : celui de la définition du fibré tangent d'une variété). Les variétés compactes deviennent des incarnants plus satisfaisants de la "localité" kantienne de la figure, dès lors qu'ils sont toujours "bien abritants" en un sens adéquat (celui d'une carte locale), alors qu'un

compact général peut posséder des points isolés. Les problèmes "globaux" (problèmes de recollement) se posent et sont traités selon des voies où les techniques de l'analyse interviennent.

Tant que, donc, le concept d'espace topologique n'intervient que par le truchement des variétés différentiables, la géométrie reste liée au continu, elle manifeste seulement au moyen du langage topologique ce que la pensée de la localité a de propre et d'excédentaire par rapport à la pure pensée du continu. La pensée de l'espace comme chargé de la localité reste fondée sur celle du continu.

Du Concept de Faisceau à l'Idée d'une Reconstruction Catégorique des Notions d'Ensemble et d'Espace

Mais le concept de faisceau, surtout au vu de la fonction décisive qu'il remplit en géométrie algébrique, ouvre la possibilité d'une rupture, d'une prise d'autonomie de la pensée de l'espace par rapport à celle du continu. Ce dont il s'agit peut être rapidement résumé comme suit : le *plus* nécessaire pour que le topologique vaille comme géométrique, et que nous décrivions comme une *présence locale du nombre* tout à l'heure, en des termes un peu vagues et mystérieux, le point de vue faisceautique en donne une nouvelle interprétation. En substance, il y aura présence locale d'*opérations*, et les opérandes seront conçus selon l'analogie fonctionnelle (comme des "fonctions" définies localement).

Le concept de variété différentiable requiert que la variété soit localement *paramétrée* par des nombres réels. Le concept de faisceau implique des éléments pris dans une structure opératoire générale (groupe, anneau, algèbre) et non pas dans celle, paradigmatique et fixant la signification moderne du continu, qu'on nomme $\mathbb{R}$; il les implique de plus non comme *paramètres*, en termes desquels se laisse intégralement lire et reconstruire localement l'espace topologique, mais comme entités "au-dessus" de cet espace, analogues de fonctions. Les axiomes définissant le faisceau assurent essentiellement un "suivi" ou une surdétermination du diagramme de la localité par les "fonctions" : un ensemble de "données fonctionnelles" locales provient par une opération fondamentale dite de "restriction" d'une donnée fonctionnelle relativement globale si et seulement si elles satisfont le critère naturel de compatibilité sur les intersections (reformulé en termes des flèches de "restriction"). Par ailleurs, il est possible de construire les germes du faisceau en chaque point, et de les agglomérer en un objet topologique qui se présente comme une fibration de l'espace de base ("l'espace étalé" associé au faisceau). Il s'ensuit que l'on récupère le problème du rapport entre fonctions locales et fonctions relativement globales, ainsi que celui du rapport entre fonctions locales et fonctions infinitésimales (les éléments de l'espace étalé). Le concept de faisceau, bien entendu, n'intervient pas seulement en géométrie algébrique. Mais cette discipline a ceci de particulier que ses objets fondamentaux selon l'approche actuelle, les *schémas*, sont des faisceaux, dont l'espace topologique sous-jacent n'est pas du tout nécessairement en rapport avec le continu : dans la situation de référence la plus banale, le substrat est un ensemble algébrique affine muni de sa topologie de Zariski, sur un corps k absolument quelconque. Il s'en suit, de telles topologies n'étant pas séparées, que le rôle du compact dans le diagramme de la localité sera joué par des "quasi-compacts" (dont le concept entre lui-même en résonance avec la notion algébrique d'anneau noetherien).

Ce qui nous intéresse est que, de la sorte, s'élabore un objet théorique qui est *plus que topologique* : il vaut comme motif d'une théorie véritablement *géométrique*, on ne dit pas les spécialistes de la géométrie algébrique géomètres de manière homonyme et par abus de langage. Et pourtant, l'objet schéma n'entretient pas une dépendance de principe à l'égard du continu. La démarche de la géométrie algébrique "consomme" donc quelque chose comme une émancipation de la pensée de l'espace à l'égard de la pensée du continu.

Bien sûr, nous ne voulons pas, disant cela, prendre une position absolue et maximaliste sur la teneur du discours de la géométrie algébrique contemporaine, à la fois parce que nous n'en avons pas la compétence, et parce que nous pressentons qu'elle abrite une multiplicité d'approches. En particulier, nous ne prétendons pas nier le rôle que persistent à jouer les variétés algébriques complexes (et peut-être réelles), rôle que la généralité des objets grothendickiens est susceptible de "masquer" ; par ailleurs, au simple degré du lecteur de traités classiques qui est le nôtre, il apparaît avec une certaine évidence que subsiste, dans le cadre des schémas, un rapport de "mimétisme" de la géométrie algébrique à l'égard de la géométrie différentielle, à travers lequel l'idée du "substrat continu" reste sans doute régulatrice, à défaut d'intervenir de la même façon dans le "préjugé de l'espace".

Dans un autre ordre d'idées, il serait assez naturel, ici, d'évoquer certaines tendances encore plus radicales suscitées par les formalismes de la géométrie algébrique : on sait que le concept de *site* permet la généralisation de la notion de faisceau à une situation où n'est même pas donné un espace topologique substrat. Ceci constitue un pas vers l'abandon de plus que le continu comme "socle" de la pensée géométrique : vers une rupture avec la représentation ensembliste elle-même, soit sans doute nécessairement avec la juridiction de la "totalité ensembliste infinie" sous laquelle se place ordinairement non seulement la géométrie, mais la mathématique en général. A ce qu'il nous semble, certaines des démarches intérieures à la géométrie algébrique grothendickienne vont dans le sens d'une telle destitution de la perspective ensembliste au profit de la perspective catégorique (nous citerions volontiers la définition catégorique des "points" comme exemple). Mais il s'est par ailleurs constitué un courant *fondationnel* à partir de ces "audaces" radicales de la géométrie algébrique, à l'origine duquel se trouvent essentiellement les idées de Lawvere. A l'heure actuelle, il semble que se poursuive en parallèle un travail de reformulation *catégorique* – en fait de reformulation dans le cadre de la théorie des topoi, qui se porte de plus en plus comme candidate à la place de langage des fondements – de la théorie des ensembles, et un travail de mise en évidence des liens que le langage des topoi entretient avec le thème de la constructivité (notamment, la logique intuitionniste et le lambda-calcul) : nous renvoyons à [Goldblatt 1984] et [Lambek-Scott 1986]. Ces travaux confirment le sentiment que nous exprimions, selon lequel le nouveau langage, né au sein de la géométrie algébrique, ouvre la voie à une relance non ensembliste de l'herméneutique de l'infini, soit à une refonte du dispositif classique plus radicale que la rupture de la géométrie avec le continu.

Nous voudrions terminer cette intervention par l'évocation d'un autre cas, relevant également de la mathématique actuelle, dont il faut certes reconnaître le caractère jusqu'ici "mineur", comparé à ce "continent" de la mathématique moderne qu'est la géométrie algébrique, envisagée à l'instant (mais qui, cela dit, s'inscrit dans un continent de même envergure, celui de la théorie des systèmes dynamiques) : nous voulons parler des travaux non standard sur les "canards". Il nous paraît tout à fait intéressant, en effet, de voir que ces travaux tendent à transformer la pensée du *local*, et donc la pensée de l'espace, tout en restant superlativement fidèles à la hiérarchie herméneutique explicitée par nous à partir de Kant.

Modulation Non Standard du Lieu : son Classicisme

L'essentiel consiste en ceci : dans les $\mathbb{R}^n$, et à vrai-dire plus généralement dans les espaces métriques, le "diagramme de la localité" peut être jusqu'à un certain point contrôlé de manière quantitative, en prenant en compte la base d'ouverts constituée des boules ouvertes et en se référant à leurs rayons. Seulement, dans ce repérage ou ce contrôle, le local-infinitésimal n'a pas de répondant, pas plus que le local-intégral, ou plus généralement ce qui excède le local-compact, qu'on peut quant à lui associer au fini, à l'instar du local-indéterminé (la "distinction" entre les deux n'étant pas exprimable sur le plan numérico-quantitatif). En autorisant l'objectivation de l'infinitésimal, l'analyse non standard modifie cette situation. Le local infinitésimal, pour commencer, pourra être associé aux rayons infinitésimaux de boules, mais à la vérité, l'incidence la plus importante du non standard est l'effet de bord de cette objectivation : le local-infinitésimal se démultiplie, se ramifie, et ce d'une façon que le numérique contrôle ; il est clair que cet effet de bord est lié au caractère algébriquement intégré de l'infinitésimal réel. Si a est un point d'un espace métrique, et ϵ un infinitésimal positif, on pourra définir, outre le halo de a (ensemble (externe) des éléments à distance infiniment petite de a), son ϵ-galaxie, son ϵ-halo, son ϵ-microhalo et son ϵ-microgalaxie :

$$\begin{aligned}
\epsilon - gal(a) &= \{x \in X | \exists^{st} n \in N \quad d(a,x) \leq \epsilon\} \\
\epsilon - hal(a) &= \{x \in X | \forall^{st} n \in N \quad d(a,x) \leq \frac{\epsilon}{n}\} \\
\epsilon - microhal(a) &= \{x \in X | \forall^{st} n \in N \quad d(a,x) \leq \epsilon^n\} \\
\epsilon - microgal(a) &= \{x \in X | \exists^{st} n \in N \quad d(a,x) \leq e^{\frac{1}{n\epsilon}}\}
\end{aligned}$$

Il est facile de voir que ces différents ensembles externes définissent des ordres de proximité toujours plus resserrée avec a, un "abîme" séparant chaque ordre de l'ordre suivant (dans le sens où il contient des éléments dont la distance à a est infiniment grande par rapport à celle des éléments de l'ordre suivant).

Cette gradation ou ramification existe de manière symétrique dans l'infiniment grand, et peut être considérée comme un repérage de l'écart *intégral/local compact* (repérage qui est ainsi susceptible d'être pris en charge par le *local indéterminé*, conformément à la réversibilité d'insertion entre les deux pôles que nous avons mise en avant). De la sorte, le *local compact* codé par le "à distance finie", c'est-à-dire par la galaxie principale constituée des éléments *limités*, prend un sens distinctif numérique à l'égard de l'*intégral* (de la fuite du *local intéderminé* vers l'intégral), en même temps que le concept de réel limité (dans la mesure où il coïncide avec celui de réel *presque-standard*) reflète quelque chose du concept *topologique* de compacité (alors que celui de réel fini, dont l'extension est la même que celle du concept de réel, dans le discours classique, n'a pas ce contenu).

Cette surdétermination métrico-numérique du diagramme de la localité, qui va de pair avec sa diversification-ramification, est exposable de loin, sans entrer dans les travaux sur les canards que nous avons nommés plus haut. Mais ces travaux montrent l'importance et la pertinence de cette surdétermination, manifestent le rôle décisif qu'elle peut jouer dans une étude essentiellement *géométrique*. Le concept même de canard, la détection informatique du phénomène, et le développement d'un intérêt mathématique à son endroit relèvent des instruments non standard de "cadrage" de la localité : alors que, pour le discours classique, les cycles canards de l'équation de Van der Pol sont des formes curieuses observées lorsque le paramètre décisif n'est pas assez près de 1 pour que l'on ait un cycle de taille prévue par le théorème de Hopf, et donc correspondent à une localité indéterminée en quelque sorte "empirique", à laquelle ne s'intéresse ordinairement pas la mathématique – dont la tâche serait uniquement de décrire ce qui se passe "en fin de compte" – l'objectivation infinitésimale regarde les canards empiriques comme expression de quelque chose qui a lieu canoniquement lorsque la proximité infinitésimale du paramètre à 1 est corrélée d'une certaine manière algébrique à la proximité infinitésimale du coefficient de $\ddot{x}$ à 0, et par conséquent, elle fait des cycles canards un thème mathématique.

Plus techniquement, il est observable que les théorèmes de base sur les canards s'expriment tous en termes de la "nouvelle" possibilité d'évaluation de la localité offerte par l'analyse non standard : nous pensons essentiellement au théorème qui dit que deux solutions canards, sur un intervalle où elles longent toutes deux la courbe lente, sont dans une proximité mutuelle "micro-halienne" (puisqu'elles ont même développement en ϵ-ombres) et à celui qui dit qu'une valeur canard du paramètre étant donnée, toute autre valeur (nécessairement infiniment proche) est à une distance microgalactique de la première [3].

Du point de vue du thème principal de cet article, c'est-à-dire de l'éventualité d'un conflit de préséance et de fondamentalité entre le continu et l'espace, il est intéressant d'observer que ce complément à la pensée de la localité, et donc à l'herméneutique de l'espace, est proposé par l'ANS dans une fidélité absolue à la hiérarchie herméneutique signalée par nous à partir de Kant : la surdétermination quantitative-numérique du diagramme de la localité est fondée sur la disponibilité d'un "modèle" du continu incorporant l'infiniment petit et grand dans un cadre algébrique commun avec le limité ; ce modèle du continu est lui-même fondé, comme on peut s'en convaincre en revenant à la source (au travail de Robinson), sur la découverte que l'opposition fini/infini avait changé de sens (s'était relativisée, notamment) sous l'influence du formalisme moderne. En fait, les présentations ultérieures de l'analyse non standard ont approfondi sur ce point l'apport de Robinson : les éléments théoriques touchant le fini, l'infini et leurs rapports sur lesquels repose l'ANS y sont formulés dans des termes indépendants de la théorie des ensembles et du théorème de Löwenheim-Skolem, et plutôt liés au théorème d'incomplétude de Gödel, c'est-à-dire à la tension qui s'établit nécessairement entre toute formalisation et le niveau "constructif"[4].

Il apparaît donc que la structure liant les trois questions "Qu'est-ce que l'espace?", "Qu'est-ce que le continu?" et "Qu'est-ce que l'infini?" est ici exactement la structure classique, selon laquelle l'espace renvoie au continu et le continu à l'infini, le progrès dans une herméneutique étant à chaque fois conditionné par le progrès dans l'herméneutique "inférieure", c'est-à-dire plus "fondamentale". En particulier, l'espace est bien pris, dans cet affaire, comme indéchirablement solidaire d'un continu sur lequel il est fondé, alors que, nous l'avions vu, la géométrie algébrique moderne ouvre une perspective selon

---

[3] Dans [Zvonkin-Schubin 1984] ce sont respectivement les théorèmes 3.1 et 3.3.

[4] Cf. à ce sujet [Reeb 1979] et [Nelson 1986]

laquelle l'herméneutique géométrique de la localité, soit en fait la pensée de l'espace, prendrait son indépendance à l'égard de la question du continu ; peut-être même, à la limite, les rapports de fondamentalité entre les trois questions, et notamment la valeur de présupposition décisive de l'infini ensembliste, seraient-ils réévalués dans une perspective fondationnelle inspirée par la théorie des topoi.

Conclusion

Disons pour conclure que, en tout état de cause, la façon de concevoir philosophiquement le conflit du continu et de l'espace que nous avons trouvée chez Hegel n'est pas homogène avec le style et les modalités de la pensée mathématique :

1) d'une part, le lien classique, le lien de référence entre continu et espace en mathématiques n'est pas que l'espace serait premier et "privé de pensée", et le continu second, venant dissoudre l'espace en apportant la qualité, l'infini et la pensée, mais tout au contraire, il consiste en ce que l'espace est fondé sur l'abîme infinitaire du continu ;

2) d'autre part, l'éventuel divorce entre l'espace et le continu dans l'aire mathématique n'est pas celui d'une réflexivité purement conceptuelle du continu avec un positivisme géométrique, n'équivaut pas à une "rupture disciplinaire" ; il est plutôt le symptôme de la dérive d'une herméneutique à l'égard d'une autre, au sein d'un continent juridique commun définissant la discipline (la mathématique ensembliste), cette dérive pouvant, à la limite, induire une refonte de ce sol juridique, sans que jamais il soit question pour autant de nier "l'unité des mathématiques", à comprendre ici comme l'unité ultime de responsabilité de la communauté mathématique à l'égard des trois questions "Qu'est-ce que l'espace?", "Qu'est-ce que le continu?" et "Qu'est-ce que l'infini?".

Bibliographie

Goldblatt R. [1984] *Topoi, the categorial analysis of logic*, Amsterdam : North-Holland.
Hegel G.W.F. [1807] *Phénoménologie de l'Esprit*, Paris : Aubier, traduction Hyppolite.
Hegel G.W.F. [1812] *Science de la logique (L'Être)*, Paris : 1972 Aubier.
Hegel G.W.F. [1813] *Science de la logique (La doctrine de l'essence)*, Paris : 1976 Aubier.
Hegel G.W.F. [1816] *Science de la logique (La logique subjective, ou doctrine du concept)*, Paris : 1981 Aubier.
Hegel G.W.F. [1830] *Précis de l'Encyclopédie philosophique*, Paris : 1967, Vrin.
Kant E. [1787] *Critique de la Raison pure*, Paris : 1971 P.U.F..
Heidegger M. [1954] Bâtir, habiter, penser in *Essais et conférence*, Paris : 1958 Gallimard, p. 170-193.
Lambek J. & Scott P.J. [1986] *Introduction to higher order categorical logic*, Cambridge : Cambridge University Press.
Nelson [1986] *Predicative arithmetic*, Princeton : Princeton University Press.
Poincaré H. [1913] *Dernières Pensées*, Paris : Flammarion.
Reeb G. [1979] *La mathématique non standard, vieille de soixante ans?*, Strasbourg : IRMA.
Zvonkin A.K. & Schubin M.A. [1984] Non standard analysis and singular perturbations of ordinary diffenrential equations, in *Russian Math. Surveys* **39** : **2**, p.69-131.

5

Some Insights About Modern Physics

Geometrical Concepts in Quantum Physics

Gilles Cohen-Tannoudji

Service de Physique des Particules. CE Saclay F-91191 Gif sur Yvette Cedex[1]

The Quantum Horizon

The concept of horizon, which is a genuine geometrical concept has emerged in the controversies raised by quantum theory, in 1948, in the review *Dialectica* directed by Ferdinand Gonseth. An issue of this review was devoted to the idea of complementarity. Several contributions of the prominent physicists at the origin of quantum theory (like N. Bohr, A. Einstein, W. Pauli, W. Heisenberg or L. de Broglie) made of this issue a major contribution to the history and the philosophy of sciences.

As a philosopher, F. Gonseth drew a few philosophical conclusions of the debate, in an article intitled *"Remarques sur l'idée de complémentarité"*. Starting from the observation that "le dévoilement progressif de la réalité par la connaissance scientifique déçoit progressivement l'espérance d'aller toucher l'ultime réalité des choses" and that "l'idée de réalité et, derrière celle-ci, l'idée même de l'être se trouvent mises en question", Gonseth proposes what he calls the dialectical conception, which, according to him allows to avoid "l'antinomie du commencement". According to this conception, "la connaissance scientifique commence par les expériences et s'ancre dans les évidences de notre *horizon naturel de réalité*". The idea of complementarity arises in a rather natural way as a relation between two subsequent reality horizons : "l'horizon quantique se dévoile comme un approfondissement de l'horizon apparent, assimilé à l'horizon classique ou naturel".

Gonseth then introduces the "apparent" and "profound" horizons which are such that "Un événement de l'horizon profond n'est connu expérimentalement que par ses traces phénoménales dans l'horizon apparent". But he adds that "s'il est vrai que l'expérimentation finit toujours par avoir notre monde propre pour siège, le travail théorique semble avoir constamment eu l'édification d'un nouvel horizon de réalité pour but."

This long quotation allows to appreciate how the concept of reality horizon is suited to the problems of quantum theory. Strictly speaking, a horizon is *subjective* as well as *objective*: if there were no world, there would be no horizon, but a horizon exists

[1] Le texte présenté ici n'est que le résumé de la conférence donnée par l'auteur. Toute personne désireuse d'en savoir plus, peut se reporter à Cohen-Tannoudji, "La pertinence du concept d'horizon de réalité en physique théorique contemporaine", *Dialectica*, **44**, 3-4, 1990, 323-332; et à *L'horizon des particules. Complexité et Elémentarité dans l'Univers Quantique.* Gallimard, Paris, 1989.

always with respect to some observer. Now the main problem raised by the existence of a quantum of action, is that, in contradiction with the dogma of classical physics, one cannot any more neglect the conditions of observation in the descrition of microscopic reality. Whereas classical physics could, at least in principle, ignore the *subject* of scientific knowledge, quantum physics has to include the subject in the very content of the concepts. This feature was one of the main point of the argumentation of N. Bohr. According to him one has to keep in mind that quantum concepts are not intended to describe directly the physical reality by itself, but only what he called *phenomena*, i.e. elements of reality as it appears under some conditions of observation which are as well defined as possible.

I would even go farther and say that quantum concepts are *"horizontal"*, in the sense that they are not intended to decribe the reality, but only reality horizons, which are, to use a geometrical concept some *apparent contours* of reality.

The Real-Potential-Actual Triad

Thanks to their horizontality, the quantum concepts take into account the dialectics of the subject and the object. About reality, the quantum theory rejects any *a priori* criterion (it has drawn the lesson of the failure of the EPR criterion), any assumption but the mere existence independant of observation. All other determinations of reality are to be uncovered *a posteriori*, according to a methodology which is completely open to experiment. Reality is thus considered as a pole of a triad: the *real* pole is thought of in terms of the dialectics of the two other poles of the triad, namely the *potential* and the *actual* poles.

Now a reality horizon is precisely a borderline between actuality and potentiality. The actual pole is essentially the apparent horizon in which necessarily occur the concrete (i.e. experimental and phenomenological) traces of reality, whereas the potential pole, or profound horizon, allows to think theoretically of the possible modes of existence of reality.

On the other hand, whereas it cannot be eliminated, a horizon can always be moved. To move a horizon it is enough to move oneself. But since a horizon is always in contact with reality, one can think of reality as *the locus of all possible horizons.* It is precisely this idea which underlies the quantization method of Feynman, known as the *"path integral method"*. According to this method, the transition probability amplitude of a given quantum process is written as a functionnal integral over a phase involving the classical lagragian in which are taken into account all the symmetry properties of the considered interaction.

It is this method which has allowed to perform spectacular progresses in elementary particle physics (which is the most advanced part of quantum physics). The so-called *standard model* in which is embedded all our knowledge about elementary particles and fundamental interactions consists on the discovery of a new elementarity level, the level of *quarks*, on the unification of electromagnetic and weak nuclear interactions, and the establishment of a theory of the strong nuclear interaction, *Quantum Chromodynamics (QCD)*. The standard model owes the path integral method its main lines of force which are a unifying principle, the *principle of local gauge invariance*, a principle of differentiation, *the phenomenon of spontaneous symmetry breaking* and a practical criterion[2], *the renormalizability criterion* which allows to choose, among candidate theories the most convenient ones.

The Geometry of Matter-Space-Time

The above mentionned lines of force correspond to the relationships between the three poles of the real-potential-actual triad. They can be expressed in a quantitative, mathematical way, by means of the basic concepts of differential geometry. One knows that quantum relativistic processes are described in terms of *quantum fields*, which are fields of probability amplitudes defined for each point of space-time. The field generally has internal degrees of freedom. The structure of a quantum field is thus the one of a *fiber bundle*, that is a product of a *base*, namely the Minkowski space-time, by, at each point of the base, a *fiber*, namely the space of the internal degrees of freedom. The true space of representation of quantum field theory is thus not space-time, but rather what I call "matter-space-time", in which the "material dimensions" are just the internal degrees of freedom of the field.

The concept of quantum field is horizontal in the sense we have introduced above. The modulus of the probability amplitude refers to the actual pole, since it is strictly related to the probability which is accessible in actual experiments. The phase in turn, refers to the potential pole since it allows to think of a class of amplitudes, differing only by their phases, and thus giving rise to the same probability.

On the other hand, it is known, even in classical electrodynamics, that the theory of the interaction can be formulated in terms of the *potential four-vector*. This formulation is known to be redundant: for a given electromagnetic field, there exists a whole class of potentials giving rise to the same field, these are potentials which differ by the divergence of an arbitrary scalar function. This "horizontal invariance" is known as *gauge invariance*.

In quantum electrodynamics, gauge invariance is fundamental because it matches the phase invariance of the matter field. In fact, it turns out that the dynamics of a matter field cannot be invariant under the change of the origin of phases, at each point of space-time, without the existence of an elctromagnetic field, obeying gauge invariance. In quantum electrodynamics, thus, the interaction results from the matching of two horizontal invariances, phase invariance of the matter field and gauge invariance of the interaction field. It appeared, in the seventies, that such a matching occurs in *all fundamental interactions*, gravitational, electromagnetic, nuclear strong and weak interactions. Which seems to me a major triumph of geometry in modern quantum physics.

---

[2] Certainly not an *a priori* criterion!

Physics and differential Geometry

Tullio Regge

Dipartimento di Fisica Teorica, Universit di Torino, Turin, Italie

In adressing this audience I am aware that the subject is extremely interesting but also rather old and dating back to the celebrated works of Einstein on General Relativity (GR). Some comments may be however useful in bringing up to date the relationship between Physics and Geometry.

We know that Einstein tried in vain in his later years to reach the final unified theory in which gravitation and electromagnetism appeared as a single entity.

Let meremember that unification does not simply mean peaceful cohexistence of different theories but rather a state of mutual necessity. Special relativity unified electromagnetism in that it made the existence of the magnetic field a necessary corollary of Lorentz invariance and of the existence of the electric field and viceversa. None of the unifications proposed in the post-Einstein era really went close to the goal but many of them contain interesting proposals which have reappeared in later attempts. For these reasons I fell compelled to discuss them.

A first class idea came from H. Weyl who proposed the first gauge theory.

Weyl tried to generalize the non integrable parallel transportation by introducing entities which would scale up and down in a non integrable manner around a close circuit. A Weyl rod could change its lenght if moved around.

Weyl scaling is not a feature of the laws of physics at the microscopic level and this is the main reason as to why the attempt had no lasting success. If however we move from the non compact group R/0 of dilations used by Weyl into the compact SO(1) or U(1) at the price of minor and obvious changes in the formalism we see that indeed we arrive at the modern form of the EM field as a gauge theory. There is no difficulty in writing this form on a Riemannian manifold and it is also possible to write, a la Cartan, general relativity as a SO(3,1) gauge theory. In any case we do not obtain a true unified theory.

By introducing torsion we may broaden the scheme and consider tetrads as particular components of a Poincar group ISO(3,1) gauge theory. As I said earlier there is however no obligation to introduce an EM field in general relativity. The proposal of Weyl was reaching in that it shed light on important similaries and structural analogies between apparently very different theories but it did not unify them.

Einstein himself did not react favorably to Weyl's proposal nor did he pay any notice to the other paper by Kaluza and Klein based on a different idea. In the K&K paper the authors proposed a scheme based on a 5 dimensional space having 1 time and 4 space dimensions. In fact their space-time was a fiber bundle whose base was ordinary 1+3 spacetime and whose fibre was S^1 which, not by accident, can be endowed with the group structure of U(1).

The authors noticed that an ordinary pure gravitational theory in 1+4 dimensions may appear as gravitation + electromagnetism if viewed by a 1+3 observer in the base space. This requires that the extra space dimension along U(1) of the fibre should be given and extremely short length of the order of Planck's length, about 10^{-33} cm, if we want to have any contact with a realistic physical theory.

Unfortunately both the fibering and this metrical constraints do not follow from any higher principle and must be assigned from the outside on phenomenological grounds. In this sense not even the K&K theory is unified.

Elie Cartan was largely responsible for the introduction of the concept of torsion in Physics and had on this point a lively correspondence with Einstein. It is difficult to immagine a higher level of exchange of ideas but in spite of this these two great protagonist of 20th Century science did not apparently understand each other and got bogged down to endless counting of components of some impressive array of tensors.

Torsion remains a very interesting idea. We need to use it, even by just declaring it to vanish, if we want to write GR as a gauge theory in which all fields and not only the spin connection appear a gauge potentials.

The interesting feature of GR is that the associate curvature of the vierbein, i.e. torsion, vanishes as a consequence of the variational princple of Hilbert-Einstein-Cartan. And in fact the Lagrangian density is not invariant under all gauge transformations of the Poincar group but only under those of the Lorentz subgroup. Although nature has prepared the gauge potentials for the full group it end up by requiring invariance under a subgroup only. This feature has prevented torsion from becoming popular.

A world with torsion would appear inescapable if we had around enough densith of high spin particles which act as sources, but this density seems at the moment well below the limit of observability.

A final comment on torsion regards the kind of space in which it is supposed to appear. It would not be any more a Riemannian manifold or, rather, none of the Riemannian structures existing on the manifold would be directely related to Physics and the theory would not be a geometrical theory in the sense envisaged by Einstein.

One way to see this point more clearly is to consider GR as GL(4,$\mathbb{R}$) theory with the Christoffel connection playing the role of a Yang-Mills potential. If the torsion vanishes it follows that the Christoffel symbol is symmetrical into the 2 lower indices whose role is however quite different. The first index is a GL(4,$\mathbb{R}$) gauge index, the second labels instead the differentials on spacetime. We may relate them because of the accidental and marvellous fact that the jacobian group of derivatives on a differentiable manifold is isomorphyc to GL(4,$\mathbb{R}$) and that we use the same indexing for differentials and vectors in GL(4,$\mathbb{R}$).

Once the symmetry is established the theory becomes almost by definition geometrical. If there is no symmetry but we can control torsion by introducing suitable norms and bounds then we may still speack of an almost geometrical theory whose exact mathematical definition is still lacking.

The K&K theory has received a good deal of attention in recent years in connection with string theory. String theorist live in spacetimes having tipically 1+10 dimensions and physicists need some technique to reduce this enormous number of extra dimensions to 1+3. The idea is to consider fiber bundles whose base is ordinary 1+3 dimensional space-time but whose fibre is compact and accomodates the extra 7 odd space dimensions.

The trouble with these schemes is that current particle phenomenology is yet unable to provide the necessary contact with reality and to decide among different models.

It is not unconceivable that future advances in the theory and experiment may find the necessary point of contact and open new vistas in the microcosmos. I feel however that early hopes created an overoptimistic sense of expectancy and that we are currently experiencing a feeling of anticlimax.

Tied up with these topics is the problem of describing the early universe. If spacetime has really more that 1+3 dimensions then it could happen that in the very early stages of the expansion there existed a time in which the ordinary universe, as simbolized by the base space 3 space dimension, was so small as to be comparable with the remaining 7 compactified ones of the fiber. In this case the effective dimension of the universe would be 10.

This means that what we call number of space dimensions is really a function of the energy scale and may have been much higher in the early stages, possibly even unbounded, depending on the final theory, if one exists at all.

Related to these topics is the interesting idea repeatedly proposed by Hawking about the quantum theory of the early universe. According to this point of view the Wick rotated universe should be a regular manifold in the 4 resulting space dimension with no particular singularity remaining in the rotated time direction. This proposal was clearly proposed for the standard GR theory and does not apply directly to exotic dimensions or to torsion. It is however conceivable that it may be broadened in scope in order to accomodate higher schemes.

In this essence it contains the not so hidden message that the true universe is a locally euclidean one, i.e. our universe is a fictitius appearance or an antropomorphyc artifact created by a counter Wick rotation from the true universe which appears here a late version of Parmenides cosmos. I hope that someone will someday put together all these ingredients and produce a really revolutionary cosmos, in any case Geometry and Physics have still a long way to go together.

Actuality of Transcendental Æsthetics for Modern Physics

Jean Petitot

Séminaire d'Epistémologie des Mathématiques
Ecole des Hautes Etudes en Sciences Sociales. 54, Bd Raspail, 75006 Paris

Contents

Introduction

In 1935 the 15th-21th September, near the Institut Henri Poincaré, in the Sorbonne, Hans Reichenbach opened the first "International Congress for the Unity of Science" – Congress organized by Carnap, Frank, Neurath, Rougier and himself under the french title "Congrés International de Philosophie Scientifique" – by giving a conference about the theme "L'empirisme logistique et la désagrégation de l'a priori". In this celebrated talk, he explained why logical empiricism must constitute henceforth the only acceptable scientific philosophy. And concerning the concept of *a priori*,concept "qui englobe lui-même toutes les difficultés d'une théorie de la science empirique *et* mathématique", he explained why and how "le développement de la science, peut être considéré comme une décomposition constante des fondements du rationalisme et de l'*a priori* ". As is well known, his essential argument was that mathematical propositions being purely analytical (and then tautological) according to Russell's and Wittgenstein's thesis, they cannot provide any cognitive content for the empirical sciences: "tout ce que nous savons du monde est tiré de l'expérience, et les transformations des données empiriques sont purement analytiques". Knowledge can grow only through induction. Consequently, any transcendental perspective had to be drastically banished.

In spite of the disagreement of eminent philosophers and mathematicians such as Enriques, Cavaillès and Lautman, this dogmatic and militant point of view was accepted almost unanimously. Since these heroic times, modern epistemology has been dominated by these empiricist dogma, even if they have been progressively "liberalized" (cf. e.g. Hempel's works).

I want to take the view opposite to Reichenbach's verdict and explain why, contrary to accepted opinions, *the transcendental point of view about objectivity provides the effective physico-mathematical sciences with an adequate and natural epistemology.*

In fact, I want to show that transcendental epistemology is nothing else than the rigourous philosophical thematization of a very special type of objective determination which is essential and pervasive in fundamental physics. Of course, such a thesis can be defended only if transcendentalism is no longer identified with its kantian foundations. In the same manner as we can do axiomatics without confining ourselves to Euclide, or be empiricists without confining ourselves to Hume, we can be transcendentalists without confining ourselves to Kant.

1. Motivations

1. The Lack of an Objective Epistemology

But *why* do we need again the *transcendental* (neither logical, nor innatist) concept of *a priori* ?

There is of course no necessity dictating a philosophical investigation about physics. Physics and Mathematics don't need epistemology. But *if* we try nevertheless to work out an epistemology, then our categorical imperative must be to work out a *plausible* one. I call an epistemology plausible if :

(i) it concerns the *contents* – the internal, specific, technical and mathematical contents – of physical theories, and not exclusively the general structures (abstraction, induction, denotation, verification, etc.) of physical knowledge ;
(ii) it provides a *philosophical* concept of knowledge which is *descriptively adequate* for the *effective* structures of *physical* knowledge ;
(iii) it leads to select as main *philosophical* problems, problems which are also essential *physical* problems.

Now, it is easy to verify that the dominant trends in present-day epistemology are not plausible in this sense. I think that the reason for this strange state of affairs is that :

(i) what we need for an epistemology to be plausible is a very particular and specific doctrine of *objectivity*, and
(ii) such a doctrine is *not* compatible with the tenets and the dogma of empiricism and positivism.

But why a plausible epistemology of physical objectivity had to be a transcendentalist one? How can we reinstate transcendentalism in its rights after more than seventy years of banishment?

The reason is that there exist two main ways for epistemology :

(i) on the one hand, the *objective* way which concerns for instance the question of the reality of mathematical entities (platonism, intuitionism, constructivism, etc.) or the question of the status of reality in Quantum Mechanics (strong objectivity VS weak objectivity, etc.);
(ii) on the other hand, the *cognitive* way which concerns the mental acts and cognitive processes which are characteristic of scientific activity (classification and categorization, abstraction, induction, denotation, relations between syntax and semantics, etc.).

According to the positivist thesis, the objective way of epistemology is reducible to the cognitive way. According to the rationalist thesis, such a reduction is impossible.

Now, the point is that, as we shall see in a moment, the very nature of physical objectivity imposes drastic constraints to the objective way. In fact, Kant was the first philosopher who has elaborated a plausible epistemology (for rational mechanics). He discovered the *constitutive* nature of objectivity – "discovery of the constitutive element" which Hans Reichenbach himself called an "eminent philosophical result" – : objective principles are *prescriptive* – not descriptive – principles which are *constitutive* of physical reality. But in Kant, this objective component was *founded* in a *cognitive* basis (representational theory, doctrine of faculties, etc.). As was stressed by Schlick : in Kant, the constitutive principles are characteristic of our representational consciousness. So, a *transcendental subjectivism* became the foundational basis for the objective way.

The developments of physics (General relativity and Quantum Mechanics) have made the objective component and the cognitive component enter in conflict. The result has been a major Krisis of epistemology.

The most serious response to this Krisis has been that of logical positivism (Carnap, etc.) which changed the foundational basis of objective epistemology : transcendental subjectivism was substituted for by a radically anti-kantian logical objectivism (Bolzano, Frege, Russell, etc.). But, in spite of its successive "liberalizations", logical positivism, and its project of "logical reconstruction" of reality, failed. As we shall see, the reason for this failure is that, in physics, the relation of mathematics to empirical reality is *not* analogous with the relation of a "language" to an empirical "world". It is not a denotative relation between a symbolic syntax and an objectal semantics. For this reason, logicism cannot serve as a foundational basis for objective epistemology.

Two consequences have followed from such a failure :

(i) the sceptic dissolution of objective epistemology (methodological anarchism, consensual, hence sociological, theory of truth, thesis of the unreachability of objective reality, etc., etc.);

(ii) the fall back of the hard core of epistemology on purely *cognitive* epistemology (AI models, expert systems, computational baconism, etc.).

Delenda ratio! In this context it is a categorical imperative to defend objectivity.

So, we want to rectify the positivist orientation and argue that there existed an alternative response to the main conflict opposing the objective and the cognitive components of transcendental epistemology. Logical positivism reduced, as we have seen, the constitutive element (and in particular the synthetic a priori) to a cognitive innatism, a logical necessity or a consciousness of evidence. It is *this* reduction that led it to sacrify the synthetic a priori constitution of objectivity. But there existed clearly *another* response to the main conflict. It was to methodologically *uncouple* the objective and the cognitive components *and to ask whether the objective component was descriptively correct for modern physics*. Such a methodological revival of transcendentalism was initiated by the neo-kantian School (the Marbourg School : Herman Cohen, Paul Natorp, and over all Ernst Cassirer). But, due to well known political diseases, this revival was not carried through.

The conclusion of this complex story has been the lack of a truly philosophically based objective epistemology, at least in the anglo-saxon area and in France. In Germany, the situation is different, due in particular to Kantian tradition and Weizsäcker's school. [1]

[1] Cf. for example Scheibe [1981]. Cf. also for instance rare works as those of H.J. Folse "Kantian Aspects of Complementarity" (Folse [1978]) or of J. Honner "The Transcendental Philosophy of Niels Bohr" (Honner [1982]).

2. The Need for an Objective Epistemology

We *need* an objective epistemology if we want to work out a plausible one.

Consider for exemple the discussion about quantum reality. In Quantum Mechanics (QM) there exists a non marginal and a non eliminable interaction between the microphysical objects and the measure apparatus. This basic physical fact implies the basic philosophical fact that *physical objectivity cannot be an ontology.* Physics cannot describe an ontological reality. It can only *objectively determine* phenomena. As was stressed by many physicists (from Bohr and Heisenberg to B. d'Espagnat and G. Cohen-Tannoudji, cf. also the celebrated Max Jammer's "The Philosophy of Quantum Mechanics" [2]), physical theories cannot refer to a reality "an sich" but only to phenomena.

As G. Cohen-Tannoudji writes: "Toute la problématique de la physique des particules est d'atteindre l'objectivité scientifique (matérialiste) malgré le caractère insécable de l'intéraction entre l'objet et l'appareil de mesure".[3] In QM, the experimental apparatus allow the manifestation (the phenomenalization process) of microphysical phenomena. *This implies the disjunction between these phenomena and any underlying reality : microphysical elementary phenomena lack any underlying ontology.* "C'est cette nouvelle conception des phénomènes qui est peut-être l'innovation la plus importante apportée par la théorie quantique. Les concepts quantiques ne se rapportent plus à l'objet en soi, mais ils se rapportent à des phénomènes"[4]. But how is it then possible to reach objectivity? What are the conditions of possibility of physical objectivity ? In a same vein, B. d'Espagnat opposes two concepts of reality : the reality "an sich", which is an independent ontological reality, and the observable reality, which is empirical and phenomenal. "Ma thèse est que c'est la science elle-même qui – s'autocorrigeant sur ce point – fournit aujourd'hui au penseur de pressantes raisons d'accepter la dualité (philosophique) de l'être et du phénomène".[5] We must go beyond the dogmatic conflict opposing, on the one hand, the "physicalist a priori" according to which physics possess an ontological content, and, on the other hand, the "mentalist a priori" according to which physics is a mere cognitive construction.

So, objective physical reality is by no means an independant ontological reality. It is nevertheless a true objectivity. Physics is undoubtedly much more than a simple nominalistic systematization of phenomenal appearances. But what can then be the philosophical status of such an objectivity wich is neither an ontology nor a mere conceptual systematization ? *It is clear that this problem is exactly the transcendental one.* To solve it we need at least three things.

1 A Non Empirical Definition of the Concept of Phenomenon

As says G. Cohen-Tannoudji, microphysical phenomena are "horizons of reality". According to the non commutativity of the algebras of observables (Heisenberg's relations) there exist absolute limits imposed upon the epistemic accessibility of physical reality. These limits are absolute but not rigid. They constitute an horizon of discernability, of predictability and of reality.

[2] Jammer [1974].

[3] Cohen-Tannoudji and Spiro [1986], pp. 80-81.

[4] *Ibid.*, p. 141.

[5] D'Espagnat [1985], p. VII.

2 The Integration of the Horizon Status of Phenomena to the Content of Theoretical Objectivizing Concepts (System, State, Causality, Interaction, etc.)

This constraint is, in fact, a tremendous one. Actually, the content of theoretical concepts must be *mathematically* interpreted : in physics, there exists a sort of *mathematical "hermeneutics"* which constitutes the basis of the theories. Now, the constraint is that *this mathematical interpretation must derive from an adequate mathematical formulation of the horizon status of phenomena.*

3 The *Relativization* of the Concepts of Truth and Existence, and also of the *Modal* Categories of Possibility, Reality and Necessity, to such a Process of Objectivation

The *ontological* (metaphysical, dogmatic) use of these modalities leads unavoidably to *dialectical antinomies.* Cf. for instance the typically dialectical debate between Bohr and Einstein, and, more generally, the debate about hidden variables (from von Neumann's theorem to Aspect's results) : as it was stressed by Jerry Marsden : "Indeed, the entire point of the negative results concerning "hidden variables" is that there is no "objective underlying state" [in the sense of an ontological reality behind phenomena] of the system!".[6]

Then, we can say that QM has conquered a new concept of objectivity which is in fact of a constitutive and critical nature. *"Weak" objectivity* (as opposed to the "strong" objectivity of an independent reality) *is no more than objectivity in the transcendental sense !* As was claimed by Weizsäcker: "Bohr was essentially right, but he himself did not know the reasons why". [7] These reasons are transcendental. Someone has spoken with humour of the "Königsberg's interpretation" of QM ...

We shall also see (sections IV to VI) that the symmetry groups of relativity allow to make the *mathematical* descriptions *independent* of the *conventional epistemic* elements (coordinate frames, etc.) which overdetermine them. But these groups possess a rich geometry and this geometry becomes a *determining* component of the *physical* models. The fact that epistemic constraints on mathematical descriptions can become constitutive and determining for the explanation of physical contents implies that physical objectivity *cannot* be thought of as a relation between a cognitive epistemology and a substantialist ontology of things, events and states of affair. We need therefore a specific and adequate *objective* epistemology. The relevance of the transcendental approach is *to provide the philosophical tools which allow us to apply the traditional philosophical concepts of reality, truth, necessity, essence/existence, a priori/ a posteriori, substance, causality, etc. to these very special type of objective determination.* We must convince ourselves that theses concepts *do not possess a scientific meaning antecedent to strategies of objective determination.* It is only in their dogmatic metaphysical applications that they can function in an absolute manner. But in science, *they are always relative to a previously constructed concept of objectivity.* Transcendental philosophy can be defined as the philosophical reflexion of this specific type of relativity.

---

[6] Marsden [1974], p. 187.

[7] Weizsäcker [1979], p. 229.

2. The Actualization of Transcendental Philosophy

Our purpose here is not to comment Kant himself. But let us nevertheless emphasize the main points of an actualization of transcendental philosophy.

1. The Uncoupling of the Objective Component of Transcendental Epistemology from its Foundational Cognitive Basis

We think that transcendental epistemology is essentially descriptively correct for physics, at least in what concerns its general *moments* (Transcendental Aesthetics, Categories, Principles, etc.). This descriptive adequacy does not need to be rooted in a transcendental subjectivism, for exemple in a cognitive innatism. As we shall see the *synthetic a priori* has nothing to do with such a psychological basis.

2. The Analogy between the Axiomatic Method and the Transcendental Method

We think that the transcendental method is analogous, for the explanation in objective sciences, to the axiomatic method for the definition in semantic analysis. Let us make this parallel more explicit.

1 The Axiomatic Method

In the semantic analysis of concepts, what we seek is a *definition* of the concepts. There exist many levels of concepts, hierarchically organized, and definitions allow the *reduction* of higher levels to lower ones. This definitional reduction leads necessarily to *primitive* concepts (of lowest rank) which are *undefinable*. The main problem is then to go *beyond* this limit of definability. We meet here an alternative.

(i) We can admit the existence of *innate* primitive concepts which are originally given. This is a classical metaphysical trend.

(ii) But we can also change our strategy in a radical manner and *substitute* to the undefinable content of primitive concepts *syntactic rules* which determine normatively their use. It is such "implicit definitions" which are provided by the axiomatic method.

2 The Transcendental Method

In objective sciences, what we seek is an "ontological" explanation of the empirical phenomena. There exist many levels of reality, hierarchically organized, and "ontological" explanations allow the *reduction* of higher level to lower ones. Lower levels operate as a *foundation* – a ground of reality and of explicative causality – for the higher ones. In that sense, causal reductionism is perfectly justified. The consequence of this is that at the *higher* levels, the underlying being ("l'être", "das Sein", the "esse") and the observed phenomena *are unified in a substantial "ontology" of things, events and states of affairs.* But this explicative reduction leads necessarily *to a fundamental level* which is no more explainable using such a strategy. At the fundamental level, *there operates a disjunction between the underlying being and the observed phenomena.* The underlying being becomes a reality "an sich", a pure ground of reality ***which lacks any causal explicative power***. The main problem is then to go beyond this limit of explanation. We meet here an alternative.

(i) We can admit fundamental elementary phenomena as irreducible data which are originally given and support *as phenomena* all the causal explicative power of physical theories. Such a phenomenist empiricism leads to a conception of the theories wich is essentially "semiotic", in fact *nominalistic*.

(ii) But we can also change our strategy in a radical manner. We can elaborate an epistemology which *starts from* the disjunction "underlying being vs phenomenon" and concerns *the conditions of possibility of objectively determining phenomena with no underlying ontology*. This is the transcendental strategy.

We can therefore say that transcendental objective epistemology *is required for fundamental physics and elementary phenomena*.

3. The Evaluation and the Rectification of the Standard Positivist Critiques against Kantian Doctrine

We give very briefly some examples.

(a) The confusion between the Kantian opposition between phenomenon and noumenon and the opposition between psychological internal representations and ontologically independent (external) entities. We agree with Henry Allison's and Gerold Prauss' "two aspect vew" for asserting that transcendentalism is methodological and concerns the epistemological conditions of objectivation, conditions which are neither internal (psychological) nor external (ontological).

(b) The thesis that Euclidean geometry is *cognitively* a priori. Kant admitted the logical possibility of non Euclidean geometries. For him, the primate of Euclidean geometry derived from the inertia principle. [8] The true point is not this one but that in physics the mathematical "hermeneutics" of theoretical concepts must be geometrical.

(c) The confusion between *transcendental* necessity and *logical* necessity. In Kant, necessity – the crowning category of Transcendental Analytic – is *relative to the radical contingency of possible experience*. (Cf. the third Postulate of empirical thinking: "Dessen Zusammenhang mit dem Wirklichen nach allgemeinen Bedingungen der Erfahrung bestimmt ist, ist (existirt) nothwendig.").

(d) The thesis that the synthetic a priori characterizes some class of propositions. It is false. Synthetic judgments are not some special sort of propositions. The synthetic a priori is a *function* in an objective process of constitution (cf. Cassirer).

(e) The claim that axiomatics renders geometry purely analytic and that, consequently, the synthetic a priori does not exist. As was stressed by Michael Friedman in his "Kant's Theory of Geometry" [9], the fundamental properties of the continuum are not expressible in a monadic first order logic such as that used by Kant. "So, for Kant, one cannot represent or capture the idea of infinity formally or conceptually"

---

[8] Cf. for instance J.E. Wiredu "Kant's Synthetic a priori in Geometry and the Rise of non Euclidean Geometries" (Wiredu [1970]), and R.J. Gomez "Beltrami's Kantian View of Non-Euclidean Geometry" (Gomez [1986]).

[9] Friedman [1985]. Cf. also Brittan [1991].

and then there exists a clear distinction between "intuitive" and "discursive". Of course, with polyadic logic and Skolem's functions we can express purely logically "intuitive" properties like the property of density:

$$\forall x \, \forall y \, \exists z \quad (x < y \Rightarrow x < z < y)$$

or the property of completion of $\mathbb{R}$ (convergence of Cauchy sequences):

$$\forall \varepsilon \, \exists N \, \forall m \, \forall n \, (m, \, n > N \Rightarrow |s_m - s_n| < \varepsilon) \Rightarrow$$
$$\exists s \, \forall \varepsilon \, \exists N \, \forall n \, (n > N \Rightarrow |s_n - s| < \varepsilon).$$

But it would be dogmatic to conclude that the heterogeneity between intuition and discursivity has disappeared. Actually:

(i) such quantifications on infinite sets raise all the difficult problems emphasized by Brouwer, Weyl or Wittgenstein;
(ii) non standard Analysis (from Veronese to Nelson, Reeb and Harthong) has shown that it is impossible to completely determine the continuum in a logical analytic manner.

This basic fact provides the modern sense of kantian intuition. As was emphasized by the second Gödel, the theorems of limitation (Löwenheim-Skolem, incompleteness, indecidability, etc.) show that the continuum is an objective informationally infinite reality which transcends its symbolic (logical analytic) determination. For the continuum to possess a "good" structure in a model of ZFC set theory, we must introduce new axioms, and in particular existence axioms for very large cardinal (measurable cardinals, etc.). As Gödel said, these axioms are like "physical hypothesis". They concern an "external" and "intuitively given" reality, and not a conceptual one.[10]

4. The Actualization of the Main Constitutive Moments of the Transcendental Doctrine

1. Transcendental Aesthetics

If phenomena are not ontologically based, how can we avoid psychologizing them as mere appearances ? The response is : *Transcendental Aesthetics.* The essential relevance of transcendental arguments is not of a "syntactic" nature. It does not concern the trivial fact that a linguistic framework is an a priori necessity for talking about knowledge and experience. It concerns above all Transcendental Aesthetics.

In Kant's "Kritik der reinen Vernunft" (KRV) Transcendental Aesthetics contains two "expositions" (Erörterung; exposition = clear representation of what belongs to a concept) : the metaphysical one and the transcendental one.

The metaphysical exposition explains that phenomena appear in a *specific medium of manifestation* (space and time for sensible phenomena). This phenomenological medium gives *forms of intuition.* "Intuition" (Anschauung) has nothing to do here with a *cognitive* evidence (as in Descartes). It is not an intuitive *knowledge.* It expresses the fact that phenomena (Erscheinungen) are *given* in some receptive way as the trace of some unknown – and unknowable – transcendent reality, and that there do exist *forms* of

[10]Cf. Petitot [1989], [1991a], [1991b]. For philosophical comments about non standard Analysis, cf. Salanskis [1989], [1991].

this receptivity. We meet here the fundamental Kantian distinction between "gegeben" and "gedacht". Space is an actual infinite which is originally given. Its function is to "depsychologize" the psychological content of the concept of phenomenon.

In his article "The Metaphysical Exposition : An Analysis of the Concept of Space"[11] P.M. Mc Goldrick has analysed the status of space as a *singular* concept possessing a reference which is *unique in all its genus*. As the universal substratum of sensible intuitions, the concept of space is neither a category subsumming the empirical diversity under the unity of aperception nor an empirical (classificatory) concept. It cannot be abstracted from sensible experience. Using the Kripkean concept of "rigid designator", we can say that space is a rigid designator, "a term which designates the same object in all possible worlds". So, if space was equivalent to a leibnizian order of co-existence between external objects, this equivalence would have to be necessary and, consequently, the concept of space would imply the concept of object. But this is false. Therefore, space is effectively an intuition. This argument agree with Gordon Brittan's definition of the synthetic a priori as "true in all real possible worlds" (as opposed to "true in the actual real world" and "true in all logical possible worlds"). [12] In that sense the synthetic a priori concerns effectively the conditions for assertions about the actual world to be true or false.

Transcendental exposition concerns the *mathematical* determination of the forms of manifestation that is to say the conversion of pure intuitions into what Kant called *formal intuitions* (cf. the celebrated note of the §26 of the KRV). The main fact is that such a mathematical determination discloses an absolute limit of the *conceptual* description in physics, description linked to the existence of *symmetries* (relativities) (Kant was the first philosopher who has philosophically understood and formulated the basic role of symmetries and relativities in physical theories).

If one imposes cognitive foundations to objective epistemology, one is necessarily led to the conclusion that transcendental exposition must be cognitively determined and, therefore, that the mathematical determination of pure intuition must be *univoque*. And this is false. But if one frees objective epistemology from its cognitive basis, then one can easily accept that the mathematical determinability is actually *under-determined* and *depends upon the mathematical possibilities of determination*. In that sense *conventionalism* (in Poincaré's sense) possesses a deep transcendental meaning. It concerns the *transcendental ideality* of the forms of intuition.

2. Transcendental Analytic

The possibility of philosophically working out a concept of objectivity wich is neither an ontology nor a mere systematization of empirical data, relies upon the "copernicean revolution" – the critical "turning point" – that is to say upon the conversion of a descriptive conception of objectivity *into a prescriptive one* (it is exactly the same critical "turning point" that was accomplished by the second Wittgenstein for the symbolic objectivity of mathematics).

Even if phenomena do not possess an underlying ontology, they can nevertheless be *legalized*. Objectivity is legality of experience. It is a juridical concept. But, for legalizing phenomena we need normative, prescriptive, "grammatical" (in the Wittgensteinian

[11] McGoldrick [1985].

[12] Brittan [1978].

sense) concepts. We need *forms of legality*. These are provided by the *categorial system* that is to say by a *transcendental Analytic*. The main categories define *positions* (Setzung) *of existence*.

3 Transcendental Schematism and the Construction of Categories

But a position of existence is not still an objective determination. To objectively determine phenomena, we need a link *between forms of manifestation* (what is "gegeben") *and forms of legality* (what is "gedacht"). At the level of the KRV such a link is provided by the transcendental schematism which converts the categories into *principles* (Grundsätze). But in fact, the true link is given by what Kant called the *construction* ("Konstruktion") of the categories. The construction is a mode of presentation ("Darstellung"). It means that one can *interpret mathematically the schematized categorial contents (i.e. the categorial semantics) using mathematics coming from the transcendental exposition of Transcendental Aesthetics.* We think that it is *in this very special sort of "mathematical hermeneutics"* – not only for the forms of manifestation but also for the forms of legality themselves – *that the synthetic a priori finds its true and deep transcendental meaning.* We agree with Henry Allison's and Gerold Prauss' interpretation [13] according to which :

(i) transcendental schematism must be not only temporal but also spatial (in fact spatio-temporal) and that it is much richer than Kant believed,
(ii) synthetic a priori judgments refer to transcendental schemata.

When they are so constructed, the categories and principles of the Analytic become *generators* of mathematical *models* for the considered phenomena. So the "construction" unifies :

(i) forms of manifestation, that is to say intuitive modes of phenomenological giving;
(ii) forms of legality, that is to say categorial modes of existential position ;
(iii) basis for mathematical modelling.

This point is scientifically *and* philosophically fundamental. To understand the implication of mathematics in physical reality, we must avoid two dead ends.

(i) A purely denotative point of vew according to which mathematics provide a syntax and empirical phenomena provide a semantics. The relation between syntax and semantics in logical model theory *cannot be used* as a metaphor for the relation between mathematics and the world.
(ii) An obscure platonist "participation" of mathematics to reality.

We can avoid these two dead ends if we note :

(i) that mathematics determine not the reality itself but *forms* of reality and that these forms are *ideal* components of reality ;
(ii) that mathematics allow also a constructive interpretation of the categories of existence;
(iii) that through such a mathematical hermeneutics, we can construct diversified and falsifiable models of empirical phenomena.

Our hypothesis is that it is exactly this type of strategy of determination which has become more and more systematic, sophisticated and pervasive in modern physics, even

[13] Cf. for instance Allison [1981], [1983] and Prauss [1980], [1981a], [1981b].

if it is not philosophically thematized by physicists themselves. It is clear that it is not compatible with a positivist (logical, predicative and denotative) conception.

3. The "Metaphysische Anfangsgründe der Naturwissenschaft"

In the *Anfangsgründe*, Kant applies his transcendental point of vue to newtonian mechanics (that is to say to the regional object "movement" (trajectory)). [14] In modern language his interpretation can be summarized in the following manner. We must remind here the fundamental difference between the "mathematical" categories concerning *essence* and the "dynamical" categories concerning *existence*.[15]

1. Phoronomy (Kinematics)

Phoronomy applies the "mathematical" categories of quantity and the principles called "Axioms of Intuition" (Axiomen der Anschauung) governing the "extensive" magnitudes. It concerns :

(i) The problem of the *metric* of space (the link between the informal phenomenological space of the metaphysical exposition and the measure of distances by numbers). It clarifies the link between the transcendental exposition and the conversion of space as form of intuition into space as formal intuition (determined as object). It explains also the inertia principle (geodesics are straight lines).

(ii) The galilean relativity and the symmetry group which expresses it (of course the concept of group was not known to Kant).

2. Dynamics

Dynamics applies the "mathematical" categories of quality and the principles called "Anticipations of Perception" (Anticipationen der Wahrnehmung) governing the "intensive" magnitudes. It concerns :

(i) The debate with Leibniz and the metaphysical tradition concerning the "substantial" *interiority* of matter. For Kant, such an interiority is "noumenal" (an sich) and therefore must be excluded from mechanics (but this very difficult problem is reopened in the *Opus Postumum* where Kant speculates about the system of the fundamental primitive internal forces governing the derivative mechanical forces and, therefore, the external movements of matter).

(ii) In space-time, matter expresses itself through intensive magnitudes as velocity and acceleration. These intensive magnitudes are *differential* data. And to be compatible with Phoronomy, they must possess an *intrinsic* meaning, that is to say they must be *covariant*. Kant was also the first philosopher who understood clearly – even if it was at an elementary level – the constitutive role of invariance and covariance principles in physical objectivity.

---

[14] Cf. for instance Howard Duncan "Inertia, the Communication of Motion, and Kant's Third Law of Mechanics" (Duncan [1984]).

[15] Cf. for instance Jules Vuillemin's classical "Physique et Métaphysique kantiennes" (Vuillemin [1955]).

3. Mechanics

Mechanics applies the "dynamical" (physical) categories of relation (substance = Inhärenz und Subsistenz, causality = Causalität und Dependenz, community, reciprocity and interaction = Gemeinschaft) and the principles called "Analogies of Experience" (Analogien der Erfahrung). We meet here the true sense of the critical turning point : the separation of physical objectivity from any (aristotelian) substantial ontology. The *construction* of the category of substance (schematized as transcendental principle of temporal permanence : "die Beharrlichkeit des Realen in der Zeit") leads to its complete reinterpretation : *substance is the general principle underlying the conservation laws of physical magnitudes.* We see very well here why physics cannot be a predicative description of some empirical states of affairs. We see also how a priori principles can become generators of a great diversity of mathematical models : covariance principles and conservation laws express themselves through fundamental *equations* whose *solutions* are models for phenomena.

In Mechanics, causality – i.e. the *force* – is expressed by Newton's law and reciprocity by the law of equality of action and reaction.

4. Phenomenology

Phenomenology applies the categories of modality and the principles called "Postulates of empirical thinking" (Postulate des empirischen Denkens überhaupt). According to the principle of relativity, movement *cannot* be a *real* (wirklich) predicate of matter. It is only a *possible* (möglich) predicate. Absolute spatial and temporal positions or absolute velocities are not observable. It is for this reason that the "empirical realism" of space and time cannot be ontologically strengthened up to a "transcendental realism". The celebrated *"transcendental ideality"* of space and time is no more than the correct philosophical formulation of the fact that *principles of relativity are, and must be, constitutive of physical objectivity.*

But forces (causality) are real and are governed by necessary laws. Necessity is not here a logical modality but a transcendental one. It is a conditional necessity, relative to the radical contingency of possible experience.

The *Anfangsgründe* show that the transcendental structure of rational mechanics – as science determining the regional object "trajectory" – *coincides* with its fundamental physical characteristics and problems. It is in that sense that kantian objective epistemology is plausible.

Our research program is to work out the equivalent *Anfangsgründe* for *other* physical theories. Our leading idea is that there exists some historical permanence of the transcendental *structure* of physical objectivity, even if the mathematical and physical *contents* of the transcendental moments develop themselves historically. As we have seen, transcendental moments are in fact *"hermeneutical"* ones and this mathematical hermeneutics is of course an historical one. We shall now give some examples.

4. Nœther's Theorem and Symplectic Mechanics

In classical mechanics, an extraordinary confirmation and deepening of the Kantian linkage between Transcendental Aesthetics and the reinterpretation of substance by conservation laws is provided by the linkage ***between relativity groups, symmetries and conservation laws.*** In that sense Nœther's theorem possesses an eminent transcendental meaning. In fact, we think that the *direct* legacy of Transcendental Aesthetics is to be found :

(i) in the constitutive role of global geometrical symmetry groups (cf. the works of Lie, Poincaré, Cartan, Weyl)[16] ;

(ii) in Nœther's theorem ;

(iii) more generally in the possibility of reducing conservation laws to purely geometrical a priori. ***This possibility of reduction provides one of the true modern physical meanings of the synthetic a priori.***

1. Nœther's Theorem

Lagrangian Version

Let S be a mechanical system, M its configuration space (M is a differentiable manifold) and $L : TM \rightarrow \mathbb{R}$ its Lagrangian. The action integral is

$$\phi(\gamma) = \int_{t_0}^{t_1} L(q, \dot{q})dt$$

γ being a path from (q_0, t_0) to (q_1, t_1).

The least action principle $\delta\phi = 0$ implies the well known Euler-Lagrange equations:

$$\frac{d}{dt}\left(\frac{\partial L}{\partial \dot{q}}\right) - \frac{\partial L}{\partial q} = 0$$

which are Newton's equations $\dot{p} = \partial L/\partial q$ for the generalized moments $p = \partial L/\partial \dot{q}$ and the generalized forces $\partial L/\partial q$.

Let φ_s be a one parameter subgroup of diffeomorphisms of M (i.e. a flow) and suppose that the φ_s are symmetries of L, that is to say that L is φ_s-invariant for all the φ_s: $L(v) = L(D_q\varphi_s(v))$ for every tangent vector $v \in T_qM$ (T_qM is the tangent vector space of M at q and $D_q\varphi_s$ is the tangent linear map of φ_s at q). Then Nœther's theorem asserts that S admits *a law of conservation* (a first integral):

$$I : TM \rightarrow \mathbb{R},\ I(q, \dot{q}) = \frac{\partial L}{\partial \dot{q}} \cdot \left.\frac{d\varphi_s(q)}{ds}\right|_{s=0} = p \,.\, X_\varphi(q)$$

where X_φ is the vector field generating the flow φ_s.

---

[16]For a transcendental approach of these fundamental geometric traditions, cf. for instance Boi [1989].

Hamiltonian Version

Let $H : T^*M \rightarrow \mathbb{R}$ be the hamiltonian of S, where T^*M is the phase space of S, that is to say the cotangent bundle of the configuration space M. T^*M is a symplectic manifold endowed with the canonical (closed and non degenerate) 2-form $\omega = \omega_0$ (in canonical local coordinates q for M and p for T^*_qM, ω_0 has the Darboux form $dq \wedge dp$). Let $F : T^*M \rightarrow \mathbb{R}$ be an observable such that H is invariant under the flow $\exp(tX_F)$ generated by the hamiltonian field X_F derived from F (X_F is defined by $i_{X_F}\omega = dF$, where $i_X\omega$ is the 1-form derived from the 2-form ω by contraction with the field X). Then F is a first integral of the hamiltonian flow X_H, that is to say the Poisson bracket of F and H vanishes : $\{F, H\} = 0$ (where $\{F, H\} = \omega(X_F, X_H) = i_{X_F} i_{X_H}\omega$).

The three classical examples of Nœther's theorem are provided by the geometrical relativity group of classical mechanics. They are summarized in the following table.

Non observability of some absolute entities	*Symmetry groups*	*Conservation laws*
Origin of time	Temporal translations	Energy
Origin of space	Spatial translations	Impulse (momentum)
Privileged direction	Rotations	Angular momentum

The transcendental meaning of Nœther's theorem is to *demonstrate* the *correlation* between two things :

(i) some mathematical (non physical, non observable) entities which are necessary for the mathematical description of physical phenomena : these entities are conventional epistemic elements submitted to relativity principles (transcendental ideality),

(ii) physical magnitudes which are observable and measurable and constitute the "substantial" content of the theory.

This correlation between the impossibility of observing certain absolute entities and the possibility of observing conserved correlated magnitudes is mediated here by the Lagrangian or the Hamiltonian. But it is clear that its philosophical meaning is typically transcendental. As Gilles Cohen-Tannoudji says, "la relativité signifie l'impossibilité d'effectuer des mesures *absolues*" and Nœther's theorem establishes "la correspondance entre *relativité* (non observabilité de certaines entités absolues), *symétries* (invariance par transformation de symétrie) et *lois de conservation* (conservation et donc observabilité de certaines quantités)". [17] Actually this confers its true physical meaning to Transcendental Aesthetics.

A phenomenon is something which is conditionned by forms of manifestation. Therefore it is a *relational* entity relative to its means of observation. To describe it we must mathematically determine the forms of manifestation. But this introduces some conventional (epistemic) elements. Objectivity essentially means that we can eliminate these over-determining elements. But such an elimination is expressed by new mathematical structures (in particular by symmetry groups). These structures are therefore constitutive of objectivity.

---

[17] Cohen-Tannoudji and Spiro [1986], pp. 106-107.

2. The Moment Map

In fact the transcendental nature of Nœther's theorem can be mathematically deepened using for instance the works of B. Kostant, J.M. Souriau, V. Arnold, A. Weinstein, R. Abraham and J. Marsden. [18] With the ***moment map*** formalism, it is possible to derive ***directly*** first integrals from relativity groups ***without specifying any Lagrangian.***

The phase space T^*M of S is a symplectic manifold (P, ω) (endowed with the canonical, closed and non degenerate, 2-form $\omega = dp \wedge dq$). Let G be a Lie group operating symplectically on P. So, the action $\Phi : G \times P \to P$ is such that $\Phi_g(x)$ belongs to the group $\mathrm{Sympl}(P, \omega)$ for each g.

There are essentially two manners for generating vector fields on P : from symmetries and from observables.

(i) Let $\mathcal{G} = T_eG$ be the Lie algebra of G and let $\xi \in \mathcal{G}$. Through the action Φ, the one parameter subgroup $\exp(t\xi)$ acts on P as the flow $\varphi_\xi(t) = \Phi_{\exp(t\xi)}$. Let $\xi_P = \frac{d}{dt}\varphi_\xi(t)|_{t=0}$ be the infinitesimal generator of the flow $\varphi_\xi(t)$. $\xi_P \in \mathcal{X}(P)$, $\mathcal{X}(P)$ being the algebra of vector fields on P, and $\xi \to \xi_P$ is a morphism of Lie algebras $\mathcal{G} \to \mathcal{X}(P)$.

(ii) Let $F : P \to \mathbb{R}$ be an observable. F generates the hamiltonian field X_F such that $i_{X_F}\omega = dF$.

The idea is to associate with every $\xi \in \mathcal{G}$ – in an automatic and natural manner – an observable F_ξ such that :

(i) $X_{F_\xi} = \xi_P$,

(ii) the F_ξ are first integrals for every hamiltonian H which is G-invariant (i.e. for which $H(x) = H(\Phi_g(x))$, $\forall g \in G$).

Let us suppose that such an association exists. If we fix some $x \in P$, we can consider the system $J(x)$ of the values $F_\xi(x)$ (x constant, ξ variable) of the first integrals F_ξ. $J(x)$ is a *linear form* on $\mathcal{G}$, i.e. an element of the dual $\mathcal{G}^*$. So we get a map – ***the moment map*** :

$$\begin{aligned} J : P &\to \mathcal{G}^* \\ x &\mapsto J(x) : \mathcal{G} \to \mathbb{R} \\ &\qquad\qquad \xi \mapsto J(x)(\xi) = F_\xi(x) \end{aligned}$$

Nœther's theorem. If $H : P \to \mathbb{R}$ is G-invariant, J is a set of first integrals for X_H.
Remark.
If $(P, \omega) = (T^*M, \omega_0)$ and if G acts *on* M, then we have

$$\begin{aligned} J &: T^*M \to \mathcal{G}^* \\ &\alpha \in T_q^*M \to J(\alpha) : \xi \to J(\alpha)(\xi) = F_\xi(\alpha) = \alpha(\xi_M(q)). \end{aligned}$$

We have therefore $F_\xi = P(\xi_M)$ where the map $P(X)$ is defined, for $X \in \mathcal{X}(M)$ by

$$\begin{aligned} P(X) &: T_q^*M \to \mathbb{R} \\ &\alpha \in T_q^*M \mapsto \alpha(X(q)). \end{aligned}$$

[18] Cf. for instance Abraham-Marsden [1978], Arnold [1976], Souriau [1975], Weinstein[1977].

3. The Symplectic Structure of the Orbits of the Co-adjoint Representation

We have just seen that first integrals can be directly associated to a relativity group. But the synthetic a priori reduction of the conservations laws can be taken even further. In fact, as was shown by Kirillov, Souriau, Kostant, Weinstein and Arnold, it is possible to apply this construction *to the Lie groups themselves* and *to derive hamiltonian systems from the symmetries alone.*

Let G be a Lie group. G acts on itself by left translations $L_g : h \rightarrow gh$. Let us consider the adjoint and co-adjoint representations of G.

The *adjoint representation* is given by :

$$\begin{aligned} Ad \; &: G \rightarrow \mathrm{Aut}\mathcal{G} \\ & g \mapsto Ad_g = T_e(R_{g^{-1}} \circ L_g) \end{aligned}$$

where $R_{g^{-1}} \circ L_g$ is the interior automorphism of G associated with g. It is the infinitesimal expression of the non commutativity of G. If we take the linear tangent map $T_e Ad$ of Ad at e, we get the map :

$$\begin{aligned} ad \; &: \mathcal{G} \rightarrow \mathrm{End}\mathcal{G} \\ & \xi \mapsto ad_\xi \; : \; \eta \mapsto ad_\xi(\eta) = [\xi, \eta] \end{aligned}$$

which return the Lie bracket of G.

The *co-adjoint representation* is given in a dual manner by:

$$\begin{aligned} Ad^* \; &: G \rightarrow \mathrm{Aut}\mathcal{G}^* \\ & g \mapsto Ad_g^* \; : \; \alpha \mapsto Ad_g^*(\alpha) \; : \; \eta \mapsto Ad_g^*(\alpha)(\eta) = \alpha(Ad_g \eta). \end{aligned}$$

If we take the linear tangent map $T_e Ad^*$, we get the map:

$$\begin{aligned} ad^* \; &: \mathcal{G} \rightarrow \mathrm{End}\mathcal{G}^* \\ & \xi \mapsto ad_\xi^* \; : \; \alpha \mapsto ad_\xi^*(\alpha) \; : \; \eta \mapsto ad_\xi^*(\alpha)(\eta) = \alpha(ad_\xi(\eta)) = \alpha([\xi, \eta]). \end{aligned}$$

To go further, we need the concept of a *Poisson structure* (A. Weinstein). If (P, ω) is a symplectic manifold, the Poisson bracket:

$$\{f, g\} = \omega(X_f, X_g) = \sum_i \left(\frac{\partial f}{\partial q_i} \frac{\partial g}{\partial p_i} - \frac{\partial f}{\partial p_i} \frac{\partial g}{\partial q_i} \right)$$

defines a Poisson structure (PS) on P. This means that $\{f, g\}$ is a Lie bracket on the algebra $\mathcal{F}(P)$ of the differentiable functions on P, which is related to the algebra product in $\mathcal{F}(P)$ by a derivation law $\{fg, h\} = f\{g, h\} + \{f, h\}g$. These particular PS are called symplectic.

It is possible to reformulate all hamiltonian mechanics only in terms of PS. If $H \in \mathcal{F}(P)$ is an hamiltonian on P, the hamiltonian field X_H (considered as a derivation on $\mathcal{F}(P)$) is defined by $X_H(f) = \{f, H\}$. Likewise, if $\theta : \mathcal{G} \rightarrow \mathcal{F}(P)$ is a morphism of Lie algebras, the moment map is given by

$$\begin{aligned} J \; &: P \rightarrow \mathcal{G}^* \\ & x \mapsto J(x) \; : \; \xi \mapsto J(x)(\xi) = \theta(\xi)(x) \end{aligned}$$

Let us then consider the commutative diagram of fibrations:

$$\begin{array}{ccc} T^*P & \xrightarrow{\Pi} & TP \\ dH \uparrow & & \uparrow X_H \\ P & \xrightarrow{Id_P} & P \end{array}$$

where Π is the morphism defined by $X_H = \Pi \circ dH$. The symplectic PS are ***non degenerate*** in the sense that Π is an ***isomorphism*** (of fibrations). This is not the case in general. Nevertheless, in general, the image of Π is an ***integrable foliation*** S on P. It can be shown that the leaves $S(x)$ of S are symplectic manifolds and that the canonical inclusions $S(x) \to P$ respect the PS. ***Weinstein theorem*** asserts that any PS is ***locally*** the direct product of a symplectic PS and a "transversal" PS which is tangent to the PS of a Lie algebra $\mathcal{G}^*$.

The PS on the Lie algebras $\mathcal{G}$ are those which transform all the moment maps $J : P \to \mathcal{G}^*$ in Poisson maps. They are ***degenerate***. Their symplectic leaves are the orbits of the co-adjoint representation of a Lie group G with Lie algebra $\mathcal{G}$. These orbits are endowed with the Kirillov symplectic structure defined in the following manner. Let $\alpha \in \mathcal{G}^*$, and $G_\alpha = \left\{Ad_g^* \alpha\right\}_{g \in G}$ its orbit. Let $\theta \in T_\alpha G_\alpha$. We can suppose that $\theta \in \mathcal{G}^*$ for, $\mathcal{G}^*$ being a vector space, we have $T_\alpha \mathcal{G}^* \cong \mathcal{G}^*$. When θ is considered as a velocity vector at α, it is identified with the $\eta \in \mathcal{G}$ such that $ad_\eta^* \alpha = \theta$. The Kirillov structure is then given by the Lie bracket $\sigma(\theta_1, \theta_2) = \alpha([\eta_1, \eta_2])$. Using the moment map $J : T^*G \to \mathcal{G}^*$ of the action of G on itself by left translations, it can be shown that this symplectic structure comes, by means of an operation of "reduction", from the canonical symplectic structure ω_0 on T^*G.

So it is possible to generate hamiltonian systems directly from symmetries. This fact allows the reduction to a priori principles of entire parts of mechanics.

4. Example : the Reduction to A Priori Principles of the Eulerian Movement of a Non Ponderous Solid around a Fixed Point (Non Commutative Kinematics)

A beautiful example of the previous construction was given by Arnold. [19] Let S be a non ponderous solid moving around a fixed point. The configuration space of S is the rotation group $G = SO(3)$. We have $\mathcal{G} = so(3) \cong \mathbb{R}^3$ with $[x, y] = x \times y$ (wedge product). Through this identification, the adjoint representation becomes simply the rotations in $\mathbb{R}^3$. Let R be the rotation of S ($x = RX$, with x (resp. X)= position of S in a fixed (resp. moving) coordinate frame e (resp. E)), ω and Ω the rotation velocities relative to e and E ($\omega, \Omega \in \mathcal{G} = \mathbb{R}^3$ and $\omega = R\Omega$), and (m, M) the angular moment relative to e and E ($m, M \in \mathcal{G}^* \cong \mathcal{G} \cong \mathbb{R}^3$ and $m = RM$). If we express the conservation law $\dot{m} = 0$ in the frame E, we get the *Euler equation* $\dot{M} = [M, \Omega]$ which is the Euler-Lagrange equation for the momentum $p = \partial L/\partial \dot{q} = M$ and the force $\partial L/\partial q = [M, \Omega]$. But, v being the velocity of a point x and μ its mass, we have $v = [\omega, x]$, $m = [x, \mu v] = \mu[x, [\omega, x]]$ and then $M = \mu[X, [\Omega, X]] = -\mu(adX)^2(\Omega) = A\Omega$. The linear operator A is associated with the kinetic energy $T = 1/2(A\Omega \mid \Omega) = 1/2(M \mid \Omega)$ (scalar product). It is the inertia tensor at X. By summation over X, we get the inertia tensor I. By diagonalisation, we

[19] Arnold [1976]. Cf. also Abraham-Marsden [1978].

get the inertia axis A_1, A_2, A_3 and the eigenvalues I_1, I_2, I_3. The Euler equations are then $I_i \dot{\Omega}_i = (I_j - I_k)\, \Omega_j\, \Omega_k$, with (i, j, k) a cyclic permutation of (1, 2, 3). It is easy to deduce from them the Poinsot description of the movement of S : the ellipsoid of inertia $\mathcal{E} = \{\Omega \mid (I\Omega \mid \Omega) = 1\}$ rolls without sliding on a fixed plan orthogonal to m.

As was shown by Arnold, this classical theory is completely reducible to a priori principles. The Lagrangian of S is given by the kinetic energy T. It is defined on TG and it is left invariant. According to Nœther's theorem, there exist 4 first integrals : the energy T and the three components of M. If S does not possess particular symmetries (i.e. if the I_j are different from one another), then these first integrals are *independent.* For given T and M, the invariant manifolds are therefore *surfaces* Σ. But $SO(3)$ is a *parallelizable* manifold (its tangent fibered bundle $TG \cong \mathbb{R}^3 \times SO(3)$ is globally trivial), hence *orientable.* It is also a *compact* manifold. Then, every Σ is also compact. But if $T \neq 0$, the velocity vector field is everywhere $\neq 0$. It is well known that the only compact and orientable surfaces where such vector fields exist are the torus. Then the Σ are invariant torus (it is a particular case of the Liouville-Arnold theorem on the integrable hamiltonian systems) and the rotation of S is the composition of two periodic movements (Poinsot movement). If the periods are non commensurable, it is a *quasi periodic* movement.

In fact, this movement is *a purely inertial one.* Its trajectories are the *geodesics* of G for the Riemannian structure derived from the kinetic energy. Let $g(t)$ be the movement of S in the configuration space $G = SO(3)$. The velocity vector $\dot{g}(t) \in T_g G$. We can transport $\dot{g}$ in $\mathcal{G} = T_e G$ by a right translation or by a left translation. We get $\omega = R^*_{g^{-1}} \dot{g}$ and $\Omega = L^*_{g^{-1}} \dot{g}$ (where R^*_h (resp. L^*_h) is the linear tangent map of the right (resp. left) translation R_h (resp. L_h)). We then have $\omega = R^*_{g^{-1}} L^*_g \Omega = Ad_g \Omega$: ω and Ω are related by the *adjoint representation.* The angular momentum (m, M) is the moment associated to (ω, Ω). Therefore m, $M \in \mathcal{G}^*$ are related by the *co-adjoint representation.* The kinetic energy $T = 1/2(A\Omega \mid \Omega) = 1/2M(\Omega)$ defines a metric on $\mathcal{G}$ which can be transported on G by left translations. This metric μ on G is, by definition, left invariant. Let H : $T^*M \to \mathbb{R}$ be the hamiltonian on G defined by T. Euler equations $\dot{M} = [M, \Omega]$ are the geodesic equations for μ. They are the equations of the hamiltonian systems defined by T on these symplectic manifolds that are the orbits of the coadjoint representation of $G = SO(3)$ in $\mathcal{G}^* \cong so(3)^* \cong \mathbb{R}^3$ (these orbits are the spheres $||M|| =$ constant). It is in that sense that, as was stressed by R. Abraham and J. Marsden, "the Euler equations (...) are purely geometrical or kinematical".[20]

This shows that there exist hamiltonian systems which are completely derivable from symmetry (relativity) groups. Such a reduction constitutes a reduction of the transcendental moment of Mechanics to the transcendental moment of Kinematics or, conversely, *a transformation of relativity principles into dynamical ones.*

[20] Abraham, Marsden [1978], p.327.

4. General relativity, the Shift of Transcendental Æsthetics and the Construction of the Category of Force

It was asserted by nearly all philosophers of science that General relativity (GR) marked the definitive collapse of the transcendental approach in modern physics. Our point of view is the opposite of such a claim. In fact, we think that GR supplies an extraordinary confirmation, deepening and sophistication of the transcendental approach, provided we accept the idea that the constitutive moments of objectivity can modify their content. *With GR the synthetic a priori shifts from the metrical level to the differentiable one* (that is to say the symmetry group of the theory changes radically). The geometrical a priori no longer lie at the metrical level but at the underlying differentiable level (level which was rendered autonomous by the pioneering works of Riemann and Clifford). It is only if we believe that the synthetic a priori is cognitively rooted that we are committed to reject a transcendental interpretation of GR. If, on the contrary, we seek for a descriptively correct epistemology of GR then we are irresistibly led to a transcendental one (cf. for example the works of Ernst Cassirer and Oscar Becker).

1. The Shift of Transcendental Moments in GR

Let $\mathcal{E}$ be space-time endowed with a locally minkowskian metric $g_{\mu\nu}$. If e_α is a basis of the tangent space $T_x\mathcal{E}$ and if ω^α is the dual basis of $T_x^*\mathcal{E}$, the Riemann curvature tensor $\mathbf{R}$ is given by $R^\alpha{}_{\beta\gamma\delta} = \langle\, \omega^\alpha,\ [\nabla_\gamma,\ \nabla_\delta]e_\beta\, \rangle$ (where ∇ is the covariant derivative and, for $\alpha \in T_x^*\mathcal{E}$ and $v \in T_x\mathcal{E}$, $\langle\, \alpha,\ v\, \rangle = \alpha(v)$). For $\alpha \in T_x^*\mathcal{E}$ and $u,\ v,\ w \in T_x\mathcal{E}$, $\mathbf{R}(\alpha,\ u,\ v,\ w) = \langle\, \alpha,\ R(v,\ w)u\, \rangle$ with $R(v,\ w) = [\nabla_v,\ \nabla_w]-\nabla_{[v,\ w]}$.

By contraction, we obtain the Ricci curvature tensor $R_{\mu\nu} = R_\mu{}^\alpha{}_{\alpha\nu}$. By a second contraction , we obtain the scalar curvature $R = R_\mu{}^\mu$. Einstein curvature tensor is then given by $G = \text{Ricci}-1/2gR$. It satisfies the contracted Bianchi identities $\nabla G \equiv 0$.

As was shown by Hilbert (1921) and developped by Arnowitt, Deser and Misner (1962), it is possible to derive the metric $g_{\mu\nu}$ from a variational principle using the scalar curvature as Lagrangian density. The action is then :

$$S = \frac{1}{16\pi} \int R\, \sqrt{|g|}\, d^4x$$

We consider the tridimensional geometry g^3 of space-like sections Σ of $\mathcal{E}$ (hypersurfaces of simultaneity). The configuration space is the space $\mathcal{M}$ of metrics – in fact the space of embeddings $\Sigma \to \mathcal{E}$ – , and a metric on $\mathcal{E}$ is a path in $\mathcal{M}$. Such a path is solution of an (infinite) hamiltonian system. [21]

More precisely we can compute g^4 and R^4 in a slice Σ_t–Σ_{t+dt} if we know g^3, the distance $(N,\ Ndt)$ between Σ_t and Σ_{t+dt}, R^3 and the extrinsic curvature K of the embedding $\Sigma \to \mathcal{E}$. We get in that way a (very complex) hamiltonian system on $T^*\mathcal{M}$.

$$\begin{cases} \frac{\partial g}{\partial t} &= \frac{\delta}{\delta\pi}(N\mathcal{H} + N\mathcal{J}) \\ \frac{\partial\pi}{\partial t} &= -\frac{\delta}{\delta g}(N\mathcal{H} + N\mathcal{J}) \end{cases}$$

[21] Cf. Misner-Thorne-Wheeler [1973] and Abraham-Marsden [1978].

where $\pi^{ij} = \delta S/\delta g_{ij}$ are the conjugate moments of g_{ij} and where :

$$\begin{cases} \mathcal{H}(g,\,\pi) = & \frac{1}{\sqrt{g}}\,(Tr\,\pi^2 - 1/2(Tr\,\pi)^2) - \sqrt{g}\,R^3 \\ \mathcal{J}^i(g,\,\pi) = & \nabla^3_j\,\pi^{ij} \end{cases}$$

(If $F(f) = \int \mathcal{F}(f(x))dx$ is a functional, the functional derivative $\delta F/\delta f$ is given by : $F(f_0 + g) = F(f_0) + \int \delta F/\delta f(x)\,|_{f=0}g(x)\,dx$).

This shows that in GR, *metric is no longer an a priori component but on the contrary a physical phenomenon which has to be determined.* It is for this reason – as it was deeply anticipated by Clifford – that metric can absorb the forces. In transcendental terms, we can say that in GR there is a conversion of the *kinematical* moment concerning metric – and therefore the category of quantity and the Axioms of intuition – into the *dynamical* moment concerning forces – and therefore the category of causality and the Analogies of experience. But this *does not mean* that GR falsifies transcendental epistemology. In fact, the equivalence principle and the general covariance principle possess an eminent transcendental meaning. Philosophically, the two most important points are :

(i) That the Transcendental Aesthetics *shifts from the metrical and global level to the local and differentiable one* : the relativity group of the theory is no more the Galilean or the Poincarean group but the group $Diff(\mathcal{E})$ of the diffeomorphisms of space-time. This implies that the synthetic a priori concerns henceforth this level.

(ii) Such a shift of Transcendental Aesthetics allows *the construction of the category of force*, and, even more, *its reduction to the a priori.*

2. The A Priori Determination of Einstein Equations Proposed by Wheeler

Actually, in *Gravitation* John Archibald Wheeler (with Charles Misner and Kip Thorne) has proposed an *a priori determination* of Einstein's field equations. Of course, such a determination is not a complete reduction. But nevertheless it reduces drastically the contingency of the empirical content of mechanics.

Wheeler starts from the equivalence and locality principles according to which the movement of matter expresses the space-time curvature. Then, the source of the curvature must be the tensor T of energy-momentum. This source must satisfy a *physical* principle of conservation $\nabla T \equiv 0$. Wheeler requires that this conservation be "automatic", that is to say due to purely *mathematical* reasons. We must therefore put $G' = kT$ where G' is a tensor of the same type as T, constructed in a purely geometrical way and satisfying *for a priori reasons* the identity $\nabla G' \equiv 0$. If we require that $G' \equiv 0$ when $\mathcal{E}$ is flat (nullity of the cosmological constant) and that G' be a linear function of the Riemann curvature tensor $\mathbf{R}$, then the Einstein tensor G is the only solution. The analysis of examples and of the newtonian limit shows that $k = 8\pi$. So we get Einstein's equations $G = 8\pi T$.

Now, $\nabla G \equiv 0$ is the Bianchi identity which is a consequence of the *cohomological* theory of differential forms ($d^2 = 0$, $\partial^2 = 0$ and Stokes' theorem). This means that *the synthetic a priori belongs henceforward to the local and differentiable level and, in particular, to the cohomological one.*

Wheeler is very explicit (quite lyric) about this fundamental possibility for geometrodynamics to derive physical equations from purely mathematical a priori. "This conservation is not an accident. According to Einstein and Cartan, it is "automatic" ; and automatic, moreover, as a consequence of exact equality between energy-momentum and an automatically conserved feature of the geometry". "Thus simply is all of general relativity tied to the principle that the boundary of a boundary is zero. No one has ever discovered a more compelling foundation for the principle of conservation of momentum and energy. No one has ever seen more deeply into that action of matter on space, and space on matter, which one calls gravitation. In summary, *the Einstein theory realizes the conservation of energy-momentum as the identity, "the boundary of a boundary is zero"*."[22] Here again, conservation laws (which guarantee the physical accessibility to physical magnitudes) are correlated with purely geometrical a priori.

3. The Critique of Wheeler by Adolf Grünbaum [23]

Philosophically, it is crucial to understand this *shift* of Transcendental Aesthetics in GR. Otherwise we are led to absurd conclusions. A good example is given by Grünbaum's critique of Wheeler.

In his "Philosophical Problems of Space and Time" – chapter 22 : General relativity, Geometrodynamics and Ontology – Grünbaum tries to show that there exists a *contradiction* between, on the one hand, GR as it is treated by Wheeler's geometrodynamics (GMD) according to Clifford's slogan "Physics is Geometry" and, on the other hand, the "Hypothesis" of Riemann ("Über die Hypothesen, welche der Geometrie zugrunde liegen"). According to him, GMD is "an all-out geometrical reductionism and absolutism" which *ontologizes* GR and transforms metric into an *intrinsic* property of space-time. Now, Grünbaum wants to maintain the thesis of the *conventionality* of metric. According to Riemann's metric hypothesis (RMH), we choose the metric in such a manner that :

(i) the class U_p of photon trajectories is composed with metrically null curves,
(ii) the class U_m of free fall movements is composed with geodesics.

"An indefinite Riemann metric (tensor) is to be SO CHOSEN that with respect to *one and the same such metric* ds

(a) The photon trajectories at any given world point are to BECOME *metrically null*.
(b) The Space-Time-trajectories belonging to the proper subclass U_m of U are to be TURNED INTO time-like GEODESICS via the intra-theoretic defining equation $\delta \int ds = 0$."[24]

These two classes U_p and U_m are empirically given. It is only the conventional choice of the metric which *qualifies* them geometrically. "It follows that if RMH is true, then the metrical structure of space-time – and thereby the very constitution of the only autonomous substance in the GMD monistic ontology – depends crucially for being what it is not only on the physically-determinate membership of U, but also on the *intrinsically*-UNFOUNDED, humanly-stipulated ascriptions of metrical *geodecisity* and *nullity* to the appropriate members of U !!" . "Granted RMH, human stipulations enter ontologically – not just verbally ! – into making the metric geometry of space-time be what it is, and thereby these stipulations paradoxically generate the character of the only autonomous

[22]Misner-Thorne-Wheeler [1973], p. 380.
[23]Cf. Grünbaum [1973], Chap. 22.
[24]Grünbaum [1973], p. 757. The emphasis is Grünbaum's.

"physical" substance recognized by GMD. For on the assumption of RMH, human conventions are indispensable to the GMD metric geometry".[25]

It is clear that Grünbaum wants to go beyond the dogmatic conflict between an ontological realism and a nominalistic ("verbal", non geometrical) conventionalism. His conception is clearly objective in the transcendental sense. According to him, Wheeler's GMD must be criticized because it is "ontological" and runs into a contradiction : "there is the outright inconsistency that space is and also is not intrinsically metric". But he misses the point. In GR, the dialectic between "is" and "is not" is exactly the same as the "is" and "is not" in the sentence "a body is and also is not moving" in classical mechanics. As Einstein said , acceleration (and therefore force) becomes as relative as velocity. In GR, according to the fact that metric has become ***a physical phenomenon***, the modality of metric ***is no longer the modality of reality but the modality of possibility.*** The shift of Transcendental Aesthetics renders irrelevant the classical problem of the conventionality of metric. It is the *differentiable* structure (with its underlying *topological* stucture and its cohomological correlates) which is now conventional (non physical) and nevertheless physically determinant, that is to say synthetic a priori.

6. Quantum Mechanics and the Generalized Transcendental Æsthetics

Quantum Mechanics (QM) was also used by positivists as a weapon against transcendentalism. But we think that, in fact, there exists also a clear and deep transcendental structure of QM.

We have seen at the begining of this work that a plausible epistemology of QM must be based upon a non ontological concept of "weak" objectivity which is of a transcendental nature. Bohr's operationalism is in fact transcendental . This thesis can be developed in a rather technical way.

1. The Quantum Transcendental Aesthetics

In an objective process of constitution, the function of Transcendental Aesthetics is characterized by four requisites :

(i) to determine *forms* of manifestation which permit to delete the *subjective* contents of the concept of phenomenon;

(ii) to manifest *a relativity* violating the principles of substantial ontology ;

(iii) to provide a mathematical basis for the construction of the dynamical (physical) categories of substance, causality and interaction ;

(iv) to lead to a non metaphysical – but only physical and specifically relative to the process of constitution – interpretation of the modal categories of possibility (virtuality), reality (actuality) and necessity.

For classical mechanics, special relativity and general relativity, Transcendental Aesthetics is spatio-temporal. But, according to the previous definition, ***it can perfectly exist non spatio-temporal generalized Transcendental Aesthetics.*** It is the case in QM.

[25]Ibid., p. 758.

If we try to recognize in QM such an instance associated with the specific quantum concept of phenomenon, we are immediately led to the concept of ***probability amplitude***. As was emphasized by G. Cohen-Tannoudji, "the recourse to statistics permits to take into account in quantum phenomena the inseparability between the object and the measure apparatus, *whithout explicitely introducing the measure apparatus*". [26]Quantum states are sets of ***possibilities*** – a "potentia" – which are ***actualized*** by measure operations. This ***relational*** status of the concept of quantum state (often stressed by Bohr) and its interpretation as a new type of relativity, were very well formulated by Vladimir Fock[27]: "The probabilities expressed by the wave function are the probabilities of some result of the interaction of the micro-object and the instrument (of some reading on the instrument). The wave function itself can be interpreted as the reflection of the potential possibilities of such an interaction of the micro-object (prepared in a definite way) with various types of instruments. A quantum mechanical description of an object by means of a wave function corresponds to the relativity requirement with respect to the means of observation. This extends the concept of relativity with respect to the reference system familiar in classical physics".

The axioms governing the use of probability amplitudes give an example of a radically non classical Transcendental Aesthetics. Of course, one of the main problems is to establish the link with the spatio-temporal Transcendental Aesthetics. It is a very difficult one. It is well known (Gelfand's spectral theory) that if $\mathcal{A}$ is a commutative C^*-algebra (that is to say a Banach algebra over $\mathbb{C}$ which is involutive and where the norm is $||x||^2$ = spectral radius of (x^*x)), then $\mathcal{A} \cong C(X)$ the algebra of continuous complex functions on the space $X = \mathrm{Sp}(\mathcal{A})$ of characters of $\mathcal{A}$: $\{\chi : \mathcal{A} \to \mathbb{C} \mid \chi \neq \imath\}$ (X is also the space of maximal ideals of $\mathcal{A}$). So, in the classical case, there exists a sort of duality between space and observables. In QM it is no longer the case and we must use ***non commutative geometry*** (in Alain Conne's sense) to find again a duality between the (non commutative) algebras of observables and some sort of "underlying space".[28]

2. Theories with Hidden Variables

The basic idea of the theories which accept hidden variables is to interpret a quantum – an then ***dispersive*** – state ψ as a statistic set of ***non dispersive*** ideal states. If A is an observable and $\psi(A) = \langle A\psi \mid \psi \rangle$ [29], we try to express $\psi(A)$ by an integral of the form:

$$\psi(A) = \int_{\Omega} f_\omega(A)\, d\mu(\omega)$$

where Ω is a space of hidden variables, μ a probability measure on Ω and the f_ω are ideal non dispersive states. [30] So, we try to associate to ψ a probability measure ν_ψ on Ω and to A a function $\underline{A} : \Omega \to \mathbb{R}$ in such a way that, if $\mu_{\psi,A}$ is the distribution on $\mathbb{R}$ for the measure of A when the system is in the state ψ, then we have for every $E \subset \mathbb{R}$, $\mu_{\psi,A}(E) = \nu_\psi(\underline{A}^{-1}(E))$.

[26]Cohen-Tannoudji and Spiro[1986], p. 152.
[27]Cf. Jammer [1974], p. 202.
[28]Cf. Connes [1990].
[29]$\langle \psi \mid \varphi \rangle$ is the scalar (hermitian) product in the Hilbert space $\mathcal{H}$ of the system.
[30]Cf. Marsden [1974].

But a celebrated theorem of von Neumann shows that this is essentially impossible:

Von Neumann's theorem. If a C^*-algebra of observables (self-adjoint operators) admit hidden variables then it is commutative (i.e. classical).

Mackey, Gleason, Bell, have refined von Neumann's theorem. They have shown that some too strong hypothesis could be removed. But the negative result remains: as is now well known, there exist no *local* theories admitting hidden variables. We can evoque here again Marsden's verdict (cf. §I.2.(iii)): "The entire point of the negative results concerning "hidden variables" is that there is no "objective underlying state" [in the ontological sense of the term "objective"] of the system !". It is henceforth the science itself which imposes an objective epistemology which can philosophically thematize the fact that a "weak" objectivity is nevertheless a true (but non ontological) objectivity.

3. Gauge Theories and the Construction of the Category of Interaction

Philosophically, one of the main achievement of quantum field theory is to have *constructed* (in the transcendental sense) the third dynamical category of *interaction* (GR constructed the second dynamical category of causality i.e. the force) using a typically transcendental principle, *Weyl's gauge principle converting gauge invariances into dynamical principles.*

In quantum field theories we start with Lagrangian densities $\mathcal{L}(\varphi,\ \partial_\mu\varphi)$ where the fields $\varphi(x,\ t)$ may possess *internal symmetries.* The action functional is :

$$S(\Gamma) = \int_\Omega \mathcal{L}\ d^4x = \int_{t_1}^{t_2} \int_{\mathbb{R}^3} \mathcal{L}(\varphi,\ \partial_\mu\varphi)\ d^3x\ dt$$

computed along the paths $\Gamma = \varphi(x,\ t)\ :\ \varphi_i = \varphi(x,\ t_1) \rightarrow \varphi_f = \varphi(x,\ t_2)$.
The axioms of QM governing probability amplitudes lead to Feynman's path integrals :

$$\langle\ \varphi_f\ |\ \varphi_i\ \rangle\ =\ \int_\Gamma \exp(iS(\Gamma)/\hbar)\ \mathcal{D}\Gamma$$

It is well known that this formula (which is analogous with the partition functions Z in statistical mechanics) encodes a considerable amount of information. It is possible to derive from it inumerable explicit, quantitative and predictive models, using appropriate mathematical tools as for example:

(i) perturbative developments ;
(ii) Wick's theorem : all the moments of a gaussian probability can be expressed as functions of its two order moments ;
(iii) stationary phase theorem : an oscillatory integral $e^{i\tau\varphi(x)}$ concentrates, for $\tau \rightarrow \infty$, on the critical points of the phase φ ;
(iv) methods of the renormalisation group.

We find here a splendid example of an objective determination leading from constitutive principles to very diversified and very specific models : the constitutive principles (relativity, symmetry) provides Lagrangians, wich in turn provide Feynman's integrals, wich provide themselves the models.

Constitutive a priori Relativities Symmetries	Lagrangian Action	Path integrals	Specific models for phenomena

Gauge theories have shown that if we *localize the internal symmetries* and if we impose the invariance of the Lagrangians for these supplementary symmetries, we can reconstruct *in a purely mathematical manner* the *interaction* Lagrangians (reduction to a priori principles). [31]

The standard example is that of the (minimal) coupling between an electron and the electro-magnetic field. We have :

The Dirac Lagrangian : $\mathcal{L}_D = \bar{\psi}(x)(i\gamma^\mu\partial_\mu - m)\,\psi(x)$;
The Maxwell Lagrangian : $\mathcal{L}_{EM} = -1/4F^{\mu\nu}F_{\mu\nu} - J^\mu A_\mu$;
The Interaction Lagrangian : $\mathcal{L}_{Int} = -e\bar{\psi}(x)\gamma^\mu A_\mu\psi(x) = -J^\mu A_\mu$ (where $J^\mu(x) = e\bar{\psi}(x)\gamma^\mu\psi(x)$ is the electro-magnetic current generated by the field ψ).

• The Dirac Lagrangian $\mathcal{L}_D$ is invariant under the internal and global symmetry: $\psi \to e^{-ie\theta}\psi$.

• The Maxwell Lagrangian $\mathcal{L}_{EM}$ is invariant under a gauge transformation: $A \to A + d\Lambda$ (for Maxwell's equations are $F = dA$, $dF = 0$, $d^*F = 4\pi^*J$ and we have $d^2 = 0$).

•If we let $\theta = \theta(x)$ depends on the spatial position x, then $\mathcal{L}_D$ is no longer invariant. But the supplementary term $e\bar{\psi}\gamma^\mu\partial_\mu(\theta(x))\psi$ can be exactly balanced by the gauge transformation : $A \to A + d\theta$. This is equivalent to the substitution of ∂_μ by the covariant derivative: $D_\mu = \partial_\mu + ieA_\mu(x)$. The vector potential A_μ is geometrically interpreted as a *connexion* on a vector bundle over space-time. The EM field is then identified with the *curvature* of this connexion.

In the non abelian case, let G ($G \neq U(1)$, $G = SU(2)$, $SU(3)$, etc.), be the non abelian Lie group of internal symmetries. Let ξ_a be the generators of the Lie algebra $\mathcal{G}$ of G. The Lagrangians are invariant under the $U(g) = \exp(-i\theta_a\xi_a)$. If we let the $\theta_a = \theta_a(x)$ depend on x, they are no longer invariant. We recover invariance if we introduce gauge fields $A^\mu_a(x)$. We get in that way a field $\mathcal{A}^\mu(x)$ with values in the Lie algebra $\mathcal{G}$.

• The gauge transformations $A_\mu \to A_\mu + \partial_\mu\theta$ are generalized by :

$$A^\mu_a \to c_a{}^{bc}\theta_b A^\mu_c + \partial^\mu\theta_a$$

i.e.

$$\mathcal{A}'_\mu = U(g)\,[\mathcal{A}_\mu + iU(g)^{-1}\partial_\mu(U(g))]\,U(g)^{-1}$$

• The covariant derivative $D_\mu = \partial_\mu Id + ie\mathcal{A}_\mu$ is transformed into $D'_\mu = U(g)D_\mu$ which balances the term $\partial_\mu(U(g))\varphi$ in $\partial_\mu\varphi \to U(g)\partial_\mu\varphi + \partial_\mu(U(g))\varphi$.

• The field $F^{\mu\nu} = \partial^\mu A^\nu - \partial^\nu A^\mu$ is generalized in $\mathcal{F}^{\mu\nu} = \partial^\mu\mathcal{A}^\nu - \partial^\nu\mathcal{A}^\mu + i[\mathcal{A}^\mu, \mathcal{A}^\nu]$.

• The default of commutativity introduces non linearities and a self-coupling of the gauge fields.

[31] For a presentation of Gauge Theories, cf. for instance Itzykson-Zuber [1985], Le Bellac [1988], Manin [1988], Quigg [1983].

Using Feynman's algorithm it has been possible to work out a true *ontogenesis* of physical reality. Such an ontogenesis converts the synthetic a priori in explicit models. The mathematical constraints are so strong (renormalizability, elimination of anomalies, Higgs mechanism of symmetry breaking for confering a mass to the gauge bosons, etc.), that it is possible to infer from very few (but crucial) empirical data to the choice of a symmetry group.

4. Superstring Theory

In superstring theory this fact is even more evident.[32]

One supposes that elementary particules are non punctual entities (strings). Let $\sigma \in [0, \pi]$ be a parametrization of the string. If τ is its proper time, the parametrization of its world leaf is $X_\mu(\sigma, \tau)$ endowed with the metric $g_{ab} = g_{\mu\nu}\, \partial_a X^\mu \partial_b X^\nu$ $(a, b = \sigma \text{ or } \tau)$. This leads to the introduction of new Lagrangians, for instance the Polyakov Lagrangian :

$$L = -\sqrt{g}\, g^{ab}\, \partial_a X_\mu \partial_b X^\mu$$

with $g = |det(g_{ab})|$. In that way, we get *a new symmetry group.*

Feynman's interaction graphs are substituted by Riemann surfaces (which are topological configurations of interaction). One has then to compute functionnal integrals of the following type :

$$Z = \sum_{\text{topologies}} \int_{\text{metrics}} \mathcal{D}g_{ab} \int_{\text{leaves}} \mathcal{D}X^\mu \, \exp(iS/\hbar)$$

To avoid redondancies corresponding to gauge invariances, we must know the spaces and the measures occuring in the integration. For doing that, we need Riemann surfaces theory. For example we need Teichmüller theory of moduli spaces for knowning exactly what are the automorphisms of a Riemann surface (what are its diffeomorphisms which are not isotopic to the identity, what are the complex structures compatible with a given differentiable structure, etc.). We need also the solution of the Schottky problem. Let S be a Riemann surface of genus g. It is well known that it is possible to find a basis (a_i, b_i), $i = 1, ..., g$ of the homology of S and a basis (ω_j), $j = 1, ..., g$ of the space of differentiable 1-forms which are the simplest possible, that is to say which satisfy: $\int_{a_i} \omega_j = \delta_{ij}$ and $\int_{b_i} \omega_j = \Omega_{ij}$, the matrix $\Omega = (\Omega_{ij})$ of periods being symmetric and of imaginary part positive definite: $Im\Omega > 0$. But if $g > 3$, the space of the matrices Ω which are symmetric and of imaginary part > 0 has a dimension $1/2g(g+1)$ which is greater than the dimension $3g$–3 of the moduli space of S. Therefore we must characterize the Ω which can be the period matrices of Riemann surfaces. This is the Schottky problem. It has been solved only in 1984.

All these sophisticated and deep mathematical results are necessary to express the independence of objectivity relatively to the new conventional elements introduced in superstring theory.

Renormalization constraints and the elimination of anomalies impose for instance the dimension of space-time (10 or 26) (elimination of the *conformal* anomaly) and the gauge group ($O(32)$ or $E_8 \otimes E_8$) (elimination of the *chiral* anomaly).

[32] For a presentation of Superstring Theories, cf. for instance Green-Schwarz-Witten [1987] or Kaku [1988].

The chiral anomaly is defined in the following manner. On a spin manifold where the Dirac equation is defined, the chirality corresponds to the invariance under the transformation $\psi \rightarrow e^{i\gamma_5\varepsilon}\psi$. This invariance implies the conservation law of the axial current $\partial_\mu J^{\mu,5} = \partial_\mu(\bar{\psi}\gamma^5\gamma^\mu\psi) = 0$.

For the dimension $D = 4$, it can be shown that:

$$\begin{cases} \partial_\mu J^{\mu,5} &= -\frac{1}{16\pi^2}\, \varepsilon^{\mu\nu\sigma\rho}\, Tr\,(F_{\mu\nu}F_{\sigma\rho}) \\ J^{\mu,5} &= -\frac{1}{4\pi^2}\varepsilon^{\mu\alpha\beta\gamma}\, Tr\,(A_\alpha\partial_\beta A_\gamma + \frac{2}{3}A_\alpha A_\beta A_\gamma) \\ &= \bar{\psi}\,\xi_a\gamma^5\gamma^\mu\,\psi \end{cases}$$

where F is the field and A the potential vector. More intrisically we have:

$$\partial_\mu J^{\mu,5} = -\frac{1}{16\pi^2}\, Tr\,(F \wedge F^*)$$

where F (the field) is the curvature 2-form $F = dA + A \wedge A$ and A the 1-form $A = A_\mu^a\, \xi_a\, dx^\mu$ (the ξ_a are the gauge fields). The 4-form equal to the square of this curvature 2-form is an exact differential : $Tr\,(F \wedge F^*) = d\omega$, ω being the Chern-Simons 3-form $\omega = Tr\,(AdA + \frac{2}{3}A^3)$.

When one introduces the corrections wich are necessary to make the Feynman integrals converge, one *brokes* this symmetry. This produces an anomaly which is deeply linked with the index of the Dirac operator (which is a *topological* number). In fact the anomaly is a cohomology class. Let S be the spin bundle of the manifold M and D the Dirac operator on the smooth sections of S. The symmetry group G is here the group of diffeomorphisms of S (considered as a fiber bundle). Let Det : $G \rightarrow \mathbb{C} - \{0\}$ be the *determinant* of the operator D. The anomaly α is the pull-back by Det of the generator dz/z of the first integral cohomology group $H^1(\mathbb{C} - \{0\},\ \mathbb{Z})$ of $\mathbb{C} - \{0\}$. We have by definition $\alpha \in H^1(G,\ \mathbb{Z})$. It can be shown that the elimination of the chiral anomaly imposes the dimension $n = 496$ for the gauge group, and this imposes $O(32)$ or $E_8 \otimes E_8$.

We must return in dim 4 if we want to recover the phenomenology. For instance – using a Kaluza-Klein device – we can compactify 16 dimensions (starting from $D = 26$) using the lattice of the roots of the Lie gauge group $E_8 \otimes E_8$ and then compactify again 6 dimensions : $M^{10} \rightarrow M^4 \times K^6$. Physical constraints of preservation of supersymmetry impose for example that there exists on K a spinor field ξ which is constant for the covariant derivation (i.e. $D_i\xi = 0$). This fact imposes drastic constraints upon the geometry of K^6 : the Ricci curvature must be $= 0$, the holonomy group must be $= SU(3)$ (and non $O(6) \cong SU(4)$), the first Chern class $c_1(K)$ of K must be $= 0$, there must exist a Kähler metric on K, etc.(In fact, according to a celebrated theorem of Calabi and Yau, a Kähler manifold K^{2n} with $c_1(K) = 0$ admits necessarily a Kähler metric with holonomy group $SU(n)$ (and not $O(2n)$).

To relate this compactification $M^{10} \rightarrow M^4 \times K^6$ to the compactification $M^{26} \rightarrow M^{10}$ using the lattice of $E_8 \otimes E_8$, the most direct method consists in identifying a part of the gauge fields of $E_8 \otimes E_8$ to the spin connexion of M^{10}. This relate the gauge fermions to the geometry of K^6. More precisely, we start with the spin connexion $\omega = \omega_\mu^{ab}\, dx^\mu\, M_{ab}$ of K^6 which is given by the covariant derivative of spinors $D_\mu\psi = (\partial_\mu + \omega_\mu^{ab}M_{ab})\psi$ (where M_{ab} is a representation of the generators of the Lorentz group) and we embed ω in the gauge connexion of the Yang-Mills theory with gauge group $E_8 \otimes E_8$. This allows to satisfy the *Bianchi identities* :

$$Tr\,(R \wedge R) - \frac{1}{30} Tr\,(F \wedge F)$$

which relate the Riemann curvature tensor **R** to the Yang-Mills tensor **F**. As for the spin connexion the holonomy group is $SU(3)$, we must consider $SU(3)$ as a sub gauge group of $E_8 \otimes E_8$. The simplest manner for doing this is to consider $SU(3) \otimes E_6 \subset E_8$.

If we consider the decomposition of the adjoint representation of E_8 (which is of dimension 248) relatively to the subgroup $SU(3) \otimes E_6$ we find :

$$248 = (3,\ 27) \oplus (\overline{3},\ \overline{27}) \oplus (8,\ 1) \oplus (1,\ 78)$$

that is to say 27 sets of fields which transform themselves according to the representation $3 \oplus \overline{3}$ of $SU(3)$ and a set of octets which transform themselves according to the adjoint representation 8 of $SU(3)$. But such an octet must be counted 3 times relatively to $3 \oplus \overline{3}$. There exists therefore in the representation $3 \oplus \overline{3}$ of $SU(3)$ a total redondancy equal to $27 + 3 \times 1 = 30$. The Bianchi identities can therefore be satisfied.

We see perfectly well with this technical and recent example how general a priori principles lead to very precise models and predictions which can be experimentally falsified. The epistemological point is that such falsifiable models and predictions *are remote mathematical consequences of a transcendental type of objective determination.*

Conclusion

1. The more mathematics and physics unify themselves in the physico-mathematical modern theories, the more an objective epistemology becomes necessary. Only such a transcendental epistemology is able to thematize correctly the status of the mathematical determination of physical reality.
2. There exists a *transcendental history* of the synthetic a priori and of the construction of physical categories.
3. The transcendental approach allows to supersed Wittgenstein's and Carnap's antiplatonist thesis according to which pure mathematics are physically applicable only if they lack any descriptive, cognitive or objective, content and reduce to mere prescriptive and normative devices. In fact, pure mathematics are prescriptive-normative in physics because :
   (i) the *categories* of physical objectivity are prescriptive-normative, and
   (ii) their categorial content is mathematically "constructed" through a Transcendental Aesthetics. Only a transcendental approach make compatible, in the one hand, a grammatical conventionalism of Wittgensteinian or Carnapian type and, on the other hand, a platonist realism of Gödelian type. *Mathematics are not a grammar of the world but a mathematical hermeneutics of the intuitive forms and of the categorial grammar of the world.*
4. The transcendental approach allows also *to reconcile the objective validity and the historical value* of scientific theories. It allows to go beyond the epistemological antinomy opposing dogmatic positivism (there exists an absolute value of objective truth) and

sceptic post-positivism (there exists an historico-anthropological relativity of truth). As we have seen, truth, reality, necessity are *moments* of the procedures of constitution and determination. They are *relative* to them, and therefore relative to an *historical* (non cognitive) synthetic a priori.

References

Abraham, R., Marsden, J. 1978. *Foundations of Mechanics*, Benjamin Cummings, New-York, Reading.

Allison, H.E., 1981."Transcendental Schematism and the Problem of the Synthetic a priori", *Dialectica*, 35, 1-2, 57-83.

Allison, H.E., 1983. *Kant's Transcendental Idealism. An Interpretation and Defense*, New-Haven, Yale University Press.

Ameriks, K., 1982. "Current German Epistemology : The Significance of Gerold Prauss", *Inquiry*, 25, 1, 125-138.

Arnold, V., 1976, *Méthodes mathématiques de la mécanique classique*, Moscou, Mir.

Bohr, N., 1935. *Physical Review*, 48, 696.

Boi, L., 1989. "Idéalisation et objectivation, ou des rapports entre géométrie et physique", *Fundamenta Scientiae*, 10, 1, 85-114.

Bouveresse, J., 1987. *La Force de la Règle*, Paris, Editions de Minuit.

Brittan, G., 1978. *Kant's Theory of Science*, Princeton University Press.

Brittan, G., 1991. *Algebra and Intuition*, Department of Philosophy, Montana Sate University, (to appear).

Cassirer, E., 1910. *Substanzbegriff und Funktionbegriff*, Berlin. *Substance et Fonction* (trans. P. Caussat), Paris, Editions de Minuit, 1977.

Châtelet, G., 1985. "Le retour de la monade", *Fundamenta Scientiae*, 6, 327- 345.

Cohen-Tannoudji, G., Spiro, M., 1986. *La Matière - Espace - Temps*, Paris, Fayard.

Connes, A., 1990. *Géométrie non commutative*, Paris, InterEditions.

d'Espagnat, B. 1985. *Une incertaine réalité*, Paris, Gauthier-Villars.

Duncan, H., 1984. "Inertia, the Communication of Motion, and Kant's Third Law of Mechanics", *Philosophy of Science*, 51, 93-119.

Ehlers, J., 1973. "The Nature and Structure of Spacetime", *The Physicist's Conception of Nature*, (J. Mehra ed.), 71-91, Dordrecht, Reidel.

Folse, H.J., 1978. "Kantian Aspects of Complementarity", *Kant-Studien*, 69, 58-66.

Friedman, M., 1985. "Kant's Theory of Geometry", *The Philosophical Review*, XCIV, 4, 455-506.

Gomez, R.J., 1986. "Beltrami's Kantian View of Non-Euclidean Geometry", *Kant-Studien*, 77, 1, 102-107.

Green, M. B., Schwarz, J. H., Witten, E., 1987. *Superstring Theory*, Cambridge University Press.

Grünbaum, A., 1973. *Philosophical Problems of Space and Time*, Dordrecht - Boston, Reidel.

Honner, J. 1982. "The Transcendental Philosophy of Niels Bohr", *Studies in History and Philosophy of Science*, 13, 1, 1-29.

Itzykson, C., Zuber, J.B., 1985. *Quantum Field Theory*, Singapour, Mc Graw-Hill.

Jammer, M., 1974. *The Philosophy of Quantum Mechanics*, New-York, John Wiley and Sons.

Kaku, M., 1988. *Introduction to Superstrings*, New-York, Springer.

Kant, E, 1980-1986. *Oeuvres philosophiques* (F. Alquié ed.), Paris, Bibliothèque de la Pléïade, Gallimard.

Kant, E., 1781-1787. *Critique de la Raison pure*, (trans. A.J.L. Delamarre et F. Marty), Paris, Pléïade, Gallimard, 1980.

Kant, E., 1786. *Premiers Principes métaphysiques de la Science de la Nature*, (trans. J. Gibelin), Paris, Vrin, 1971.
Kant, E., 1790. *Critique de la Faculté de Juger*, (trans. A. Philonenko), Paris, Vrin, 1979.
Kant, E., 1796-1803. *Opus Postumum*, (trans. F. Marty), Paris, Presses Universitaires de France, 1986.
Kant, I., 1781-1787. *Kritik der reinen Vernunft*, Kants gesammelte Schriften, Band III, Preussische Akademie der Wissenschaften, Berlin, Georg Reimer, 1911.
Kant, I., 1786. *Metaphysische Anfangsgründe der Naturwissenschaft*, Kants gesammelte Schriften, Band IV, Preussische Akademie der Wissenschaften, Berlin, Georg Reimer, 1911.
Kant, I., 1790. *Kritik der Urtheilskraft*, Kants gesammelte Schriften, Band V, Preussische Akademie der Wissenschaften, Berlin, Georg Reimer, 1913.
Lautman, A., 1937-1939. *Essai sur l'unité des mathématiques et divers écrits*, Paris, Bourgois, 1977.
Le Bellac, M., 1988. *Des phénomènes critiques aux champs de jauge*, Paris, InterEditions - C.N.R.S.
Manin, Y. I., 1988. *Gauge Field Theory and Complex Geometry*, Berlin, New- York, Springer.
Marsden, J., 1974. *Applications of Global Analysis in mathematical Physics*, Berkeley, Publish or Perish.
Mc Goldrick, P.M., 1985. "The Metaphysical Exposition : An Analysis of the Concept of Space", *Kant-Studien*, 76, 3, 257-275.
Misner, C.W., Thorne, K.S., Wheeler, J.A., 1973. *Gravitation*, San Francisco, Freeman.
MNS, 1989. *La Mathématique non standard*, (H. Barreau, J. Harthong, eds.), Paris, Editions du CNRS.
Paty, M., 1988. *La Matière dérobée*, Paris, Editions des Archives Contemporaines.
Petitot, J., 1979-1982. "infinitesimale", "Locale/Globale", "Unità delle matematiche", *Enciclopedia Einaudi*, VII, 443-521 ; VIII, 429-490 ; XV, 341-352 ; XV, 1034-1085, Turin, Einaudi.
Petitot, J., 1987a. "Refaire le Timée. Introduction à la philosophie mathématique d'Albert Lautman", *Revue d'Histoire des Sciences*, XL, 1, 79-115.
Petitot, J., 1987b. "Mathématique et Ontologie", *La scienza tra filosofia e storia in Italia nel Novecento*, (F. Minazzi, L. Zanzi, eds.), 191-211, Rome, Edizione della Presidenza del Consiglio dei Ministri.
Petitot, J., 1988. "Logique transcendantale et ontologies régionales", *Colloque de Cerisy : Rationalité et Objectivités*, (to appear, Editions Patiño).
Petitot, J., 1989. "Rappels sur l'analyse non standard", *MNS [1989]*, 187- 209.
Petitot, J., 1990. "Premiers principes métaphysiques d'une Science de la Forme", *Colloque de Cerisy autour de la Critique de la Faculté de Juger*, (to appear).
Petitot, J., 1990a. "Logique transcendantale, Synthétique a priori et Herméneutique mathématique des objectivités", *Fundamenta Scientiae*, 10, 1, 57-84.
Petitot, J., 1990b. "Logique transcendantale et Herméneutique mathématique : le problème de l'unité formelle et de la dynamique historique des objectivités scientifiques", *Il pensiero di Giulio Preti nella cultura filosofica del novecento*, (F. Minazzi ed.),155-172, Milano, Franco Angeli.
Petitot, J., 1990c. "Note sur la querelle du déterminisme", *La Querelle du Déterminisme*, (K. Pomian ed.), 202-227, Paris, Le Débat, Gallimard.
Petitot, J., 1991a. "Idéalités mathématiques et Réalité objective. Approche transcendantale", *Hommage à Jean-Toussaint Desanti*, (G. Granel ed.), 213-282, Mauvezin, Editions TER.
Petitot, J., 1991b. "Continu et Objectivité", *Le Continu mathématique*, (J.M.Salanskis, H. Sinaceur eds.), Colloque de Cerisy (to appear).
PQG, 1988. *Physique quantique et géométrie* (Colloque André Lichnerowicz, D. Bernard, Y. Choquet-Bruhat eds.), Paris, Hermann.

Prauss, G., 1980. *Einführung in die Erkenntnistheorie,* Darmstadt, Wissenschaftliche Buchgesellschaft.

Prauss, G., 1981a. "Time, Space and Schematisation", *The Philosophical Forum,* XIII, 1, 1-11.

Prauss, G., 1981b. "Kants Theorie der ästhetischen Einstellung", *Dialectica,* 35, 1-2, 265-281.

Quigg, C., 1983. *Gauge Theories of the Strong, Weak, and Electromagnetic Interactions,* Reading, Benjamin-Cummings.

Salanskis, J.M., 1989. "Le potentiel et le virtuel", MNS [1989], 275-303.

Salanskis, J.M., 1991. *L'Herméneutique formelle,* Paris, Éditions du CNRS.

Scheibe, E., 1981. "Invariance and Covariance", (J. Agassi, R. S. Cohen, eds.), *Scientific Philosophy Today,* 311-331, Dordrecht, Reidel.

Souriau, J.M., 1975. *Géométrie symplectique et physique mathématique,* Coll. Internat. du C.N.R.S., 237, Paris.

Stegmüller, W., 1979. *The Structuralist Views of Theories,* Berlin, New- York, Springer.

Vuillemin, J., 1955. *Physique et Métaphysique kantiennes,* Paris, Presses Universitaires de France.

Wang, H., 1987. *Reflections on Kurt Gödel,* Cambridge, M.I.T Press.

Weinstein, A., 1977. *Lectures on Symplectic Manifolds,* C.B.M.S., Conf. Series, Am. Math. Soc., 29, Providence.

Weizsäcker, C.F. von, 1979. *Die Einheit der Natur,* Munich, Hauser.

Weyl, H., 1922. *Space - Time - Matter,* New-York, Dover.

Wiredu, J.E., 1970. "Kant's Synthetic A Priori in Geometry and the Rise of Non-Euclidean Geometries", *Kant-Studien,* 61, 1, 5-27.

GPSR Compliance

The European Union's (EU) General Product Safety Regulation (GPSR) is a set of rules that requires consumer products to be safe and our obligations to ensure this.

If you have any concerns about our products, you can contact us on ProductSafety@springernature.com

In case Publisher is established outside the EU, the EU authorized representative is:

Springer Nature Customer Service Center GmbH
Europaplatz 3
69115 Heidelberg, Germany

Batch number: 10620881

Printed by Printforce, the Netherlands